Lecture Notes in Control and Information Sciences

Edited by M. Thoma and A. Wyner

For information about Vols. 1–61 please contact your bookseller or Springer-Verlag.

Lecture Notes in Control and Information Sciences

Edited by M. Thoma and A. Wyner

129

W. A. Porter
S. C. Kak (Editors)

Advances in Communications and Signal Processing

Springer-Verlag Berlin Heidelberg GmbH

Editors
William A. Porter
Subhash C. Kak

Department of Electrical
and Computer Engineering
Louisiana State University
Baton Rouge, LA 70803
USA

ISBN 978-3-540-51424-4   ISBN 978-3-540-46259-0 (eBook)
DOI 10.1007/978-3-540-46259-0

Offsetprinting: Mercedes-Druck, Berlin

2161/3020-543210   Printed on acid-free paper.

PREFACE

This volume is a collection of papers selected out of those presented at the 1988 International Conference on Advances in Communication and Control Systems. This conference was held in Baton Rouge, Louisiana during October 19-21, 1988 and it was sponsored by National Science Foundation, Louisiana State University, Southern University, and International Federation of Information Processing.

The aim of this conference was to bring together researchers in the fields of communications, control, signal processing, and computing to explore emerging common themes and report advances in research. A total of 131 papers were presented at the conference. Sixty-eight of these papers were selected for inclusion in two Springer-Verlag LNCIS volumes. These volumes are titled <u>Advances in Communications and Signal Processing</u> and <u>Advances in Computing and Control</u>.

We would like to thank A.V. Balakrishnan, Nirmal Bose, Zdzislaw Bubrucki, Romano DeSantis, Thomas Dwyer, Erol Gelenbe, Don Halverson, Manju Hegde, Edward Kamen, Hong Lee, Ruey-wen Liu, Mario Lucertini, Steve Marcus, Sanjit Mitra, Mort Naraghi-Pour, Dave Neuhoff, Roberto Triggiani and Nick Tzannes for helping with the organization of the sessions and selection of contributors.

We also thank Margaret Brewer and Rachel Bryant for their enthusiastic help with the preparation of the volumes.

William A. Porter
Subhash C. Kak
Editors

TABLE OF CONTENTS

PERFORMANCE ANALYSIS OF NON-ORTHOGONAL BLOCK DESIGNS FOR FH/MFSK SYSTEMS UNDER PARTIAL BAND INTERFERENCE

Guillermo E. Atkin   Hector P. Corrales

Department of Electrical and Computer Engineering
Illinois Institute of Technology, Chicago, Illinois, 60616

1. Introduction. Spread Spectrum Communication Systems designed to combat intentional interference or jamming have been widely investigated [1,2,3], in particular Frequency Hopped M-ary Frequency Shift Keying (FH/MFSK) System. In a FH/MFSK System, several uncorrelated frequency bands or subchannels are available to the transmitter, each one divided into M orthogonal subbands. On each symbol interval, the modulator selects one subchannel, according to a pseudorandom pattern, and transmit the MFSK symbols, selected according to the input bits. A partial band interference channel is modeled as a gaussian noise jammer that restricts its power to a fraction ρ ($0 < \rho \leq 1$) of the total spread spectrum bandwidth. To maximize its interference action, the jammer also randomly hops the location of the jammed band over the total spectrum, however the pseudorandom hop pattern is only known by the transmitter and the receiver. To simplify the analysis, it is assumed that the jammer hops are made in coincidence with the symbol transitions, and that the jammed band lies entirely inside or outside the subchannel used by the system, therefore the channel is considered stationary on each hop. The interference action is maximized considering that the jammer is able to adjust the interfered fraction of the band, such that a maximum degradation is achieved. The worst case jamming obtained with this strategy, has proven to be an effective interference for FH/MFSK, because for large signal to noise ratio, the jammer is able to concentrate its power within a small region, thus when the proper subchannel is interfered, a symbol error is more likely to occur. Time diversity is an effective countermeasure against this type of jamming. Using time diversity, several active cells or chips are transmitted on each symbol interval, and hopped randomly over separated subchannels, selected according to a pseudorandom pattern. Therefore, although the signal to noise ratio per subchannel is reduced, the probability of interference decreases, because the jammer must interfere more than one subchannel to produce a symbol error. Also, the use of coding is effective on this type of interference, since an equivalent diversity is obtained with the Hamming distance between codewords, when block codes are used, or the free distance, for convolutional codes.

In this paper, the performance of a Non-Orthogonal FH/MFSK System [4,5] is presented for the interference channel described before. In this type of system, the modulator divides its power into several tones, transmitted over separated subchannels. The selection of the subchannels used by the system, on each symbol interval, is based on a non-orthogonal combinatorial construction called t-designs. The strategy assumed for the jammer is also to divide its interference into several subchannels, however only the transmitter and the receiver know the pseudorandom pattern used to hop the multiple tones. It is shown that non-orthogonal FH/MFSK Systems based on this type of designs, achieve comparable performance to FH/MFSK with time diversity (or Fast FH/MFSK), Also, it is shown that a smaller signaling frame length is required to represent a given alphabet size, therefore the transmission rate can be increased, maintaining the same system bandwidth. Additionally, time diversity can be used on this scheme, to obtain a Fast FH/MFSK scheme with non-orthogonal signaling. Using this additional diversity, it is shown that a lower number of time repetitions per symbol, are required to obtain comparable performance to a conventional Fast FH/MFSK System. In section 2, Non-Orthogonal FH/MFSK Systems are described, along with its performance in broadband gaussian noise. Its performance on a partial band gaussian noise interference channel is presented in section 3. Finally in section 4, the transmission rate improvement is shown. A brief description of t-designs is given in the Appendix, along with its application to this particular FH/MFSK System.

2. Non-Orthogonal FH/MFSK Systems. Let $S(w,v)$ be a t-design (Appendix), used to select a signaling block of v elements, and w active cells ("ones") per block. If k is the number of input bits, grouped to form a symbol, then at least $M=2^k$ different blocks on $S(w,v)$ are required to represent all possible input symbols. As is shown in the Appendix, a total of $b=v(v-1)/w(w-1)$ blocks are obtained with a t-design with parameters (w,v), therefore v must be selected such that $M\leq b$. A Non-

FH/MFSK System based on a t-design is shown in Figures (1.a) and (1.b). The transmitter selects a block of w active elements from $S(w,v)$, according to the input data bits, and divides its energy equally into the signals corresponding to the elements forming the selected block. These signals are then transmitted over separated subchannels, chosen by independent pseudorandom patterns. The receiver forms a decision variable for each of the b blocks on $S(w,v)$. Each decision variable is then obtained combining the non-coherently detected outputs of the matched filters corresponding to the w active cells forming each signaling block. Finally, a decision is made to the largest decision variable formed in this manner.

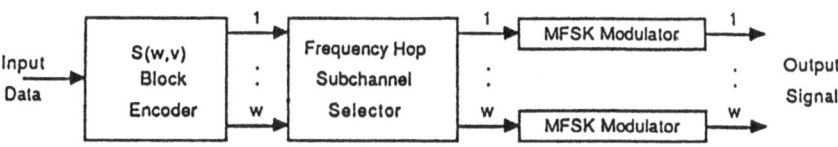

Figure (1.a). Non-Orthogonal FH/MFSK Transmitter.

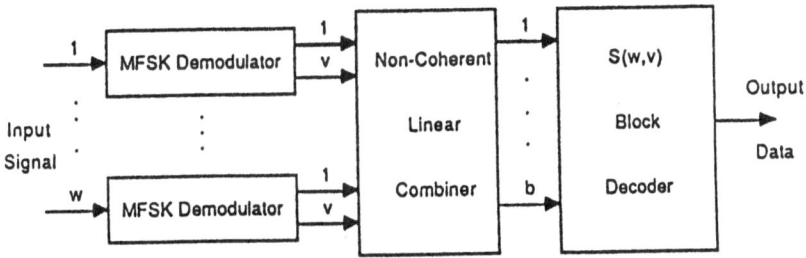

Figure (1.b). Non-Orthogonal FH/MFSK Receiver.

Let $B_j = (b_{j1}, b_{j2}, \ldots, b_{jw})$, for $j=1,2,\ldots,b$, represent the w-tuples or blocks obtained from a $S(w,v)$ t-design. The signaling blocks, formed by v elements, are obtained assigning an active cell (a "one"), on the positions corresponding to the elements on the selected w-tuple. Therefore, the signaling block corresponding to the block B_j, is represented by $X_j = (x_{j1}, x_{j2}, \ldots, x_{jv})$, where $x_{ji}=1$ if the i-th cell is an element on B_j, otherwise $x_{ji}=0$. The decoder, as was mentioned before, is based on the linear combining of the non-coherent matched filter outputs (r_{jk}), corresponding to the active cells (b_{jk}) forming each block B_j. Therefore, each decision variable is obtained as:

$$U_j = \sum_{k=0}^{w} |r_{jk}|^2 \quad \text{for } j = 1,2,...,b \tag{1}$$

Let B_i be the transmitted block, with U_i its corresponding decision variable, and consider another block B_j in $S(w,v)$, with a corresponding decision variable U_j, as it is shown in the Appendix, two cases are observed because of the intersection properties between blocks on $S(w,v)$:

    1) $B_j \in X_1(B_i)$, $(B_j$ has only one element is common to $B_i)$.
    2) $B_j \in X_0(B_i)$, $(B_j$ has no element in common to $B_i)$.

The sizes of these subsets are given respectively by $x_1 = w(v-w)/(w-1)$, and $x_0 = b - x_1 - 1$ (Appendix). If U_j^0 and U_j^1 are the corresponding decision variables for each case, then for a broadband gaussian channel, the probability of a symbol error can be upper bounded by:

$$P_e(\gamma_b, w, M) \leq x_0 P_2(U_j^0 > U_i) + x_1 P_2(U_j^1 > U_i) \tag{2}$$

The first term on equation (2), implies that no element is common to both decision variables, and then the decision error between the two variables, $P_2(U_j^0 > U_i)$, can be obtained by:

$$P_2(E_0 < 0) = P(U_i - U_j^0 < 0) \tag{3}$$

where a random variable E_0, is defined as $E_0 = \sum_{k=0}^{w}(|r_{ik}|^2 - |r_{jk}|^2)$. The second term on equation (2), implies that only one element is common to both decision variables. If $b_{iw} = b_{jw}$ is the intersecting element, then the decision error $P_2(U_j^1 > U_i)$ can be obtained as follows:

$$P_2(E_1 < 0) = P(U_i - U_j^1 < 0) \tag{4}$$

where the random variable E_1, is defined as $E_1 = \sum_{k=0}^{w-1}(|r_{ik}|^2 - |r_{jk}|^2)$. It is observed that this error probabilities are equivalent to the case of a MFSK System with diversity $L = w$, for $P_2(E_0 < 0)$, and $L = w-1$, for $P_2(E_1 < 0)$, then defining a signal to ratio per subchannel, $\gamma_c = k\gamma_b/w$, the probability of a decision error can be obtained as follows [6]:

$$P_2(E_0 < 0) = \left(\frac{1}{2}\right)^{2w-1} e^{-w\gamma_c/2} \sum_{n=0}^{w-1} \frac{1}{n!} \sum_{r=0}^{w-1-n} \binom{2w-1}{r} \left[\frac{w\gamma_c}{2}\right]^n \qquad (5.a)$$

and:

$$P_2(E_1 < 0) = \left(\frac{1}{2}\right)^{2w-3} e^{-(w-1)\gamma_c/2} \sum_{n=0}^{w-2} \frac{1}{n!} \sum_{r=0}^{w-2-n} \binom{2w-3}{r} \left[\frac{(w-1)\gamma_c}{2}\right]^n \qquad (5.b)$$

In Figure (2), results are shown for $k=5$, comparing the performance of a FH/MFSK System based on a $S(w,v)$ t-design, against a conventional FH/MFSK System with time diversity L, referred here as $O_L(k)$. A performance degradation close to 2 dB is obtained with this scheme in a broadband gaussian noise channel, however as will be shown in the next section, a performance improvement comparable to the use of time diversity is obtained for a partial band gaussian noise interference channel.

3. **Partial Band Gaussian Interference.** In this section the performance of a Non-Orthogonal FH/MFSK System is presented for an interference channel with partial band gaussian noise. Let a $S(w,v)$ t-design be used to obtain the w active elements or cells, used by the modulator. These cells are then frequency hopped over the same number of separated subchannels, the jammer also divides its energy over several subchannels. According to this model, each subchannel is interfered with probability ρ, while a noise free transmission occurs on each subchannel with probability $(1-\rho)$. Then, the probability of a simultaneous interference on j distinct subchannels is given by:

$$P_j(\rho) = \rho^j \qquad (6)$$

It is assumed that the receiver knows with certainty whether a subchannel is interfered or not. This side information can be implemented by declaring an interfered subchannel when more than one matched filter output is high. If an interference is detected, the corresponding subchannel is rejected as an element on the corresponding decision variable. A no error decision can be taken when the number of interfered subchannels is less than one half of the minimum Hamming distance between any two signaling blocks on the set. When an errorless decision is not possible, the non-coherent output of the matched filters, corresponding to the elements forming the blocks, are linearly combined, and then the largest is selected. Therefore, the symbol error probability is given by:

$$P_e(\rho\gamma_b, w, M) = \sum_{j=\left\lceil d_{min}/2\right\rceil}^{w} P_j(\rho)\, P_e(\rho\gamma_b, w, M/j) \tag{7}$$

where $P_e(\rho\gamma_b, w, M/j)$ is the symbol error probability, when j subchannel are interfered. For a $S(w,v)$ t-design, it is shown in the Appendix that two blocks intersect at most on a single element, then its minimum Hamming distance is $2(w-1)$. If $w-1$ cells are jammed, then any block in $X_1(B_i)$ (besides B_i), constitutes a possible block to be decided by the receiver, while if all w cells are jammed, any block in $S(w,v)$ is a possible block. In both cases the receiver selects the largest decision variables formed with the metric described before. Therefore, the symbol error probabilities, when $w-1$ and w cells are interfered, are upper bounded by:

$$P_e(\rho, \gamma_b, w, M/w) \le x_0\, P_2(U_j{}^0 > U_i) + x_1\, P_2(U_j{}^1 > U_i) \tag{8.a}$$

$$P_e(\rho, \gamma_b, w, M/w-1) \le x_1\, P_2(U_j{}^1 > U_i) \tag{8.b}$$

where the decision error probabilities $P(U_j{}^0 > U_i)$ and $P(U_j{}^1 > U_i)$, are computed using equations (5.a), and (5.b), with a signal to noise ratio per subchannel degraded by a factor ρ. The worst case partial band interference probability of a symbol error $P_e(\gamma_b, w, M)$, is finally obtained maximizing $P_e(\rho, \gamma_b, w, M)$ over all possible values of ρ, as follows:

$$P_e(\gamma_b, w, M) = \max_{0 \le \rho \le 1} P_e(\rho, \gamma_b, w, M) \tag{9}$$

An approximate closed expression can be obtained for this equation, considering that the major contribution to the symbol error probability is obtained from the non-orthogonal blocks (set X_1), then:

$$P_e(\rho\gamma_b, w, M) \le x_1\, \rho^{w-1} P_2(U_j{}^1 > U_i) \tag{10}$$

Using a Chernoff bound, the decision error probability, can be obtained minimizing the following expression [6]:

$$P_e(\gamma_b, w, M) \le x_1 \max_{0 \le \rho \le 1} \min_{0 \le v \le 1} \left[\frac{\rho}{1-4v^2} \exp\left(-\frac{2\rho v k \gamma_b}{w(1+2v)}\right) \right]^{w-1} \tag{11.a}$$

The following approximate bound is obtained, for $v=1/4$, and $\rho = 3w/k\gamma_b$:

$$P_e(\gamma_b, w, M) \le x_1 \left(\frac{4w}{ek\gamma_b}\right)^{w-1} \tag{11.b}$$

Results for $k=5$, are shown in Figure (3), compared to a conventional FH/MFSK System with time diversity L. A comparable performance is obtained on both systems when $L=w-1$, this result is directly related to the minimum Hamming distance of the signaling set used here. It is shown in the Appendix that this distance for a t-design is $2(w-1)$, therefore an equivalent time diversity $L=\left\lfloor d_{min}/2 \right\rfloor=w-1$, is obtained with this scheme.

Time diversity can also be incorporated to this frequency hop scheme, dividing each of the w transmitted cells into L chips, and then transmitting these Lw chips over separated subchannels. The symbol error probability here, is given by:

$$P_e(\gamma_b,w,M) \leq x_1 \left(\frac{4Lw}{ek\gamma_b}\right)^{L(w-1)}$$ (12.a)

that can be minimized, considering an optimum diversity factor (L_{opt}):

$$P_e(\gamma_b,w,M) \leq x_1 \exp\left[-\frac{(w-1)\gamma_b k}{4w}\right] \text{ for } L_{opt}=\frac{k\gamma_b}{4w}$$ (12.b)

In Table 1, the signal to noise ratio per bit required to obtain a symbol error probability of $P_e=10^{-5}$, for a non-orthogonal FH/MFSK system with optimum time diversity, referred as $S_L(w,v)$, is compared to a conventional FH/MFSK System also with optimum time diversity, referred as $O_L(k)$. A performance degradation close to 1 dB is obtained with this scheme, however the number of time repetitions required to achieve this optimum diversity is smaller, and therefore an improvement on the transmission rate is obtained, this fact will be presented on the next section.

It is also interesting to analyze the asymptotic performance of this non-orthogonal scheme, on a partial band gaussian noise channel, for a large alphabet size. Similar work has been done in [7], for a conventional FH/MFSK System, showing that the symbol error probability tends slowly to the limit $\ln 2/\gamma_b$, as $M\rightarrow\infty$, for $\gamma_b>\ln 2$. Here, the implicit diversity given to the system by the use of multiple tones modifies this limit, the exact expression for the symbol error probability has been numerically calculated, for large alphabet sizes. Table 2 shows the results obtained for each value of w, as $M\rightarrow\infty$:

Table 1. Signal to noise ratio per bit (in dB), required to achieve a symbol error probability $P_e=10^{-5}$, with optimum time diversity (shown in parenthesis).

k	$O_L(k)$	$S_L(3,v)$	$S_L(4,v)$	$S_L(5,v)$
5	10.78(15)	12.38(7)	12.00(5)	11.85(4)
6	10.19(16)	11.71(7)	11.33(5)	11.06(4)
7	9.71(16)	11.16(8)	10.69(5)	10.51(4)
8	9.31(17)	10.68(8)	10.24(5)	10.02(4)
9	8.97(18)	10.26(8)	9.82(5)	9.58(4)
10	8.68(18)	9.90(8)	9.44(5)	9.21(4)

Table 2. γ_b (in dB), at $P_e=10^{-5}$, when $M \to \infty$.

k	$S_L(3,v)$	$S_L(4,v)$	$S_L(5,v)$
5	24.03	17.60	14.19
10	24.23	17.88	14.82
15	24.48	18.01	15.02
20	24.68	18.20	15.18
25	24.88	18.39	15.48
30	24.92	18.43	15.85

According to table 2, there is evidence that the signal to noise ratio per bit approaches a constant limit, thus the symbol error probability will approach a nonzero limit, for large values of the alphabet size. As was mentioned before, similar result was suggested and proved by Hegde and Stark in [7], for a conventional FH/MFSK System on a partial band gaussian noise channel.

4. Transmission Rate Improvement. It is known that on a conventional MFSK System, orthogonality between signals is obtained for a frequency separation of $\Delta f=1/T$, when a signaling interval T is used. Consequently, the transmission rate obtained at a bandwidth $W=M\Delta f$, and a time diversity L_O, is given by:

$$R_O = W \frac{k}{ML_O} \tag{13}$$

For a t-design FH/MFSK System, w subchannels are used, however only v orthogonal frequencies are required on the signaling frame, to represent the same alphabet size. If a time diversity L_S is used, then the transmission rate, for the same bandwidth W, is given by:

$$R_S = W \frac{k}{vwL_S} \tag{14}$$

The signaling frame length required by a t-design to obtain $M=2^k$ blocks, is given by v, such that $b=v(v-1)/w(w-1)\geq M$, therefore it increases proportional to $\sqrt{Mw(w-1)}$, while for a conventional MFSK the signaling frame is equal to M. If an optimum time diversity is used on each scheme, then the improvement factor $r=R_S/R_O$, can be obtained:

$$r = \frac{ML_O}{vwL_S} \approx \frac{\gamma_{bO}}{\gamma_{bS}} \sqrt{M/w(w-1)} \tag{15}$$

where γ_{bO}, and γ_{bS}, are the signal to noise ratio per bit, required to obtain a given symbol error probability, on each case. If ϵ is this constraint, then using equation (12.b) this relation can be written approximately as:

$$r \approx \left(\frac{w-1}{w}\right) \frac{\ln(M/\epsilon)}{\ln(x1/\epsilon)} \sqrt{M/w(w-1)} \tag{16}$$

Table 3 shows, the transmission rate improvement factors, for a performance constraint $P_e=10^{-5}$, using optimum time diversity. Transmission rate improvements are achieved with this non-orthogonal scheme for $k\geq 6$, i.e. an improvement factor close to three is obtained for $S(3,v)$, and $k=6$, and increasing for a higher alphabet size. It is also noted that the best rate improvement is achieved for the lowest value of w, on the other hand the performance improvement on a partial band gaussian noise interference channel increases for a higher value of w.

Table 3. Transmission Rate Improvement Factors ($r = R_S/R_O$).

k	$S_L(3,v)$	$S_L(4,v)$	$S_L(5,v)$
5	1.48	0.97	0.61
6	2.15	1.33	1.28
7	2.95	2.55	1.74
8	4.34	3.39	2.68
9	6.68	4.95	4.24
10	9.55	7.67	6.25

5. Conclusions. In this paper we have presented the performance improvements of a FH/MFSK with a non-orthogonal signaling frame based on a combinatorial construction known as t-designs. The intersection and distance properties of the blocks obtained by this design, are such that, on a partial band gaussian noise interference channel, a comparable performance improvement to the use of time diversity, is obtained. Time diversity can be used on this non-orthogonal scheme, in that case a lower number of time repetitions are required to obtain the optimum performance. also, since the signaling frame is reduced, an improvement on the transmission rate is achieved, for alphabets higher than $M=32$, depending on the number of subchannel used. Finally, the asymptotic performance, for a large alphabet size, has been briefly described, and compared to the results obtained in [7] for a conventional FH/MFSK System, showing evidence that a nonzero performance limit is also achieved with this non-orthogonal scheme.

Appendix: t-designs A t-design [8,9,10] is a collection of b blocks, formed by the arrangement of v distinct elements, satisfying the following conditions:

i)   Each block contains w elements.

ii)   Each element occurs in r blocks.

iii)   Each subset of t elements occurs together in λ blocks.

This arrangement of elements, is also called a Balanced Incomplete Block Design (BIBD) with parameters (v, b, r, w, λ, t). The signaling set used here is taken from the set of blocks generated by a t-design with $\lambda=1$, and $t=2$, then only v, b, w and r, are to be specified. There are two basic relations that the design parameters involved on a t-design (with $\lambda=1$, and $t=2$), must satisfy:

$$bw = vr \qquad \text{(A.1.a)}$$
$$(v-1) = r(w-1) \qquad \text{(A.1.b)}$$

The first relation counts the total number of single element occurrences: b blocks of w elements, or v elements on r different blocks. The second relation counts the number of pairs containing a particular element: $v-1$ possible pairs, or $w-1$ possible pairs on a given block, repeated on r different blocks. Therefore, only two parameters are independents (w and v), and a family of code sets can be generated according to the selected values of w and v. As can be implied by these relations, it is observed that the number of codewords obtained by these designs is given by:

$$b = \frac{v(v-1)}{w(w-1)} \qquad \text{(A.2)}$$

In this paper, results will be presented for t-designs, with $w=3$, 4, and 5, the construction obtained in that case, known as a Steiner t-design, and referred here as $S(w,v)$,. To obtain integral solutions from equations (A.2.a) and (A.2.b), v must satisfy the following conditions:

$$v \equiv 1, \quad \text{or } v \equiv w \quad \text{mod } w(w-1) \qquad \text{(A.3.a)}$$
$$v > w(w-1) \qquad \text{(A.3.b)}$$

Since $t=2$, and $\lambda=1$, it is derived from the initial conditions of a t-design, that for $S(w,v)$ any pair of elements is contained in only one block, and then two different blocks can intersect at most on a single element. Then, given a block $B_i \epsilon S(w,v)$, the collection of blocks $S(w,v)$ can be partitioned into 2 disjoint subsets:

1) A subset $X_1(B_i)$, containing only non-orthogonal codewords to B_i (intersecting on one element, on the same position).

2) A subset $X_0(B_i)$, containing only orthogonal codewords to B_i (completely non-intersecting).

It can be easily proved that the sizes of the subsets $X_1(B_i)$ and $X_0(B_i)$, called intersection numbers, are given by:

$$x_1 = |X_1(B_i)| = w(r-1) = \frac{w}{w-1}(v-w) \qquad \text{(A.4.a)}$$

$$x_0 = |X_0(B_i)| = b-x_1-1 \qquad \text{(A.4.b)}$$

The first equation counts the total number of block intersections, associated to each of the w elements of B_i (excepting B_i), while the second counts the remaining blocks on $S(w,v)$. A summary of the parameters of a Steiner t-design are given in Table A.1, compared to a conventional MFSK, referred as $O(k)$.

Table A.1. Parameters of a Steiner BIB design compared to the parameters of a conventional MFSK System.

Parameter	$S(w,v)$	$O(k)$
Number of Elements	v	$n=2^k$
Number of Blocks	$b = \dfrac{v(v-1)}{w(w-1)} \geq 2^k$	$n=2^k$
Number of Bits	k	k
Minimum Distance	$2(w-1)$	2
Block Weight	w	1

Acknowledgements. This work has been supported by the Integrated, Information and Telecommunications Systems Center (IITSC), Illinois Institute of Technology, Chicago, Illinois, 60616.

REFERENCES

[1] Simon, M.K., Omura, J.K., Scholtz, R.A., and Levitt, B.K., *Spread Spectrum Communications*, Rockville MD, Computer Science Press, 1985.

[2] Houston, S.W., "Modulation Techniques for Communication, Part 1: Tone and Noise Jamming Performance of Spread Spectrum M-ary FSK and 2, 4-ary DPSK Waveforms," *National Aeronautics Conference Record*, pp.51-56, 1975.

[3] Levitt, B.K., "Strategies for FH/MFSK Signaling with Diversity in Worst-Case Partial Band Noise," *IEEE Trans. on Sel. Areas of Comm.*, vol. SAC-3, pp.622-626, 1985.

[4] Atkin, G.E., and Blake I.F., "Performance of Multitone FFH/MFSK Systems in the Presence of Jamming," *IEEE Trans. on Information Theory,*", to be published.

[5] Atkin, G.E., and Corrales, H.P., "Performance Analysis of a Multiple Tone FH/MFSK Spread Spectrum System," Conference Proceeding of the *IEEE Technical Conference on Tactical Communications*, Fort Wayne, Indiana, May 1988.

[6] Proakis, J.G., *Digital Communications*, New York, Mc-Graw Hill, 1983.

[7] Hegde, M.V., and Stark W.E., "Asymptotic Performance of *M*-ary Orthogonal Signals in Worst Case Partial-Band Interference and Rayleigh Fading," *IEEE Trans. on Communications*, vol. COM-36, pp.989-992, 1988.

[8] Hall, M.Jr., *Combinatorial Theory*, Blaisdell, Waltham, Mass, 1967

[9] Ryser, H.J., *Combinatorial Mathematics*, Mathematical Assoc. of America, Monograph no.14, J. Wiley and Sons, 1963.

[10] Blake, I.F., and Mullin R.C., *The Mathematical Theory of Coding*, Academic Press, New York, 1975.

14

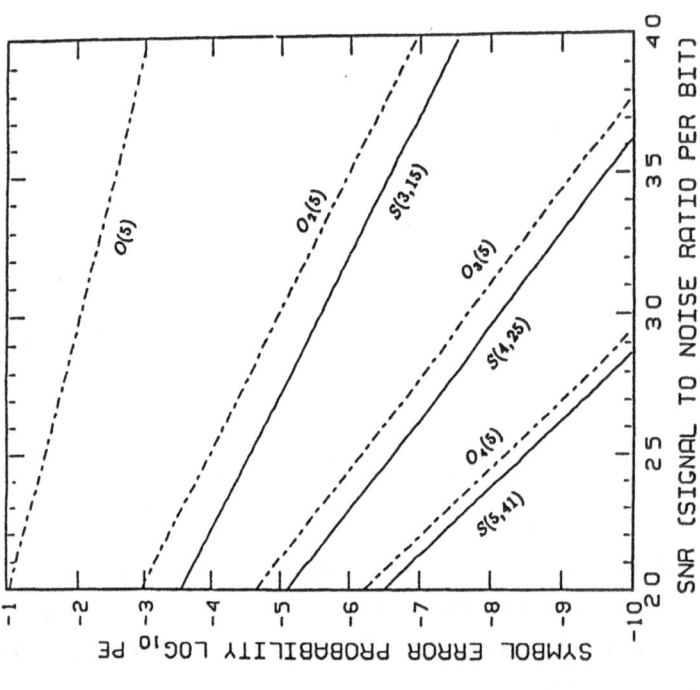

Figure (3). Non-Orthogonal FH/MFSK Performance on a Partial Band Gaussian Noise Interference Channel, compared to an Orthogonal FH/MFSK System with Time Diversity L ($k=5$).

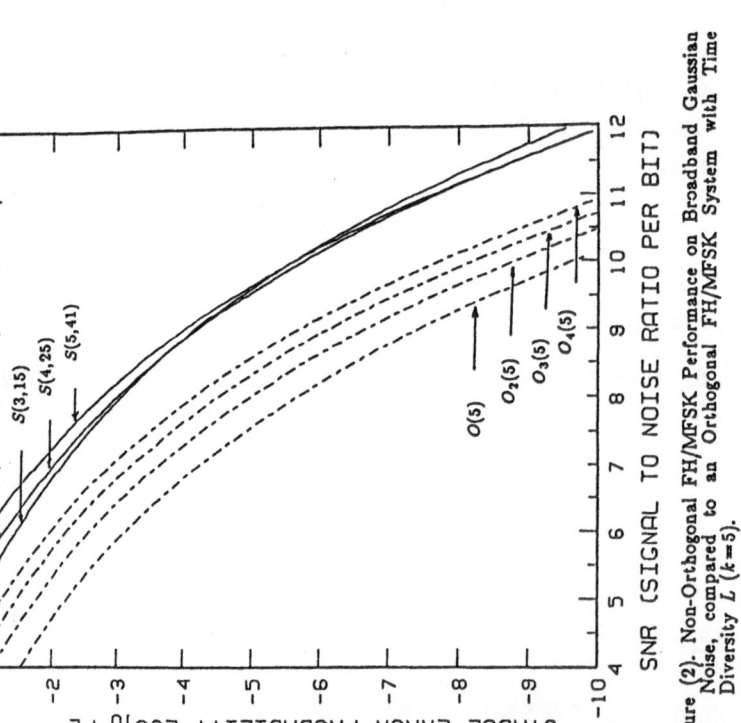

Figure (2). Non-Orthogonal FH/MFSK Performance on Broadband Gaussian Noise, compared to an Orthogonal FH/MFSK System with Time Diversity L ($k=5$).

Probability of Error Comparison of
Linear and Iterative Multiuser Detectors

Mahesh K. Varanasi and Behnaam Aazhang

Rice University

1. Introduction

In a Code-Division Multiple-Access (CDMA) system, several users simultaneously transmit information over a common channel using preassigned code waveforms. The receiver is equipped with a knowledge of the codes of some or all of the users. It is then required to demodulate the information symbol sequences of these users, upon reception of the sum of the transmitted signals of all the users in the presence of additive noise. Examples of such situations arise in a variety of communication systems such as satellite communications, radio networks and other multipoint-to-multipoint multiple-access networks.

Different approaches to the multiuser demodulation problem have been considered so far. The conventional approach consists of demodulating each signal using the corresponding single-user detector, thereby ignoring the multiple-access interference, or equivalently, ignoring the cross-correlations between the modulating signals of different users. While this approach has the virtue of being simple to implement, it has two major shortcomings. Firstly, since low cross-correlations between a given number of signals can be achieved only at the expense of an increased bandwidth, it is not surprising that reliable performance from the conventional detector has been possible only for low bandwidth efficiencies. Secondly, since error probability degrades exponentially as the interfering signal strengths increase, this detection scheme is highly vulnerable to near-far effects.

On the other hand, the study of the optimum demodulator has shown that while significantly superior performance over the conventional detector is possible, it can be obtained only at a marked increase in computational complexity which is exponential in the number of users. Since a CDMA system could potentially have a large number of users, this solution may prove to be too expensive and hence impractical to implement. Furthermore, most CDMA systems with a current need for only a small number of users may be required to have an expansion capability without needing a steep increase in computational requirements.

The primary objective of this paper is to address the need for low-complexity detectors which perform reliably in high bandwidth efficiency situations and are robust to near-far effects. Attention is focused on the symbol-synchronous CDMA system. Although this system is a special case of the important class of asynchronous CDMA systems, we shall study it for the following reasons. Firstly, this system does indeed find application in slotted channels. Secondly, the iterative detector that we will propose and analyze here, generalizes to the asynchronous situation in a natural way [4]. Therefore the synchronous problem can be construed as providing us with a simple setting to better understand the issues involved in iterative multiuser demodulation. Finally, the performance of this detection scheme is indicative of the performance of its asynchronous counterpart. It should be pointed out here that the synchronous multiuser demodulation has been previously considered by Schneider in [3] who proposed a certain linear decorrelating detector, the primary feature of which can be considered as its performance invariance to the signal energies of the interfering users. More recently,

This work was supported by the National Science Foundation under Grant NCR-8710844.

Lupas and Verdu in [1] recognized the exponential complexity of the optimum solution and proposed the *optimal linear* detector which requires a linear real-time computational complexity. However, restricting the decision statistic to be a linear function of the sufficient statistics proves to be a rather stringent requirement and hinders the achievement of near-optimum performance in a number of situations, corresponding particularly to the demodulation of relatively weak signals. A new multiuser detection strategy which is based on an iterative multiple-access interference rejection scheme is studied. Unlike the linear detectors, the decision statistics involved in this demodulation scheme are nonlinear functions of the sufficient statistics. However, the computational complexity is linear in the number of users, thereby retaining the computational advantage that the linear detectors have to offer.

2. The Iterative Multiuser Detector

Let us suppose that there are K transmitting users in the system. Assume that the modulating signal of the kth user is denoted as $s_k(t)$, which is then antipodally modulated to transmit a sequence of information bits. Let $s_k(t)$ be time-limited to a bit duration ($t \in [0,T]$) with finite energy denoted as E_k. In a synchronous CDMA system, these users maintain cooperation amongst themselves to transmit information bit sequences in time synchronism so that the signal at the receiver, denoted $r(t)$, can be written as

$$r(t) = \sum_i \sum_{k=1}^{K} b_k^{(i)} s_k(t - iT) + n_t, \tag{2.1}$$

where $b_k^{(i)}$ denotes the kth user's bit in the ith time interval and n_t represents the additive white Gaussian noise with a spectral density with height σ^2. Assuming that all possible information sequences are independent and equally likely, and defining $b^{(i)} = [b_1^{(i)}, b_2^{(i)}, \cdots, b_K^{(i)}]^T$, it is easy to see that an optimum decision on $b^{(i)}$ is a one-shot decision in that it requires the observation of the received signal only in the ith time interval. Without loss of generality, we will therefore focus attention on $i = 0$ and drop the time super-script and consider the demodulation of the vector of bits $b = [b_1, b_2, \cdots, b_K]^T$ with the observation of the received signal in the time-interval $[0, T]$.

The optimum or the maximum likelihood decision on b is chosen as $\hat{b}^* = [\hat{b}_1^*, \hat{b}_2^*, \cdots, \hat{b}_K^*]^T$ which maximizes the log-likelihood function so that

$$\hat{b}^* = \arg \left\{ \max_{b \in \{-1, +1\}^K} \left[2 \int_0^T r(t) s(t, b) \, dt - \int_0^T s^2(t, b) \, dt \right] \right\}, \tag{2.2}$$

where $s(t, b) = \sum_{k=1}^{K} b_k s_k(t)$. The optimum decision can also be written as

$$\hat{b}^* = \arg \left\{ \max_{b \in \{-1, +1\}^K} \left[2 y^T b - b^T H b \right] \right\}. \tag{2.3}$$

where $y = [y_1, y_2, \cdots, y_K]^T$ is the vector of sufficient statistics with

$$y_k = \int_0^T r(t) s_k(t) \, dt, \quad \text{for } k = 1, 2, \cdots, K. \tag{2.4}$$

Further, the matrix of cross-correlations H is such that[1]

---

[1] For the rest of the paper, we will denote the $(i, j)^{th}$ element of a matrix A with the corresponding lower-case symbols a_{ij}.

$$h_{kl} = \int_0^T s_k(t) s_l(t) \, dt, \quad \text{for } k, l = 1, 2, \cdots, K. \tag{2.5}$$

The maximization in solving (2.3) has been shown to be NP-complete [1]. That is, no algorithm which solves (2.3) in polynomial time in K is known. While the exponential complexity of (2.3) may be acceptable in CDMA systems with a small number of users, it may be prohibitive in large CDMA systems with the number of users in excess of 10 or 20, depending on available computing power and the type of coherent modulation employed.

We will now propose and study a low-complexity sub-optimum approach to the maximization in (2.3). In particular, an iterative solution is considered. Let the estimate of b in the mth stage of the iteration be denoted as $\hat{b}(m) = [\hat{b}_1(m), \hat{b}_2(m), \cdots, \hat{b}_K(m)]^T$. For $m \geq 1$, consider the $(m+1)^{st}$ stage estimate of the kth user's information bit b_k as being

$$\hat{b}_k(m+1) = \arg\left\{ \max_{\substack{b_k \in \{+1, -1\} \\ b_l = \hat{b}_l(m), \forall l \neq k}} \left[2 y^T b - b^T H b \right] \right\}. \tag{2.6}$$

It is easily shown that

$$\hat{b}_k(m+1) = \text{sgn}[z_k(m)], \tag{2.7}$$

where $z_k(m)$ is the mth stage statistic for the kth user given as

$$z_k(m) = y_k - \sum_{j \neq k} \hat{b}_j(m) h_{jk}. \tag{2.8}$$

In demodulating the information bits of all the users, the maximization of (2.6) is performed for each $k = 1, 2, \cdots, K$. The $(m+1)^{st}$ stage estimate of b can then be written as the sign of the mth stage vector decision statistic $z(m) = [z_1(m), z_2(m), \cdots, z_K(m)]^T$ so that

$$\hat{b}(m+1) = \text{sgn}[z(m)] = \text{sgn}[y - (H-E)\hat{b}(m)], \tag{2.9}$$

where the diagonal matrix E has the energies of the modulating signals for its diagonal elements so that $diag\{E\} = \{E_1, E_2, \cdots, E_K\}$. Noting from the definition of the sufficient statistic in (2.4), it is easily shown that

$$y = H b + \eta = E b + I(b) + \eta, \tag{2.10}$$

where η is a zero-mean Gaussian noise vector with a covariance matrix $\sigma^2 H$ and $I(b) \triangleq (H-E)b$ represents the multiple-access interference vector. Substituting (2.10) in (2.9), the expression for the $(m+1)^{st}$ stage estimate of b is given as

$$\hat{b}(m+1) = \text{sgn}[z(m)] = \text{sgn}\left[Eb + I(b) - I(\hat{b}(m)) + \eta \right]. \tag{2.11}$$

The result in equation (2.11) has a simple interpretation. The $(m+1)^{st}$ stage estimate of b is obtained as the sign of the mth stage statistics which in turn is obtained by *subtracting* from the sufficient statistic y, the *estimate* of the multiple-access interference based on the mth stage estimate of b.

Having described the iterative solution to the multiuser demodulation problem, we now consider the choice of an initial condition. In this paper, we will consider the initial one-stage detector as belonging to the class of linear detectors. Noting from (2.3) that y constitutes the vector of sufficient statistics, a linear detector denoted as L is such that

$$\hat{b}(1) = \text{sgn}[\mathbf{L}\mathbf{y}] = \text{sgn}[\mathbf{z}(0)]. \tag{2.12}$$

Specifically, the conventional, decorrelating and the optimum linear detectors belong to this class. These detectors have been derived and extensively studied in [1] in terms of their asymptotic efficiency and near-far resistance. The conventional approach to the demodulation in multiuser channels relies entirely on a well-designed signal constellation by *ignoring* the multiple-access interference. Under such an assumption the optimum detector corresponds to the transformation $\mathbf{L}=\mathbf{I}$, the identity matrix. It is by now well-known that this conventional detector exhibits a severe (exponential) performance degradation in its error probability performance in either one or both of the following situations of practical importance. The first corresponds to an efficient utilization of the available bandwidth so that low levels of multiple-access interference are no longer possible. The second situation corresponds to near-far environments wherein the user energies can be markedly dissimilar. On the other hand, when the modulating signals of all the users are linearly independent,[2]— a mild restriction that we will require of the signals for the rest of this paper—a decorrelating detector corresponding to $\mathbf{L} = \mathbf{H}^{-1}$—which will exist by assumption—was first proposed by Schneider in [3] as a natural strategy to eliminate the multiple-access interference from the vector of sufficient statistics given in (2.10) so that

$$\mathbf{z}(0) = \mathbf{H}^{-1}\mathbf{y} = \mathbf{b} + \mathbf{H}^{-1}\boldsymbol{\eta} \tag{2.13}$$

A treatment of the linear and the iterative detectors when the linear independence assumption is relaxed can be found in [1] and [5], respectively. It was demonstrated in [1] that the decorrelating detector corresponds to the maximum likelihood detector when the energies are not known to the receiver. From an exponentially degrading performance to increasing interfering signal strengths of the conventional detector, a significant improvement was obtained by the linear-complexity sub-optimum decorrelating detector wherein the error probability of any user remains invariant to the interfering signal strengths. However, being a *maximin* detector, the decorrelator is vulnerable to being too conservative a solution in environments where the signal energies remain fixed for a packet duration or vary slowly enough to lend themselves to easy tracking.

In the following section, the probability of bit-error for the two-stage multiuser detector is derived when the decorrelator provides the initial estimate.

3. Probability of Error Analysis

The use of a CDMA system for multiple-access communications is motivated in part by its ability to reject disturbances due to multipath and jamming—intentional or otherwise. After decoding, these disturbances are effectively of a broad-band nature in a system with well-designed code waveforms. Therefore, even in low ambient noise, the performance of CDMA receivers is of interest for high as well as low signal-to-noise ratios, depending on the levels of the external disturbances. In this sense, the probability of bit-error is a more detailed performance measure than asymptotic efficiency.

3.1. Linear Detector

Let us consider the bit error probability of the one-stage linear detector L. Without loss of generality, consider the error probability of the first user. Denoting it as $P_1^{(1)}(\mathbf{L})$, it is clear that for $\mathbf{F} \triangleq \mathbf{L}\mathbf{H}$,

$$P_1^{(1)}(\mathbf{L}) = P[\hat{b}_1(1) = 1 \mid b_1 = -1] = P[(\mathbf{F}\mathbf{b}+\boldsymbol{\gamma})_1 > 0 \mid b_1 = -1],$$

---

[2] No signal belongs to the subspace spanned by the other signals.

where the second equality follows from (2.10) and (2.12) and γ is a zero-mean Gaussian random vector defined as $\gamma \triangleq L\eta$ with covariance $\sigma^2 G$ where $G \triangleq LHL^T$. Let $\beta_1 = [b_2, b_3, \cdots, b_K]^T$ be the $(K-1)$-dimensional column vector that represents the bits that interfere with b_1. Denoting the expectation over the ensemble of independent, uniformly distributed $\beta_1 \in \{-1, +1\}^{(K-1)}$ as E_{β_1}, we have

$$P_1^{(1)}(L) = E_{\beta_1} \left[P(\gamma_1 > f_{11} - \sum_{j=2}^{K} f_{1j} b_j) \right]$$

$$= 2^{1-K} \sum_{\beta_1 \in \{-1, +1\}^{K-1}} Q\left[\frac{f_{11} - \sum_{j=2}^{K} f_{1j} b_j}{\sqrt{\sigma^2 g_{11}}} \right],$$

(3.14)

where the last equality follows from the fact that γ_1 is a zero-mean Gaussian random variable with variance $\sigma^2 g_{11}$.

3.2. Two-Stage Detector

An exact expression for the error probability of the iterative detector which uses two stages in the iteration—which will be referred to as the two-stage detector—for a K-user synchronous CDMA system with an arbitrary linear detector L serving as an initial condition, has been derived in [5]. In this paper we will focus attention on the two-stage detector with the decorrelator as an initial condition. This choice is motivated by the relative simplicity of error probability analysis that results and the fact that the decorrelating detector provides each user with a certain guaranteed error probability which decreases exponentially as the desired signal strength increases and which is invariant of the interfering signal energies.

Again, without loss of generality, consider the demodulation of the first user's bit b_1. Denoting the bit error probability for the 2-stage receiver as $P_1^{(2)}(H^{-1})$, we have that

$$P_1^{(2)}(H^{-1}) = \frac{1}{2} \left[Pr[z_1(1) \geq 0 \mid b_1 = -1] + Pr[z_1(1) < 0 \mid b_1 = +1] \right],$$

where $z_1(1)$ is obtained from (2.11) as

$$z_1(1) = E_1 b_1 + \sum_{k=2}^{K} h_{1,k} [b_k - \hat{b}_k(1)] + \eta_1.$$

(3.15)

The decorrelating detector is considered as an initial estimate so that from equation (2.10),

$$\hat{b}(1) = sgn[H^{-1} y] = sgn[b + \gamma].$$

(3.16)

Denote the first stage estimate of β_1 to be $\hat{\beta}_1(1)$ and let $\beta_1 - \hat{\beta}_1(1)$, the error vector that occurs in the decision statistic under consideration, be denoted as δ_1. Similar to the definition of β_1, let us define a $(K-1)$-dimensional Gaussian random vector as $\xi_1 = (\gamma_2, \gamma_3, \cdots, \gamma_K)^T$ where γ_k is the kth element of γ. It is clear from (3.15) and (3.16) that the decision statistic $z_1(1)$ is a function not only of η_1, b_1, and β_1 but also of ξ_1 because of its dependence on δ_1. The decision statistic can be written to show the dependence on these parameters as

$$z_1(1) = E_1 b_1 + \eta_1 + I(\beta_1, \xi_1, \delta_1).$$

(3.17)

If an arbitrary linear detector L was considered as an initial condition, it may be noted that the additive noise η_1 in $z_1(1)$ is correlated with the interference term $I(\beta_1, \xi_1, \delta_1)$ through ξ_1. In particular, $E[\xi_1 \eta_1] = \bar{f}_{c1}$, which is defined by the decomposition of F as

$$\mathbf{F} = \begin{bmatrix} f_{11} & \bar{\mathbf{f}}_{r1} \\ \bar{\mathbf{f}}_{c1} & \bar{\mathbf{F}} \end{bmatrix}, \tag{3.18}$$

so that $\bar{\mathbf{f}}_{c1}$ is a $(K-1)\times 1$ column vector and $\bar{\mathbf{f}}_{r1}$ is a $1\times(K-1)$ row vector and $\bar{\mathbf{F}}$ is a $(K-1)\times(K-1)$ matrix. However, if $\mathbf{H}^{-1}$ is chosen for the linear transformation $\mathbf{L}$, it is clear that ξ_1 is independent of η_1 since $\bar{\mathbf{f}}_{c1} = 0$. Therefore for a given $\mathbf{b}$, η_1 is independent of $I(\beta_1,\xi_1,\delta_1)$. The strategy we will therefore follow will be to first condition on $\mathbf{b}$. Note that δ_1 takes on 2^{K-1} possible values corresponding to $[\delta_1]_j \in [2(\beta_1)_j, 0]$ depending on whether $[\beta_1(1)]_j$ is in error or not. Denote the set of all possible such values of δ_1 as $S(\beta_1)$ to show its dependence on β_1. For each value of δ_1, for each possible β_1, if $b_1 = +1$, $\eta_1 < -E_1 - I(\beta_1,\xi_1,\delta_1) \triangleq L_1(\delta_1,\beta_1)$ results in an erroneous two-stage detector decision on b_1. Similarly, for a given β_1, for each possible δ_1, if $b_1 = -1$, then $\eta_1 > E_1 - I(\beta_1,\xi_1,\beta_1) \triangleq U_{-1}(\delta_1,\beta_1)$ results in an erroneous decision on b_1. Now, since η_1 is independent of ξ_1 we have that for a given $\mathbf{b}$, η_1 is independent of $\beta_1(1)$ and hence independent of δ_1. Clearly, the error probability can therefore be written as

$$P_1^{(2)}(\mathbf{H}^{-1}) = \frac{1}{2} E_{\beta_1} \left[E_{\delta_1} \left[Q \left(-\sqrt{\frac{1}{\sigma^2 E_1}} L_1(\beta_1,\delta_1) \right) \right] + E_{\delta_1} \left[Q \left(\sqrt{\frac{1}{\sigma^2 E_1}} U_{-1}(\beta_1,\delta_1) \right) \right] \right] \tag{3.19}$$

where E_{β_1} represents expectation over the ensemble of independent, uniformly distributed $\beta_1 \in \{-1, +1\}^{K-1}$. The expectation over δ_1 denoted by E_{δ_1} needs further explanation. The occurrence of every $\delta_1 \in S(\beta_1)$ corresponds to $\beta_1(1)$ being equal to $\beta_1 - \delta_1$. Now, the probability of such an event for a given $\mathbf{b}$ is equivalent to the probability that $[z_2(0), \cdots, z_K(0)]$ belongs to a certain hyperquadrant. It is clear from (3.16) that this probability can be expressed equivalently as a $(K-1)$ normal distribution function. Let us denote it as

$$P_{b_1}(\beta_1,\delta_1) \triangleq \Pr[\beta_1(1) = \beta_1 - \delta_1 \mid \mathbf{b}, \delta_1]. \tag{3.20}$$

Finally, writing the expectation over δ_1 in (3.19) explicitly, we have

$$P_1^{(2)}(\mathbf{H}^{-1}) = \frac{1}{2} E_{\beta_1} \left[\sum_{\delta_1 \in S(\beta_1)} P_1(\beta_1,\delta_1) Q \left(-\sqrt{\frac{1}{\sigma^2 E_1}} L_1(\beta_1) \right) + \right.$$

$$\left. \sum_{\delta_1 \in S(\beta_1)} P_{-1}(\beta_1,\delta_1) Q \left(\sqrt{\frac{1}{\sigma^2 E_1}} U_{-1}(\beta_1) \right) \right], \tag{3.21}$$

The evaluation of the expression in (3.20) requires the computation of a $K-1$-dimensional Gaussian distribution function, an efficient method for which can be found in [2]. As a concluding remark, it may be noted that when all the users are not linearly independent, the above derivation holds for the initial condition corresponding to any member of a certain class of linear detectors which incidentally includes also the Moore-Penrose generalized inverse. Details can be found in [5].

4. Numerical Results

The two-user example is simple to understand and is yet illustrative of the salient features of the different detectors. The simplicity lies in the fact that we need the appraisal of the detectors for each operating point; i.e., the normalized cross-correlation R_{12} denoted as ρ and the square root of the relative energy parameter $\sqrt{(E_2/E_1)}$. The analysis compares the error probability of the conventional, decorrelating, linear optimum and the two-stage detectors. The initial estimate for the two-stage detector is taken to be the decorrelating detector. The reason for this choice is two-fold. Firstly, the decorrelating detector provides a robust initial estimate which is independent of the interfering signal strength. Secondly, the error probability evaluation in this case, is simpler than it is for other initial estimates such as the conventional and in some cases the linear optimum[3] detector. The former choice was considered for asynchronous systems in [4]. The signal strength of the desired user which in our case is the first user, is fixed at 8 dB. Figures 1.a-c correspond to the cross-correlation parameter $\rho = 0.5, 0.6, 0.7$, respectively. The error probability of the first user is depicted as a function of E_2/E_1 ranging from -10dB to 10dB corresponding to the interfering signal energy being a tenth of the desired signal energy to ten times the same. It will be argued that the detection of a "strong" signal is not as critical an issue as the demodulation of relatively weak signals; that the linear approach falls short of achieving the latter task; also that the two-stage detector achieves very significant improvements over the linear detectors, specifically for the demodulation of relatively weak signals.

The explanation of the behaviour of the conventional detector is straightforward. Since it ignores the MA interference, it has an acceptable performance only for very low values of E_2/E_1 and degrades exponentially as this ratio increases. Consider the decorrelating detector. The error probability in this case remains invariant with interfering signal strength. However, note the degradation with increasing values of ρ. Meanwhile, it is seen that the optimum linear detector has a dual behaviour. If the interfering signal strength is small enough, i.e., when $E_2/E_1 < |\rho|$, some improvement over the decorrelating detector is obtained. For the two-user channel this also is the region where the optimum linear detector achieves optimum asymptotic efficiency. On the other hand, if the interfering signal is strong enough, i.e., when $E_2/E_1 > |\rho|$, the optimum linear detector coincides with the decorrelating detector. A detailed explanation of this dual behaviour is given in [1]. Notice that the two-stage detector does not necessarily perform better than the decorrelating detector for a relatively weak interfering signal. In this case, the interfering signal is not estimated well and hence is not rejected often enough for the effect of the successfully rejected interference to prevail over the effect of interference doubling corresponding to unsuccessful estimation of the interferer.

As the interfering signal strength increases, the conventional detector suffers a total breakdown whereas the decorrelator remains uneffected. The optimum linear detector coincides with the decorrelator. The two-stage detector performance improves dramatically and approaches single-user performance when the interfering signal is strong enough. It converges faster to single-user performance for lower values of ρ. However, improvements over the decorrelating (or the linear optimum) detector are more marked for larger correlation, where the decorrelator suffers from the conservativeness of a maximin solution.

The second set of examples is aimed at addressing a number of important issues that were discussed all along in this work. We will consider the modulating signals of five different users as being base-band signals derived from Gold sequences of length seven as shown in Figure 2. A justification for the choice of this signal set is now in order. As was noted earlier, the linear and the iterative suboptimum approaches lend themselves to generalizations for asynchronous CDMA systems while retaining their linear computational complexities. Therefore, a comparative performance analysis of the suboptimum detectors for synchronous channels would be indicative of results that one would expect for the corresponding suboptimum detectors for asynchronous systems, which enjoy wider application. In order to facilitate such an analysis, one would have to consider signature waveforms that are used for asynchronous CDMA systems. These waveforms are designed to keep the cross-correlations between the signals as low as the bandwidth restriction would allow for all relative delays. The

---

[3] The ith row of the optimal linear detector is obtained by maximizing the asymptotic efficiency functional of the ith user, a procedure for which is detailed in [1].

choice of the Gold sequences of Figure 2 conform to this requirement. Further, since efficient bandwidth utility is a key issue, the signature waveforms are restricted to having a limited bandwidth which along with the number of users in the system should represent high bandwidth efficiencies. Again, note that the five signature waveforms that we chose are direct-sequence waveforms corresponding to a bandwidth spread factor of seven. Finally, since there are a large number of users in a majority of CDMA systems of practical value, it is of interest to consider the performance deterioration of the suboptimum schemes as the number of users in a system of fixed bandwidth grows. In this task, however, we are restricted by the computational complexity of the numerical evaluation of the error probability of the two-stage detector. In any case, we address this issue by depicting the bit error probability of the decorrelating detector and the two-stage detector by starting out with the two-user case and then proceeding to add one user each time until we have a five-user system.

Figures 3a-d illustrate the bit error probability of the first user, the signal to noise ratio of which is fixed at 8 dB. All figures represent an increase in the strength of the interfering signals relative to the first user's signal strength from left to right. The conventional detector, the decorrelating detector and the two-stage detector based on an initial estimate from the decorrelating detector are considered for comparison. Note that as the number of users increase from Figure 3.a to 3.d, the decorrelating detector suffers a sizable degradation. Also note that as compared to this, the two-stage detector performance shows a clear improvement, which is particularly marked as the interfering signals' strength increases, a phenomenon that was also observed in the first 2-user channel examples.

5. Conclusions

Multiuser communication networks employing Code-Division Multiple-Access consist of several users transmitting information simultaneously over a common channel. The problem of demodulating every user in a synchronous network was considered. In this context, optimum demodulation is inherently a difficult problem requiring a computational complexity per demodulated bit that is exponential in the number of users. Suboptimum single-user (conventional) demodulation suffers severe degradation due to the presence of multiple-access interference when the bandwidth has to be efficiently utilized and/or when the interfering signals are strong. The class of linear detectors have the virtue of linear computational complexity per demodulated bit and are therefore simple to implement. Among these, the optimum linear approach and the decorrelating detectors alleviate the shortcomings of the conventional detector wherein their performance remains invariant to increasing interfering signals strengths. This invariant error probability is however a function of the signal cross-correlations and can be quite different from the optimum error probability. As an alternative to the linear suboptimum approach, a nonlinear detection strategy based on a linear complexity iterative multiple-access interference rejection algorithm was proposed. The class of linear detectors are considered as initial conditions for the iterative detector. A bit-error probability comparison of the linear and iterative detectors was undertaken. It was shown that the iterative detectors are capable of achieving considerable improvements over the linear detectors, particularly in near-far situations, i.e., in the demodulation of weak signals in the presence of strong interfering signals. This problem has been of primary concern for currently operational CDMA systems. It has also been shown by some representative examples that higher bandwidth utility factors are also achieved by the use of such detectors. The investigation of the iterative approach for asynchronous CDMA systems therefore require further consideration as do the use of more stages in the iteration.

Acknowledgement

The authors would like to thank Nageen Himayat for writing the computer programs to efficiently evaluate the multivariate normal distribution function.

References

[1]   R. Lupas and S. Verdu, "Linear Multiuser Detectors for Synchronous Code-Division Multiple-Access Channels," *IEEE Trans. Info. Theory, to appear.*

[2]   R. L. Plackett, "A Reduction Formula for Normal Multivariate Integrals," *Biometrika*, 1954.

[3]   K. S. Schneider, "Optimum Detection of Code Division Multiplexed Signals," *IEEE Trans. on Aerosp. Electron. Syst.*, Vol.AES-15, pp. 181-185.

[4]   M. K. Varanasi and B. Aazhang, "Iterative Multiuser Detection for Asynchronous Code-Division Multiple-Access Systems," *Submitted to the IEEE Trans. Commun..*

[5]   M. K. Varanasi and B. Aazhang, "Near-Optimum Detection in Synchronous Code-Division Multiple-Access Systems," *Submitted to the IEEE Trans. Commun..*

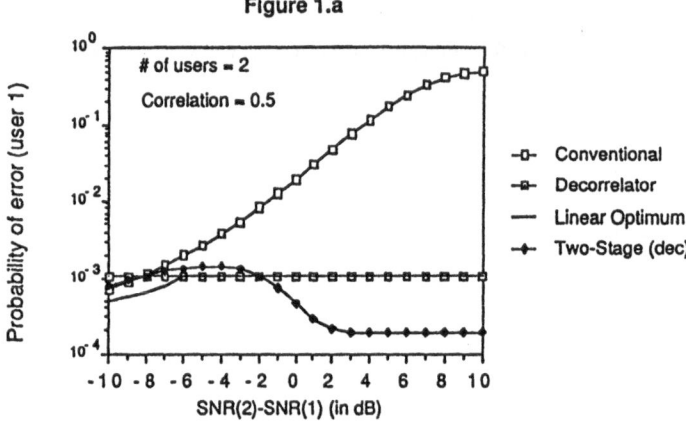

Figure 1.a

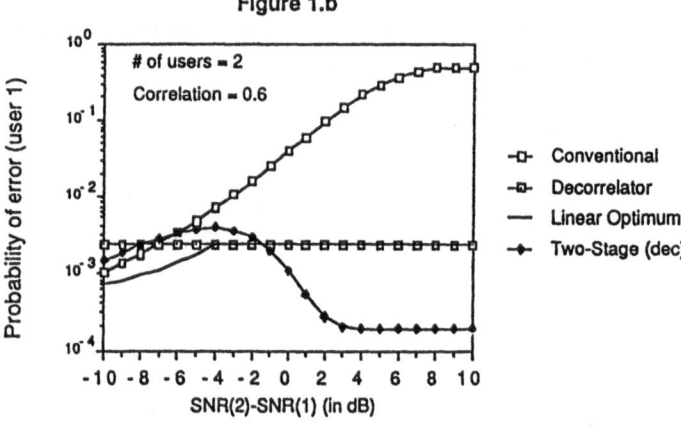

Figure 1.b

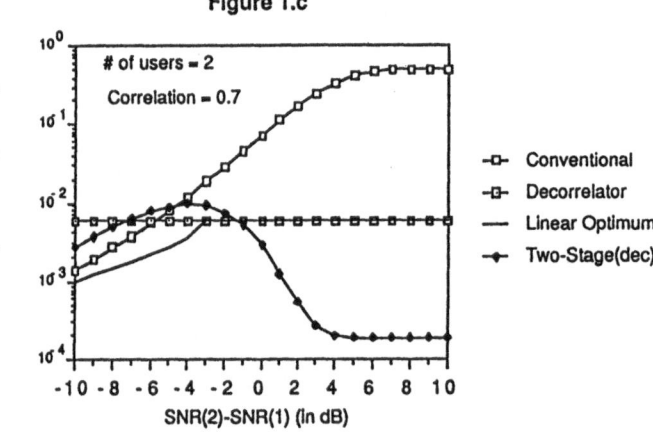

Figure 1.c

Figure 2

Base-band code signals assigned to the five users of a five-user
CDMA system derived from Gold Sequences of length 7

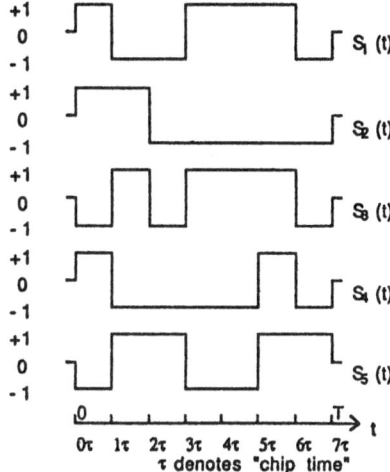

τ denotes "chip time"

Figure 3.a

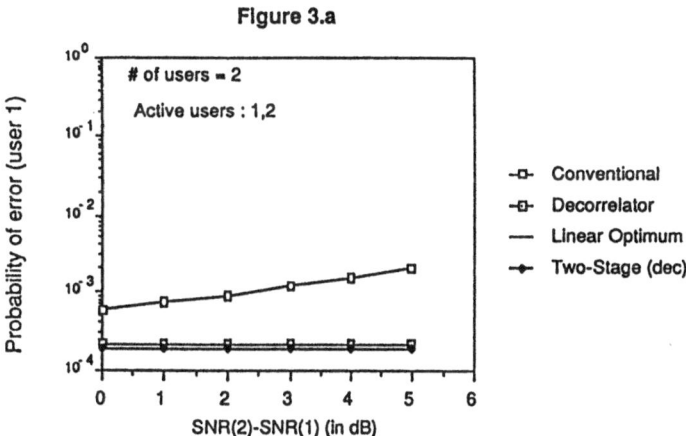

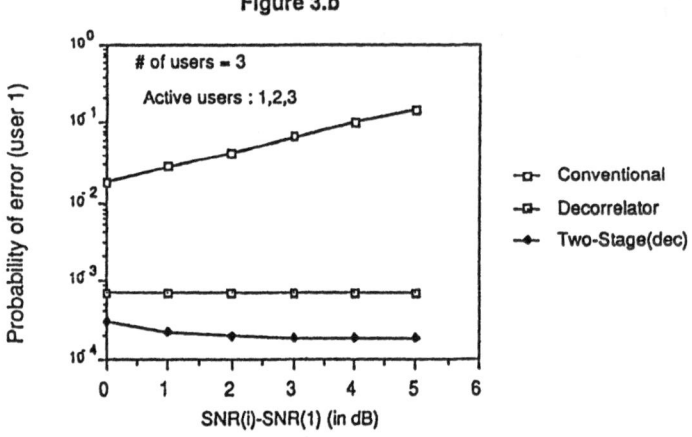

Figure 3.b

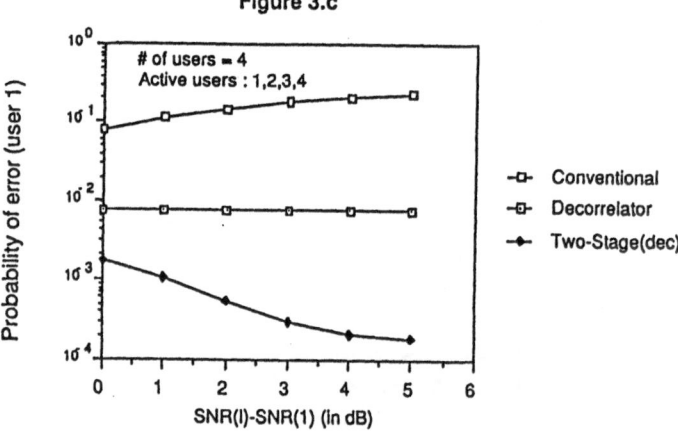

Figure 3.c

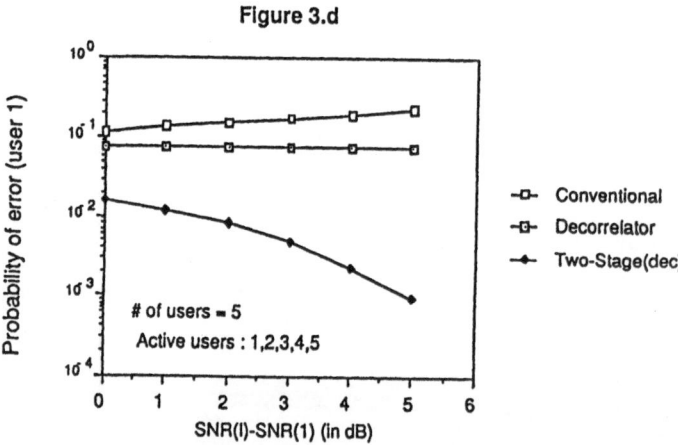

Figure 3.d

RECENT PROGRESS IN MULTIUSER DETECTION

Sergio Verdú

Princeton University

I. Introduction

Multiuser Detection is an active research area in Detection Theory. Its objective is the study of strategies to demodulate the digital information sent simultaneously by several transmitters who share a multiple-access channel. Common channels that are encompassed by this general model include up-link satellite channels, local-area networks and radio networks.

Static strategies, such as Frequency-Division Multiple-Access (FDMA) and Time-Division Multiple-Access (TDMA), whereby the multiple-access channel is effectively partitioned into independent single-user sub-channels are common in practice. The analysis of those approaches lies fully whithin the domain of classical single-user communication, and offers few, if any, conceptual challenges. Moreover, from the practical viewpoint, static strategies tend to be wasteful in applications where most users actively send information during a very small percentage of time. Current dynamic channel-sharing strategies, which allow the active users a larger share of the channel while they are transmitting, fit into two categories: random-access communication and simultaneous transmission systems. In random-access communication it is assumed that the receiver is not capable of demodulating more than one simultaneous transmission, and so the problem is to design protocols to schedule channel access at nonoverlapping times, and if collisions between messages occur (as they do when the protocol operates in a decentralized fashion) to ensure that those messages are eventually retransmitted successfully. Simultaneous transmission systems differ from static strategies and random-access protocols in that users are allowed to transmit simultaneously, asynchronously and through the same channel. The destination receives a noisy version of the superposition of all the transmitted waveforms and it is the task of the receiver to demodulate all (as in the case of the satellite channel) or a subset (as in multipoint-to-multipoint topologies) of the transmitted messages. This is the communication problem that is the subject of multiuser detection and multiuser information theory. In practice, the major multiaccess strategy using the simultaneous transmission philosophy is Code-Division Multiple-Access (CDMA). In this technique, each transmitter is assigned a fixed, distinct signature waveform which he uses to modulate his message as in single-user communication. Then the information sent by each user can be demodulated by correlating the received signal with each of the signature waveforms. This demodulator, whose use is widespread in practice, is referred to as the conventional single-user detector. As is well-known, when the channel output is corrupted by additive white Gaussian noise, the conventional single-user detector minimizes the probability of error in a single-user channel, i.e., in the absence of interfering users. The fact that this is no longer true in the multiple-access channel is the raison d'être of the area of multiuser detection.

The performance of the conventional single-user detector is acceptable provided that the energies of the received signals are not too dissimilar and that the signature waveforms are designed so that their crosscorrelations are low enough (this depends on the desired maximum number of simultaneous users). In practice, low crosscorrelations are usually achieved employing Spread-Spectrum Pseudonoise sequences of long periodicity. If the received signal energies are indeed dissimilar, i.e., some users are very weak in comparison to others, then the conventional single-user detector is unable to recover the messages of the weak users reliably, even if the signature waveforms have very low crosscorrelations. This is known as the *near-far problem* and is the main shortcoming of

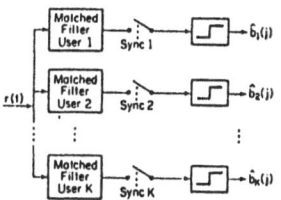

Figure 1. Conventional detector.

currently operational Direct-Sequence Spread-Spectrum Multiple-Access systems.

Due to the reduced multiple-access capability and the increased vulnerability to hostile sources caused by the near-far problem, its solution or alleviation had been a target of researchers in the area for several years. Before the emergence of solutions based on multiuser detection, success had been very limited and, essentially, the only remedies available were power control and the design of signals with even more stringent crosscorrelation properties. Unfortunately, power control (i.e., the adaptive adjustment of transmitter power depending on its location and on the received powers of the other users) dictates significant reductions in the transmitted powers of the strong users in order for the weaker users to achieve reliable communication. Thus, power control can become self-defeating since it actually decreases the overall multiple-access and antijamming capabilities of the system. Furthermore, more and more complex signature waveforms lead to rapid increases in system cost and bandwidth, and, as we have noted, do not eliminate the near-far problem. For these reasons, it can be seen why the solution to the near-far problem has been highlighted as the main achievement of multiuser detection.

The chief reason why multiuser detection did not develop until relatively recently was the belief shared by many a worker in Spread Spectrum that multiuser interference is accurately modeled as a white Gaussian random process, and thus the conventional detector is essentially optimum. This serves as a good illustration of A. N. Kolomogorov's assertion that "the Central Limit Theorem is a dangerous tool in the hands of amateurs," and it ranks along with the stability of the original ALOHA algorithm, and the independence assumption in queueing models of store-and-forward communication networks as one of the maximum exponents of the area of "voodoo multiuser communication." While it is not difficult to build an infinite population multiuser signal model which can be rigorously shown to be asymptotically Gaussian as the individual amplitudes go to zero with the appropriate speed, the number of transmitters, signature waveforms, and power levels encountered in many practical situations (e.g. in near-far environments) render the Gaussian approximation completely useless. This was recognized by H. V. Poor who proposed in 1980 [9] the use of techniques from both minimax robustness and non-Gaussian signal detection to improve the performance of the conventional single-user detector in multiuser channels. Several earlier works had already investigated receivers that used the knowledge of the interfering signature waveforms to improve the performance of Code-Division Multiplexing systems; Timor [15, 16] showed that it was possible to double the maximum number of simultaneous users achievable with the conventional noncoherent demodulator of Frequency-Hopped FSK systems, Horwood and Gagliardi [3] and Schneider [12, 13] considered multiuser receivers for synchronous systems. In particular, [12] claimed that an appropriately chosen linear transformation of a bank of matched filter outputs results in optimum decisions. Although that claim turned out to be erroneous, such a receiver and its generalization to the asynchronous case do have very desirable properties as we discuss in the sequel. In the same work, Schneider briefly considered the asynchronous channel arguing that it could be modeled as a finite state machine and conjectured that Viterbi's convolutional decoding algorithm could be used to demodulate the data. Those ideas were closely related to earlier work to combat crosstalk in single-user multichannel communication systems by Shnidman [14], Kaye and George [4] and Van Etten [1, 2] among others. Those works dealt with PAM systems through multi-input multi-output dispersive channels, which are no more than a vector generalization of the conventional scalar intersymbol interference model. In contrast, in the asynchronous multiuser channel each signal overlaps with two consecutive signals of each of the interferers; thereby introducing memory in the channel in a way which (as we will see in Section II) is akin to a scalar periodically time-varying generalization of the conventional intersymbol interference model.

Section II reviews the main strategies proposed so far in multiuser detection, and Section III summarizes the main results on the performance of the various receivers. For the sake of clarity and in order to highlight the fundamental features of the approaches reviewed in this paper, it is convenient to circumscribe the discussion to the two-user case, and the reader is referred to the corresponding treatment with an arbitrary number of users in each of the referenced sources.

II. Multiuser Detectors

The two-user white Gaussian asynchronous multiple-access channel considered in this paper is

$$y(t) = \sum_{i=-M}^{M} b_1(i)\, s_1(t-iT-\tau_1) + \sum_{i=-M}^{M} b_2(i)\, s_2(t-iT-\tau_2) + n(t), \tag{1}$$

where $n(t)$ is white Gaussian noise with power spectral density equal to σ^2. The unit-energy signature waveforms assigned to both users are $s_1(t)$ and $s_2(t)$ and have duration T. The symbol streams $\{b_1(i)\}$ and $\{b_2(i)\}$ take values on $\{-\sqrt{w_1},+\sqrt{w_1}\}$ and $\{-\sqrt{w_2},+\sqrt{w_2}\}$, respectively, where w_k denotes the received energy per bit of user k. The delays $\tau_1 \in [0,T]$ and $\tau_2 \in [0,T]$ account for the asynchronism between both transmitted streams.

The objective is to find the optimum demodulator for the streams of data transmitted by both users. This would appear to apply only to multipoint-to-point topologies where detection is centralized (e.g. the up-link satellite channel). However, in situations where each location is interested in demodulating only one of the transmitted streams, the same strategy can be used as long as the receiver knows the signature waveforms of the interfering users. (The derivation of optimum detectors without knowledge of the interfering signature waveforms is discussed later in this section.) A key point in the derivation of the optimum receiver [20] for the asynchronous channel (1) is that in order to make optimum decisions on any particular bit it is necessary to use the observation of the entire received waveform rather than just the received waveform on the interval of that particular bit. In other words, due to the asynchronism between the users, we face a problem of sequence detection rather than one-shot detection. It is easy to show [20] that the sequence of outputs of matched filters for $\{s_1(t)\}$ and $\{s_2(t)\}$ is a set of sufficent statistics to demodulate the data streams $\{b_1(i)\}$ and $\{b_2(i)\}$. Those sufficient statistics can be obtained by a receiver that knows the signature waveforms of both users upon acquisition of the timing of the bit-epochs of each stream. Assuming without loss of generality that $\tau_1 \leq \tau_2$, the sufficient statistics admit the following expression in terms of the data

$$\begin{bmatrix} y_1(i) \\ y_2(i) \end{bmatrix} = \begin{bmatrix} 0 & \rho_{21} \\ 0 & 0 \end{bmatrix} \begin{bmatrix} b_1(i-1) \\ b_2(i-1) \end{bmatrix} + \begin{bmatrix} 1 & \rho_{12} \\ \rho_{12} & 1 \end{bmatrix} \begin{bmatrix} b_1(i) \\ b_2(i) \end{bmatrix} + \begin{bmatrix} 0 & 0 \\ \rho_{21} & 0 \end{bmatrix} \begin{bmatrix} b_1(i+1) \\ b_2(i+1) \end{bmatrix} + \begin{bmatrix} n_1(i) \\ n_2(i) \end{bmatrix} \tag{2}$$

for $-M \leq i \leq M$ and where

$$\rho_{12} = \int_0^T s_1(t)\, s_2(t+\tau_1-\tau_2)\, dt, \qquad \rho_{21} = \int_0^T s_1(t)\, s_2(t+T+\tau_1-\tau_2)\, dt \tag{3}$$

are the crosscorrelations between the signature waveforms. The discrete-time random process $\{[n_1(i)\ n_2(i)]^T\}$ in (2) is Gaussian with zero-mean and covariance matrix:

$$E\left[\begin{bmatrix} n_1(i) \\ n_2(i) \end{bmatrix} [n_1(j)\ n_2(j)] \right] = \sigma^2\, H(i-j)$$

where $H(i) = 0$ if $|i| > 1$, and $H(1)$, $H(0)$ and $H(-1)$ are the matrices appearing in (2), i.e.,

$$H(0) = \begin{bmatrix} 1 & \rho_{12} \\ \rho_{12} & 1 \end{bmatrix}, \qquad H(1) = H^T(-1) = \begin{bmatrix} 0 & \rho_{21} \\ 0 & 0 \end{bmatrix}$$

Since the noise in (1) is white and Gaussian, maximum likelihood decisions are obtained by selecting the symbol sequences $\{b_1(i)\}$ and $\{b_2(i)\}$ that best explain the observations in a mean-square sense, i.e., assuming that the optimum detector knows the received energies, it should select the argument achieving

$$\min_{\substack{b_1(i) \in \{-\sqrt{w_1}, +\sqrt{w_1}\} \\ b_2(i) \in \{-\sqrt{w_2}, +\sqrt{w_2}\} \\ i = -M, \cdots M}} \int [y(t) - \sum_{k=1}^{2} \sum_{i=-M}^{M} b_k(i)\, s_k(t - iT - \tau_k)]^2\, dt. \tag{4}$$

Before considering how to obtain a detector that solves for the minimum in (4) it is instructive to consider the synchronous ($\tau_1 = \tau_2$) version of the problem for which a one-shot approach suffices:

$$\min_{\substack{b_1(0) \in \{-\sqrt{w_1}, +\sqrt{w_1}\} \\ b_2(0) \in \{-\sqrt{w_2}, +\sqrt{w_2}\}}} \int_0^T [y(t) - \sum_{k=1}^{2} b_k(0)\, s_k(t)]^2\, dt \; = \; \max_{b \in \{-\sqrt{w_1}, \sqrt{w_1}\} \times \{-\sqrt{w_2}, \sqrt{w_2}\}} 2b^T y - b^T H b \tag{5}$$

where $y = [y_1(0), y_2(0)]^T$, $b = [b_1(0), b_2(0)]^T$ and $H = \begin{bmatrix} 1 & \rho \\ \rho & 1 \end{bmatrix}$, $\rho = \int_0^T s_1(t) s_2(t)\, dt$.

The solution to the right-hand side of (5) is easy; the function therein can be computed for each of the four possible values of b and the maximizing vector gives the solution to the optimum demodulation problem. In the K-user case, there are 2^K possible values of b and no algorithm is known to solve (5) in a number of steps which is polynomial in the number of users. In fact, it has been shown that (5) is an NP-hard combinatorial optimization problem [25], and, therefore, it can admit a polynomial solution only if the same is true for longstanding combinatorial problems such as the *travelling salesman* and *integer linear programming* problems.

In the asynchronous case, it is possible to follow the same approach by viewing each symbol as transmitted by a different fictitious user over the interval $[-MT, MT+2T]$. In the K-user case, this is equivalent to having a synchronous channel with $(2M+1)K$ fictitious users. Thus, the complexity of the above brute-force method is out of the question because of the large values of M one expects in applications. Fortunately, the asynchronous problem admits a much more efficient algorithm whose complexity per demodulated bit is very similar to that of the synchronous case (exponential in K but independent of M), making it practical for situations where the number of simultaneous active users does not exceed the range 10-15. The reason for the existence of such an algorithm is that the $(2M+1)K \times (2M+1)K$ matrix H of crosscorrelations between each possible pair of symbols has a great deal of structure, as each symbol only overlaps with two consecutive symbols of the other users. For example, in the two-user case it is equal to

$$H \; = \; \begin{bmatrix} 1 & \rho_{12} & & & & & \\ \rho_{12} & 1 & \rho_{21} & & & & \\ & \rho_{21} & 1 & \rho_{12} & & & \\ & & & \cdot & & & \\ & & & & \cdot & & \\ & & & & \rho_{21} & 1 & \rho_{12} \\ & & & & & \rho_{12} & 1 \end{bmatrix} \tag{6}$$

This implies that the objective function in the right-hand side of (5) admits the following additive decomposition

$$2b^T y - b^T H b = \lambda_{-M}^{(1)}[0, b_1(-M)] + \lambda_{-M}^{(2)}[b_1(-M), b_2(-M)] + \lambda_{-M+1}^{(1)}[b_2(-M), b_1(-M+1)] + \cdots$$
$$+ \lambda_0^{(1)}[b_2(-1), b_1(0)] + \lambda_0^{(2)}[b_1(0), b_2(0)] + \lambda_1^{(1)}[b_2(0), b_1(1)] + \cdots$$
$$+ \lambda_{M-1}^{(2)}[b_1(M-1), b_2(M-1)] + \lambda_M^{(1)}[b_2(M-1), b_1(M)] + \lambda_M^{(2)}[b_1(M), b_2(M)], \tag{7}$$

where

$$\lambda_i^{(k)}[a, b] = 2 b y_k(i) - w_k - 2ab \rho_{jk}, \quad j \neq k.$$

The interesting thing to note on (7) is that the dependence on the components of b is sequential; each of the functions $\lambda_i^{(k)}$ depends on only two symbols and every symbol appears only in two consecutive functions. Therefore, the maximization of (7) is equivalent to finding the longest path in a layered directed graph (trellis) where there are two nodes in each layer (Fig. 2). (The K-user algorithm in [20] has 2^{K-1} nodes per layer and each node is connected to two nodes in the following layer.) The natural solution to this combinatorial problem is dynamic programming, which can be implemented in real time by running it in forward fashion and with decisions that look ahead over a finite window of stages, rather than waiting for all the data to be received (the Viterbi algorithm). Hence, the structure of the optimum multiuser receiver is now complete: a matched filter front-end followed by a Viterbi algorithm (Fig. 3). Due to the interference between the users, maximum likelihood is not the unique optimality criterion even if the transmitted streams are assumed independent and identically distributed. Minimum probability of error decisions can be accomplished by a backward-forward dynamic programming algorithm [19, 23]; however, the bit-error-rate of the maximum likelihood receiver turns out to be indistinguishable from the minimum achievable bit-error-rate in the region of signal-to-noise ratios of interest, so for most applications it is not worth incurring the additional implementation cost required by the backward-forward algorithm over the Viterbi algorithm.

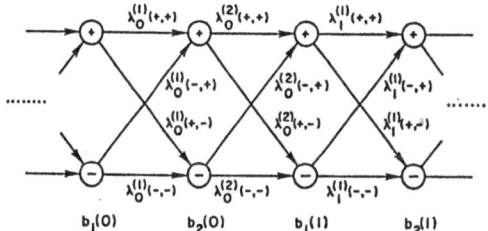

Figure 2. Trellis for two-user asynchronous demodulation.

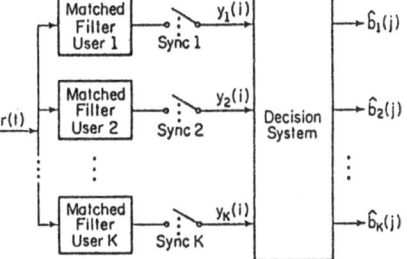

Figure 3. Optimum multiuser detector.

In the special case when the energies of the users are the same and $\rho_{12} = \rho_{21}$ (i.e. H is Toeplitz), the functions $\lambda_i^{(k)}$ in (7) no longer depend on k and i (except through the data $y_k(i)$) and they become identical to the metric functions obtained by Ungerboeck [17] for maximum likelihood sequence detection in the single-user intersymbol interference channel. This is because if the energies and signature waveforms of both users are identical and $\tau_2 - \tau_1 = T/2$, then the multiuser channel (1) is entirely equivalent to an intersymbol interference channel where each symbol overlaps with two symbols. Thus, the decomposition in (7) is akin to a periodically time-varying generalization of Ungerboeck's decomposition [17]. Similar decompositions [21] of the likelihood function are also feasible in asynchronous multiple-access channels which are much less structured than (1).

As we discuss in Section III the optimum detector affords important performance gains over the conventional single-user detector, and, in particular, it solves the near-far problem. However, the price for this is exponential complexity in the number of users, since the combinatorial optimization problem of selecting the most likely transmitted bits given the matched filter outputs is inherently hard [25]. This has spurred several works that propose lower complexity multiuser detectors whose performance lies somewhere in the big gap between optimum and conventional bit-error-rate. A natural alternative to explore is the replacement of the Viterbi algorithm by a sequential decoding algorithm. This is done in [11] where a metric is proposed for a stack sequential decoder based on the additive decomposition of the loglikelihood function in (7). At each point in the search in the binary tree of the sequential decoding algorithm the metric is the sum of two parts: one that depends on "past" symbols exclusively and the other that depends on "future" unexplored symbols. The metric proposed in [11] takes the decisions of the conventional detector in lieu of those unexplored symbols. As usual, results on the computational complexity of the resulting sequential decoding algorithm are hard since the number of steps depends on the noise realization and there is the possibility of buffer overflow. But it appears that unless the signal-to-noise ratios are extremely poor such a strategy does indeed afford important computational savings over the Viterbi algorithm when the number of users is large.

The approach taken in [5-8] to find low-complexity multiuser detectors that exhibit good performance, and, in particular, are near-far resistant, is to restrict attention to linear transformations of the bank of matched filter outputs. As in the discussion of the maximum likelihood detector, it is beneficial to consider first the synchronous version of the problem. The maximization in (5) is over a four-element set because the receiver is assumed to know the energies of both users. If those energies were unknown and we were interested in obtaining maximum likelihood estimates of the energies in addition to the maximum likelihood decisions on the transmitted streams, then the maximization in (5) would be over the whole plane, and the solution would be $H^{-1}y$, i.e., the sign of the components of $H^{-1}y$ gives the most likely transmitted bit and their absolute values give the maximum-likelihood amplitude estimates. (Note that better amplitude estimates would be possible with a sequence approach rather than a one-shot approach.) This boils down to using a simple modification of the conventional detector: instead of correlating the received signal with $s_1(t)$ and comparing the output to zero to determine the bit transmitted by user 1, we correlate the received signal with $s_1(t) - \rho\, s_2(t)$. In the hypothetical case when there is no background Gaussian noise ($\sigma^2 = 0$), we have $y = Hb$, and the bits selected by this receiver are $sgn(H^{-1}y) = sgn(b)$, i.e., the true data, regardless of the values of the received energies. A nice property, which augurs well for the performance of this receiver (at least in the high SNR region). Indeed, it is the lack of this feature that makes the conventional receiver so sensitive to the near-far problem. In the general K-user case, the essence of the $sgn(H^{-1}y)$ receiver is that instead of correlating with $s_k(t)$ it correlates with the projection of $s_k(t)$ on the subspace orthogonal to the subspace spanned by the other signature waveforms $\{s_j(t),\ j \neq k\}$. Thus, it effectively tunes out the multiuser interference, and hence its name: *decorrelating detector*.

The foregoing derivation of the decorrelating detector has highlighted that if no prior information on the transmitted energies is available, maximum likelihood decisions are obtained by a linear receiver. Note that Schneider [12] sought to minimize error probability in a synchronous multiuser channel with equal-energy signals and, erronously, arrived at the same receiver $sgn(H^{-1}y)$. At this point, it is natural to ask whether it is possible to achieve better performance by choosing a different linear transformation when the received energies are known. The answer is affirmative; for example, the optimum linear receiver (for high SNR's) for user 1 correlates the received signal with [6]:

$$
\begin{cases}
s_1(t) - \sqrt{\dfrac{w_2}{w_1}}\, sgn(\rho)\, s_2(t) & \text{if } \dfrac{w_2}{w_1} \leq \rho^2 \\[2ex]
s_1(t) - \rho\, s_2(t) & \text{otherwise}
\end{cases}
$$

However, as we discuss in Section III, the decorrelating receiver provides the same degree of near-far resistance as the optimum multiuser receiver [20] (which uses knowledge of the received energies) and it is much easier to compute than the optimum linear receiver in the K-user case.

How do we generalize the decorrelating receiver to the asynchronous case? We can take the same approach as in the derivation of the optimum receiver: each of the $(2M+1)K$ symbols can be viewed as transmitted by a different user. However, the resulting detector would be impossible to compute as it would involve the inversion of a crosscorrelation matrix of dimension $(2M+1)K$. Fortunately, as $M \to \infty$ the rule $sgn(H^{-1}y)$ is equivalent to passing the sufficient statistics $y(i) = [y_1(i)\ y_2(i)]^T$ through a two-input two-output discrete-time linear time invariant system with transfer function [7]

$$[H^T(1)\,z + H(0) + H(1)\,z^{-1}]^{-1} = \frac{1}{1 - \rho_{12}^2 - \rho_{21}^2 - \rho_{12}\rho_{21}z - \rho_{12}\rho_{21}z^{-1}}\begin{pmatrix} 1 & -\rho_{12} - \rho_{21}z^{-1} \\ -\rho_{12} - \rho_{21}z & 1 \end{pmatrix} \quad (8)$$

Notice that the matrix in (8) performs the same function as the decorrelating detector in the synchronous case, namely, it substracts from each matched filter output (say, the one corresponding to the ith bit of user 1) the matched filter outputs corresponding to the two overlapping bits of the other user (bits $i-1$ and i of user 2). However, now this is not enough because this operation introduces intersymbol interference between the symbols of user 1 as the matched filter outputs of the overlapping bits ($i-1$ and i of user 2) are contaminated by bits $i-1$ and $i+1$ of user 1. This is the origin of the recursive part of the filter (8) which acts as a linear equalizer for the intersymbol interference introduced by the nonrecursive part. Note that the impulse response of the recursive part is of the form $[\alpha(\rho_{12},\rho_{21})]^{|n|}$, i.e., it is non-causal, and therefore, in practice, it is necessary to approximate it by truncation.

An alternative to the decorrelating detector for asynchronous channels, which is presented here for the first time, can be obtained by taking a one-shot approach where each symbol interval is considered separately. Let us focus attention on bit 0 of user 1, which occupies the interval $[0, T]$ (assuming without loss of generality $\tau_1 = 0$). This bit overlaps with bit -1 of user 2, over the interval $[0, \tau_2]$ and with bit 0 of user 2 over the interval $[\tau_2, T]$. We can view this situation as a 3-user synchronous channel (see Fig. 4) with unit-energy signature waveforms $\bar{s}_1(t) = s_1(t)$; $\bar{s}_2(t) = c_2^{-\frac{1}{2}}s_2^L(t)$; $\bar{s}_3(t) = (1-c_2)^{-\frac{1}{2}}s_2^R(t)$, where

$$s_2^L(t) = \begin{cases} s_2(t+T-\tau_2) & 0 \le t \le \tau_2 \\ 0 & \tau_2 \le t \le T \end{cases} \qquad s_2^R(t) = \begin{cases} 0 & 0 \le t \le \tau_2 \\ s_2(t-\tau_2) & \tau_2 \le t \le T \end{cases}$$

and $c_2 = \int_0^{\tau_2} s_2^2(t+T-\tau_2)dt$ $(0<c_2<1)$ is the partial energy of the interfering signal over the left overlapping interval. We can now solve for a 3-user decorrelating detector with

$$H = \begin{bmatrix} 1 & c_2^{-\frac{1}{2}}\rho_{12} & (1-c_2)^{-\frac{1}{2}}\rho_{21} \\ c_2^{-\frac{1}{2}}\rho_{12} & 1 & 0 \\ (1-c_2)^{-\frac{1}{2}}\rho_{21} & 0 & 1 \end{bmatrix} \quad (9)$$

resulting in a detector for user 1 that correlates the received signal with

$$[s_1(t) - \frac{\rho_{12}}{c_2}s_2^L(t) - \frac{\rho_{21}}{1-c_2}s_2^R(t)].$$

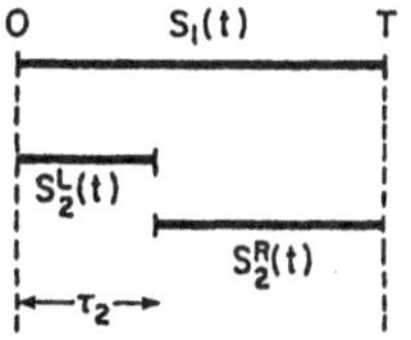

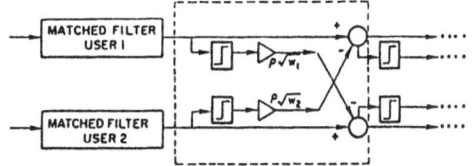

Figure 4. Intervals for one-shot decorrelating detector.

This one-shot decorrelating detector has lower complexity than the asynchronous decorrelating detector described before at the expense of some performance degradation.

Another approach to obtain suboptimum low-complexity multiuser signal detectors is proposed in [18] by Varanasi and Aazhang. Again, it is best to consider first the synchronous case. If we had an estimate, say $\hat{b}_2$, of the symbol transmitted by user 2 in the interval $[0,T]$, then we could subtract $\hat{b}_2 s_2(t)$ from the received signal and process the remaining signal as if $b_2 = \hat{b}_2$, i.e., correlate it with $s_1(t)$ and find the polarity of the output. The detector proposed in [18] is assumed to know the received energies, and the tentative decisions are obtained with the conventional single-user detector, i.e., $\hat{b}_2$ is $+\sqrt{w_2}$ or $-\sqrt{w_2}$ depending on the polarity of the output of the matched filter of user 2 (Fig. 5). If the signal of user 2 is comparatively strong, then the conventional receiver will select $\hat{b}_2 = b_2$ with high probability, in which case the decision on the bit transmitted by user 1 will noticeably improve the (poor) decision made by the conventional receiver. On the other hand, the decision on the bit of user 2

Figure 5. Multi-stage detector for two synchronous users.

will not improve the one obtained with the conventional detector since $b_1 = \hat{b}_1$ does not occur with high enough probability. However, we can now repeat the same procedure as many times as desired using the tentative decisions of the previous stage in lieu of those obtained by the conventional detector. For example, at the output of the second stage, the decisions on user 2 will be near-optimal because the available estimate of the bit of user 1 is excellent, whereas the decisions on user 1 will not improve those obtained in the first stage. Roughly speaking, the decisions put out by even [resp. odd]-numbered stages will be good for the powerful [resp. weak]-transmitters. When the energies are similar, the prerequisite for adequate operation is that the crosscorrelations and noise level be such that the initial decisions put out by the conventional detector have moderate bit-error-rate. In the asynchronous case, the principle of operation is very similar except that now we need to subtract two terms corresponding to the two symbols that overlap with each symbol. Therefore, at the output of each stage we need to store a sliding window of decisions. For example, if there is only one stage we need to store two consecutive outputs of each conventional single-user detector, whereas if there are two stages, we need two consecutive outputs of the first stage and hence, a window of three consecutive conventional decisions at the beginning. In general, if there are m stages, the kth stage needs to store $m - k + 1$ consecutive outputs for each user. As in the case of the one-shot decorrelating detector, this detector can be simplified using (partial) correlations with $s_2^L(t)$ and $s_2^R(t)$.

We now turn our attention to the derivation of demodulators for decentralized (multipoint-to-multipoint) topologies, where each receiver is interested in demodulating the information transmitted by only one of the active users. Obviously, all the foregoing detectors can still be used provided the signature waveforms of the interferers are known and provided that synchronism with each of the transmitted streams is acquired. The decorrelating detector is especially suited for this, as we can implement the linear transformation dictated by each row of H^{-1} or $[H^T(1) z + H(0) + H(1) z^{-1}]^{-1}$ separately, thus resulting in a simple decentralized version of the detector. The rest of the detectors discussed above do not lend themselves to such decentralized implementations, but in many cases it is possible to lower their complexity substantially, without compromising performance, by ignoring the comparatively weak interferers. In any case, it is important to investigate which detection structures result when the signature waveforms of the interferers are not known or when the receiver acquires synchronization only with the user of interest. The problem becomes one of single-user signal detection in additive colored non-Gaussian noise, since the equivalent noise seen by each signal is the sum of the multiple-access interference and the additive background Gaussian noise.

The solution to this signal detection problem is found in [10] and [19, Chap. 5] both in the case where the interfering signature waveforms are known (but no synchronization is available) and in the case where the modulation format is Direct Sequence Spread Spectrum and the interfering signature sequences are not known. Even though the unavailability of the timing epochs of the interfering users results in very complex nonlinear detectors, it is shown in [10] that important reductions in complexity are possible in several asymptotic cases such as when the number of chips per symbol is large and the background Gaussian noise level is low. Also, when the interfering users are weak the locally optimum detector reduces to a modification of the conventional single-user detector, whereby the received signal is correlated with a smoothed replica of the signature waveform of the user of interest. This is because for the purposes of locally optimum detection it is necessary to take care only of the nonwhiteness of the multiple-access noise.

III. Performance Analysis

We focus attention now on the bit-error-rate achieved by the detectors reviewed in Section II. We denote by $P_k(\sigma)$ the bit-error-rate of the kth user when the white Gaussian noise level is σ^2. Of particular interest is the bit-error-rate in the region of low σ^2, because that quantifies the inherent performance degradation due to the presence of other users in the channel. A convenient way to quantify bit-error-rate in the region of high signal to noise ratios is the *asymptotic efficiency* which is defined as follows [20, 22]. Define the *effective energy* of user k, $e_k(\sigma)$, as the energy that that user would require to achieve bit-error-rate $P_k(\sigma)$ in the same Gaussian channel but without interfering users, i.e., $P_k(\sigma) = Q(e_k^{1/2}(\sigma)/\sigma)$. Thus, the *efficiency* or ratio between the effective and actual energies $e_k(\sigma)/w_k$ is an alternative way to characterize $P_k(\sigma)$. The *asymptotic efficiency* is defined as $\eta_k = \lim_{\sigma \to 0} e_k(\sigma)/w_k$ and measures the slope with which $P_k(\sigma)$ goes to 0 in the high signal-to-noise ratio region. A third useful performance measure is the *near-far resistance*, which is defined as the minimum asymptotic efficiency over the relative energies of all the other users. This measure quantifies the robustness of a receiver against the near-far problem, which as we argued in the introduction is the major shortcoming of coherent CDMA systems.

The minimum probability of error achievable in a Gaussian asynchronous multiple-access channel was obtained in [20]. Prior to the appearance of that paper, works devoted to the analysis of error probability in multiuser channels had focused attention exclusively on the bit-error-rate of the conventional single-user receiver. For fixed, arbitrary offset between the signals, it is easy to see that the two-user asynchronous error probability of a conventional receiver for user 1 is

$$P_1^c(\sigma) = \frac{1}{4} Q\left[(\sqrt{w_1} + \sqrt{w_2}\,\rho_{12} + \sqrt{w_2}\,\rho_{21})/\sigma^2\right] + \frac{1}{4} Q\left[(\sqrt{w_1} + \sqrt{w_2}\,\rho_{12} - \sqrt{w_2}\,\rho_{21})/\sigma^2\right]$$
$$+ \frac{1}{4} Q\left[(\sqrt{w_1} - \sqrt{w_2}\,\rho_{12} + \sqrt{w_2}\,\rho_{21})/\sigma^2\right] + \frac{1}{4} Q\left[(\sqrt{w_1} - \sqrt{w_2}\,\rho_{12} - \sqrt{w_2}\,\rho_{21})/\sigma^2\right] \quad (10)$$

In the region of low background Gaussian noise the expression in (10) is dominated by the Q-function with the smallest argument. Hence, according to the foregoing definition, the asymptotic efficiency is

$$\eta_1^c = \max^2\{0,\ 1 - \sqrt{\frac{w_2}{w_1}}\,(\,|\rho_{12}| + |\rho_{21}|\,)\} \quad (11)$$

If the interferer is very weak $w_2 \ll w_1$, then, η_1^c is close to unity (the conventional receiver is quasi-optimal because very little penalty is paid for ignoring user 2). As w_2/w_1 grows, η_1^c decreases monotonically, until it becomes zero for

$$\frac{w_2}{w_1} \geq \frac{1}{[\,|\rho_{12}| + |\rho_{21}|\,]^2},$$

in which case, the probability of error is worse than 0.25 regardless of the background noise level. Therefore, the near-far resistance is equal to zero and there is always an interference level that renders the conventional receiver useless no matter how good the signal design.

The minimum probability of error derived in [20] exhibits a very different behavior. The main contribution of that paper was the introduction of the so-called *method of indecomposable sequences* to obtain upper bounds on the probability of error of m-ary hypothesis testing problems in terms of the error probability of binary problems. This method has been successfully used in a wide variety of channels: both single-user [24] and multiple-access channels with both synchronous and asynchronous users, and with both Gaussian and Poisson [21] observation models. Space limitations prevent us from an exposition of the method of indecomposable sequences and its application to code division multiple-access channels. The main conclusion from that analysis is that the probability of error of the maximum likelihood receiver is optimum in the region of moderate to large signal-to-noise ratios and it behaves as that of a single transmitter with reduced energy. This reduction in effective energy is quantified by the optimum asymptotic efficiency [22]

$$\eta_1 = 1 - \frac{\sqrt{w_2}}{\sqrt{w_1}} \left[\left[2\,|\rho_{12}| - \frac{\sqrt{w_2}}{\sqrt{w_1}} \right]^+ + \left[2\,|\rho_{21}| - \frac{\sqrt{w_2}}{\sqrt{w_1}} \right]^+ \right] \tag{12}$$

As in the case of the conventional receiver, η_1 is close to unity if $w_2/w_1 \ll 1$. However, unlike (11), η_1 is not monotonic in w_2/w_1. Actually, if

$$\frac{w_2}{w_1} \geq 4 \max\{\rho_{12}^2, \rho_{21}^2\}, \tag{13}$$

then $\eta_1 = 1$. Therefore, as long as the energy of user 2 exceeds the threshold given by (13) the bit-error-rate of user 1 is equivalent to the single-user case where user 2 is not active. The explanation of this behavior of the optimum receiver is that if the interfering user is sufficiently powerful, then the primary source of errors committed in the optimum demodulation of user 1 is the background Gaussian noise, rather than the randomness of the information carried by the interfering signal. Note that according to the threshold in (13), an interferer who is 3 dB weaker than the user of interest has no appreciable effect on the bit-error-rate as long as the maximum crosscorrelation is below 0.35 (a mild condition on the signal design). Interestingly, the same is true if the relative energy of the interferer is *higher* than −3 dB. The minimization of (12) with respect to w_2/w_1 yields

$$1 - \max^2\{|\rho_{12}|, |\rho_{21}|, (|\rho_{12}| + |\rho_{21}|)/\sqrt{2}\},$$

which implies that there is a nonzero level of guaranteed asymptotic efficiency regardless of the received energies (unless both signature signatures are identical and there is no offset between them). A more stringent measure of near-far resistance is obtained by minimizing the asymptotic efficiency over the energies of all the users letting those energies be time-varying; the rationale for this being that the near-far problem occurs primarily in networks with dynamically changing topologies. In such case, the near-far resistance is given by

$$\bar{\eta}_1 = \sqrt{[1 - (\rho_{12} + \rho_{21})^2][1 - (\rho_{12} - \rho_{21})^2]}. \tag{14}$$

In the general K-user case, it can be shown that $\bar{\eta}_1$ is equal to the smallest energy of the signals in Ξ_1, which is the set of multiuser signals such that $b_1(0) = 1$, i.e., $\Xi_1 = \{\sum_i \sum_{k=1}^{K} b_k(i)s_k(t - iT - \tau_k),$ $b_1(0) = 1, b_k(i) \in R, (k,i) \neq (1,0)\}$. In the synchronous case, $\bar{\eta}_k$ can be shown [6] to equal $1/(H^{-1})_{kk}$, whereas in the asynchronous case [7] we have

$$\bar{\eta}_k = \left[\frac{1}{\pi} \int_0^\pi [H(1)^T\,e^{j\omega} + H(0) + H(1)\,e^{-j\omega}]_{kk}^{-1}\,d\omega \right]^{-1}.$$

The decorrelating detector eliminates the multiuser interference from the decision statistic. Therefore, its bit-error-rate has the very desirable property that it is independent of the energy of the interfering users. It can be shown that it is equal to (in the K-user case) [7]:

$$P_k^d(\omega) = Q\left(\frac{\sqrt{w_k\bar{\eta}_k}}{\sigma} \right) \tag{15}$$

which implies that the effective energy is $e_k(\omega) = w_k\bar{\eta}_k$, and therefore, the efficiency is independent of the noise level as well as of the energies and is equal to the near-far resistance of the optimum receiver. Hence, we arrive at the interesting conclusion that the decorrelating detector achieves optimum near-far resistance. Figure 6 illustrates the near-far behavior of the conventional, optimum and decorrelating detectors, comparing their asymptotic efficiencies as a function of the relative energies. Note that there is a gap between the minimum efficiency achieved by the optimum receiver and the efficiency of the conventional detector. This is because the asymptotic efficiency of the optimum detector is the function in (12) where the energies are assumed time-invariant. The minimum of the optimum asymptotic efficiency over the amplitudes of each of the symbols does indeed coincide with the decorrelating asymptotic efficiency.

. The bit-error-rate of the one-shot decorrelating detector for asynchronous channels introduced in Section II is also independent of the energy of the interferers. Its efficiency (and near-far resistance) is given by the inverse of the first diagonal element of the inverse of (9), i.e.,

$$\eta_1^{ed} = 1 - \frac{\rho_{12}^2}{c_2} - \frac{\rho_{21}^2}{1 - c_2}.$$

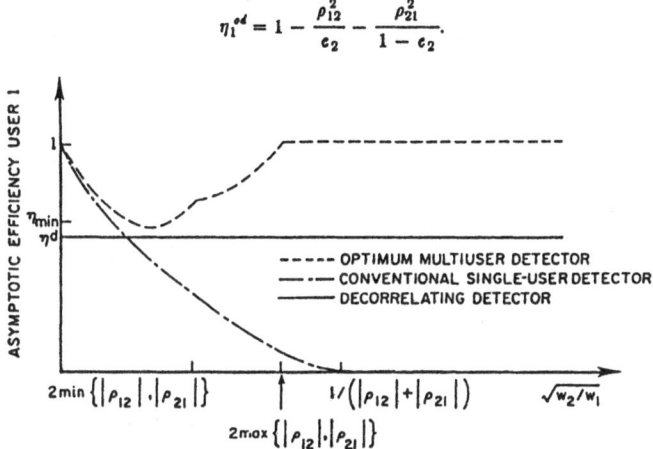

Figure 6. Asymptotic efficiencies for $\rho_{12} = 0.3$, $\rho_{21} = 0.5$. $\eta_{min} = 0.68$ and $\eta^d = 0.59$.

The performance analysis of the remaining detectors reviewed in Section II does not lend itself to similar analytical results due to the nonlinear nature of those detectors. A numerical computation of the receiver proposed in [18] shows that in the two-user case a single stage detector achieves noticeable improvements over the conventional receiver especially when the interferer is very strong and the signature waveforms have good crosscorrelation properties. No bit-error-rate analysis of the nonlinear single-user detectors obtained in [10] has been undertaken to date. Nevertheless, [10] shows the interesting result that in the absence of background Gaussian noise the transmitted bits can be demodulated perfectly even if the signature waveforms of the interferers are not known to the receiver, thus hinting that some degree of robustness to the near-far effect is also possible in that case. However, no practically implementable receiver that ignores the interfering signature waveforms is known to be near-far resistant.

References

1. W. Van Etten, "An Optimum Linear Receiver for Multiple Channel Digital Transmission Systems," *IEEE Trans. on Communications*, pp. 828-834, Aug 1975.

2. W. Van Etten, "Maximum Likelihood Receiver for Multiple Channel Transmission Systems," *IEEE Trans. Commun.*, vol. COM-24, no. 2, pp. 276-283, February 1976.

3. D. Horwood and R. Gagliardi, "Signal Design for Digital Multiple Access Communications," *IEEE Trans. Commun.*, vol. COM-23, pp. 378-383, March 1975.

4. A.R. Kaye and D.A. George, "Transmission of multiplexed PAM signals over multiple channel and diversity systems," *IEEE Trans. on Communication Technology*, vol. COM-18, pp. 520-525, Oct 1970.

5. R. Lupas-Golaszewski and S. Verdu, "Asymptotic Efficiency of Linear Multiuser Detectors," *Proc. 25th Conf. on Decision and Control*, pp. 2094-2100, Athens, Dec. 1986.

6. R. Lupas and S. Verdu, "Linear Multiuser Detectors for Synchronous Code-Divsion Multiple-Access Channels," *IEEE Trans. Information Theory*, vol. IT-34. to appear.

7. R. Lupas and S. Verdu, "Near-Far Resistance of Multiuser Detectors in Asynchronous Channels," *IEEE Trans. Communications*, vol. COM-37. to appear

8. R. Lupas and S. Verdu, "Optimum Near-Far Resistance of Linear Detectors for Code-Division Multiple-Access Channels," *Abs. 1988 IEEE Int. Symp. Information Theory*, p. 14, Kobe, Japan, June 1988.

9. H.V. Poor, "Signal detection in multiple access channels," U.S. Army Research Office Proposal (Contract DAAG29-81-K-0062 to Coordinated Science Laboratory, University of Illinois),, 1980.

10. H. V. Poor and S. Verdu, "Single-User Detectors for Multiuser Channels," *IEEE Trans. Communications*, vol. 36, pp. 50-60, Jan. 1988.

11. C. Rushforth and Z. Xie, "Multiuser Signal Detection using Sequential Decoding," *1988 IEEE Military Communications Conf.*, San Diego, Oct. 1988.

12. K.S. Schneider, "Optimum Detection of Code Division Multiplexed Signals," *IEEE Trans. Aerosp. Electron. Syst.*, vol. AES-15, no. 1, pp. 181-185, January 1979.

13. K.S. Schneider, "Crosstalk Resistant Receiver for *M*-ary Multiplexed Communications," *IEEE Trans. on Aerospace and Electronic Systems*, vol. AES-16, pp. 426-433, Jul 1980.

14. D.A. Shnidman, "A Generalized Nyquist Criterion and an Optimum Linear Receiver for a Pulse Modulation System," *The Bell System Technical Journal*, pp. 2163-2177, Nov 1967.

15. U. Timor, "Improved Decoding Scheme for Frequency-Hopped Multilevel FSK System," *The Bell System Technical Journal*, vol. 59, pp. 1839-1855, December 1980.

16. U. Timor, "Multistage Decoding of Frequency-Hopped FSK System," *The Bell System Technical Journal*, vol. 60, pp. 471-483, April 1981.

17. G. Ungerboeck, "Adaptive Maximum Likelihood Receiver for Carrier-Modulated Data Transmission Systems," *IEEE Trans. Commun.*, vol. COM-22, no. 5, pp. 624-636, May 1974.

18. M. K. Varanasi and B. Aazhang, "Near Optimum demodulation for Coherent Communications in Asynchronous Gaussian CDMA Channels," *Proc. 22nd Princeton Conf. on Information Sciences and Systems*, pp. 832-839, Mar. 1988.

19. S. Verdu, "Optimum multi-user signal detection," Ph. D. Dissertation, Department of Electrical and Computer Engineering, University of Illinois at Urbana-Champaign. Report T-151 Coordinated Science Laboratory, Urbana, IL, Aug. 1984.

20. S. Verdu, "Minimum Probability of Error for Asynchronous Gaussian Multiple-Access Channels," *IEEE Trans. on Information Theory*, vol. IT-32, pp. 85-96, Jan. 1986.

21. S. Verdu, "Multiple-Access Channels with point-process observations: Optimum demodulation," *IEEE Trans. Information Theory*, vol. IT-32, pp. 642-651, September 1986.

22. S. Verdu, "Optimum multiuser asymptotic efficiency," *IEEE Trans. Communications*, vol. COM-34, pp. 890-897, Sept. 1986.

23. S. Verdu and H. Vincent Poor, "Abstract Dynamic Programming Models under Commutativity Conditions," *SIAM J. Control and Optimization*, vol. 24, pp. 990-1006, Jul. 1987.

24. S. Verdu, "Maximum Likelihood Sequence Detection for Intersymbol Interference Channels: A New Upper Bound on Error Probability," *IEEE Trans. Information Theory*, vol. IT-33, pp. 62-68, Jan. 1987.

25. S. Verdu, "Computational Complexity of Optimum Multiuser Detection," *Algorithmica*, vol. 4, to appear.

MINIMAX CAUSAL TRANSMISSION OF GAUSSIAN STOCHASTIC PROCESSES OVER CHANNELS SUBJECT TO CORRELATED JAMMING

Tangül Ü. Başar
Dept. Electrical and Computer Engineering
Illinois Institute of Technology

Tamer Başar
Coordinated Science Laboratory
University of Illinois

1. Introduction and Problem Formulation: Consider the communication system depicted in Figure 1. A stochastic process $\{\theta_t, t \in [0, t_f]\}$, which satisfies the stochastic differential equation (SDE)

$$d\theta_t = a(t)\theta_t dt + b(t)dv_t; \; \theta_0 \sim N(0, \lambda) \tag{1}$$

where v_t is a standard Wiener process and independent of θ_0, and $a(t)$ and $b(t)$ are uniformly bounded (i.e., $|a(t)| \leq M$, $|b(t)| \leq M$), is to be transmitted through a continuous time stochastic channel which has additive white Gaussian noise and is also tapped by an intelligent

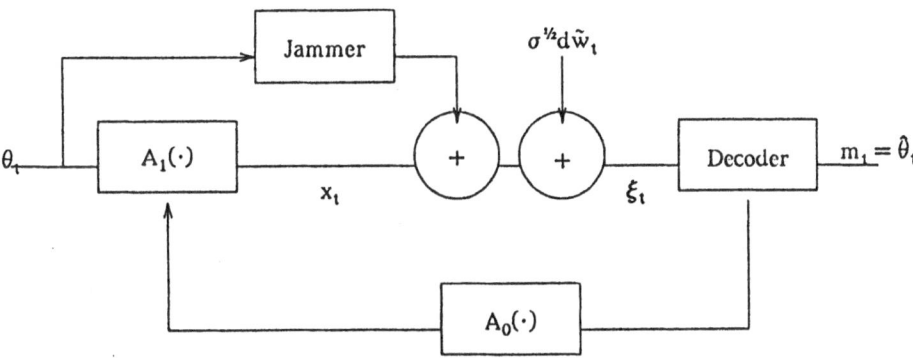

Figure 1. Schematic diagram of the communication system.

jammer. The jammer sends a stochastic process at a certain power level so as to maximize the interference at the channel. The transmission channel is to be used for exactly t_f seconds, it has an instantaneous input power constraint P, i.e.,

$$E[x_t^2] \leqslant P \tag{2}$$

and its output process ξ_t, $t \in [0, t_f]$, is described by

$$d\xi_t = x_t dt + d\beta_t + \sigma^{\frac{1}{2}} d\tilde{w}_t, \ \xi_0 = 0 . \tag{3}$$

Here $\{\tilde{w}_t, t \in [0, t_f]\}$ is a standard Wiener process, independent of $\{v_t\}$ and θ_0, and $\{\beta_t, t \in [0, t_f]\}$ is the jamming noise which is allowed to depend on the current and past values of the message process $\{\theta_t\}$ (i.e., $\beta(t, \theta_0^t)$). The jammer's policy is restricted to the class of stochastic processes given by the Wiener integral

$$\beta_t = \int_0^t B(s, \theta_s) dw_s \tag{4}$$

where $\{w_t\}$ is a standard Wiener process, independent of $\{v_t\}$, $\{\tilde{w}_t\}$ and θ_0, and B is some second-order process independent of $\{w_t\}$, and possibly dependent on θ_t, satisfying the power constraint

$$E\{ |B(t, \theta_t)|^2\} \leqslant k . \tag{5}$$

The receiver, on the other hand, applies a causal transformation on ξ_t to produce m_t, which is the desired estimate for the message process θ_t. Let the feedback transformation around the channel be $A_0(t, m_s, s \leqslant t)$, where $\{m_s\}$ is the output of the decoder, so that

$$x_t = A_0(t, \xi_s, s \leqslant t) + A_1(t, \xi_s, s \leqslant t)\theta_t \tag{6}$$

where $A_0(t; \xi_s, s \leqslant t)$ and $A_1(t; \xi_s, s \leqslant t)$ are two (nonanticipatory) coding functions.

In this framework, we address the problem of choosing the best transmitter-receiver structure for worst possible choices of jammer's policy $\{\beta_t\}$ under a quadratic distortion measure. Letting the associated probability measure for the jammer be μ, the quadratic distortion measure is given by

$$J(A_0, A_1, \theta, \mu) = E^\mu\{\int_0^{t_f} (\theta_t - \hat{\theta}_t)^2\}, \tag{7}$$

with the expectation E^μ taken over the statistics of $\{\theta_t\}$ and $\{\tilde{w}_t\}$, for a fixed probability distribution μ for $\{\beta_t\}$. Then, what we seek is a triple $\{A_0^*, A_1^*, \theta^*\}$ such that

$$\sup_\mu J(A_0^*, A_1^*, \theta^*, \mu) = \inf_{(A_0, A_1, \hat{\theta})} \sup_\mu J(A_0, A_1, \theta, \mu). \tag{8}$$

A stronger result would be the existence of a saddle-point $\{A_0^*, A_1^*, \theta^*, \mu^*\}$ satisfying

$$J(A_0^*, A_1^*, \theta^*, \mu) \leqslant J(A_0^*, A_1^*, \theta^*, \mu^*) \leqslant J(A_0, A_1, \theta, \mu^*) \tag{9}$$

for all permissible $\{A_0, A_1, \theta, \mu\}$.

An even more general formulation, that we consider, is the one where the encoder-decoder pair, together with the feedback loop, is allowed to be a mixed policy. Let η denote the corresponding probability measure, and the expected cost under η be denoted by

$$\tilde{J}(\eta, \mu) = E^\eta\{J(A_0, A_1, \theta, \mu)\} = \int_{A_0, A_1\hat{\theta}} J(A_0, A_1, \theta, \mu)d\eta(A_0, A_1, \theta). \tag{10}$$

Then we seek here a saddle-point solution (η^*, μ^*) satisfying

$$\tilde{J}(\eta^*, \mu) \leqslant \tilde{J}(\eta^*, \mu^*) \leqslant \tilde{J}(\eta, \mu^*) \tag{11}$$

for all permissible (η, μ).

A different version of this problem, where the jammer is allowed to tap the output of the encoder (instead of the input) has recently been considered in Reference [1] where it has been shown that a minimax linear causal encoder-decoder pair exists, and the least favorable probability distribution for the unknown jamming noise is Gaussian, correlated with the transmitted signal. Here, however, we will see in the sequel that the minimax encoder-decoder pair has a different structure.

2. Minimax Coding Decoding Schemes: The saddle-point solution to the problem formulated above (in the sense of (10)) is provided in the following theorem:

Theorem 1: The communication system described in Section 1 and depicted in Fig. 1 admits a unique mixed-type saddle-point solution, where

i) the corresponding transmitter-receiver pair is

$$+ (\sqrt{P/\gamma_t^*})\,(\theta_t - m_t), \ + \int_0^{t_f} [a(t)m_t^* dt + \frac{\sqrt{P\gamma_t^*}}{\sigma + k}\,d\xi_t], \quad \text{w.p.} \frac{1}{2}$$
(12)

$$- (\sqrt{P/\gamma_t^*})\,(\theta_t - m_t), \ - \int_0^{t_f} [a(t)m_t^* dt + \frac{\sqrt{P\gamma_t^*}}{\sigma + k}\,d\xi_t], \quad \text{w.p.} \frac{1}{2}$$

where

$$\dot{\gamma}_t^* = [2a(t) - \frac{P}{\sigma + k}]\gamma_t^* + b^2(t), \quad \gamma_0^* = \lambda$$
(13)

and $\{m_t^*\}$ satisfies

$$dm_t^* = a(t)m_t^* dt + \sqrt{P\gamma_t^*}\,d\xi_t, \quad m_0^* = 0$$
(14)

ii) the feedback-loop transformation $A_0(\cdot)$ is given by

$$A_0^*(t, m_s, s \leqslant t) = -A_1^*(t)m_t,$$
(15a)

and

$$A_1^*(t, \theta_t, \xi_t) = \begin{vmatrix} + (\sqrt{P/\gamma_t^*}) & \text{for +ve realization} \\ - \sqrt{P/\gamma_t^*} & \text{for } -\text{ve realization} \end{vmatrix}$$
(15b)

Hence, the encoder output is

$$x_t^* = A_0^*(t) + A_1^*(t)\theta_t,$$
(16)

and the decoder output is

$$\theta_t^* = m_t^*,$$ (17)

where the m_t^* is generated by (14);

iii) the jamming policy induced by the least favorable distribution is

$$\beta_t^* = \sqrt{k} \, w_t$$ (18)

which is independent of θ_t;

iv) the worst case distortion level is

$$\tilde{J}^* = \int_0^{t_f} \gamma_t^* dt.$$ (19)

Proof: The proof proceeds in two steps. We first establish validity of the right hand side (RHS) inequality of (11) when β^* is given by (16) and then prove the left-hand side (LHS) inequality of (11) when A_0^*, A_1^* and θ_t^* are given by (14), (15) and (12), respectively.

a) The RHS Inequality: With $\beta_t = \sqrt{k} \, \omega_t$, the channel output (3) can be written as

$$d\xi_t = x_t dt + \sqrt{k} \, dw_t + \sqrt{\sigma} \, d\tilde{w}_t$$ (20)

where

$$x_t = A_0(t) + A_1(t)\theta_t.$$ (21)

Since the noise process in (20) is equivalent to some Wiener process with incremental variance $(k + \sigma)dt$, the underlying encoder-decoder problem can be solved by basically following the steps of [Ref. 2, p. 184], to yield the optimum values

$$A_1^*(t, \xi_t) = \pm \sqrt{P/\gamma_t^*},$$

$$A_0^*(t, \xi_t) = -A_1^*(t, \xi_t)m_t^*,$$

and

$$\theta_t = m_t^*, \quad 0 \leqslant t \leqslant t_f,$$

where m_t^* is given by (14), and

$$\gamma_t^* = E[\theta_t - m_t^*]^2, \quad 0 \leqslant t \leqslant t_f$$

satisfies (13). The unique solution to (13) is

$$\gamma_t^* = \lambda \exp\{2\int\limits_0^t (a(s) - \frac{P}{2(\sigma+k)})ds\} + \int\limits_0^t b^2 \exp\{2\int\limits_s^t (a(u) - \frac{P}{2(\sigma+k)})du\}ds$$

and its integral over $(0, t_f)$ gives us the realizable minimum error (i.e, $\tilde{j}^* = \int\limits_0^{t_f} \gamma_t^* dt$). This completes the proof of validity of the RHS inequality of (11).

b) The LHS Inequality: Let A_0^*, and the decoder structure A_1^* be as given in the theorem. Then, for a "positive" realization of η^*,

$$x_t = (\sqrt{P/\gamma_t^*}) [\theta_t - m_t^+]$$

where m_t satisfies the stochastic differential equation

$$dm_t^+ = a(t)m_t^+ dt + \frac{1}{\sigma+k}\gamma_t^* A_1^+(t, \xi_t)d\xi_t, \quad m_0^+ = 0 \tag{22}$$

with A_1^+ denoting the positive realization of A_1^*.

Since

$$d\xi_t = (\sqrt{P/\gamma_t^*}) [\theta_t - m_t^+]dt + d\beta_t + \sigma^{1/2}d\tilde{w}_t \tag{23}$$

we obtain

$$dm_t^+ = a(t)m_t^+dt + \frac{1}{\sigma+k}\gamma_t^\bullet(A_1^+)^2[\theta_t - m_t^+]dt$$

$$+ \frac{1}{\sigma+k}\gamma_t^\bullet A_1^+d\beta_t + \frac{1}{\sigma+k}\gamma_t^\bullet A_1^+\sigma^{1/2}d\tilde{w}_t, ; \quad m_0^+ = 0. \tag{24}$$

Let $p_t^+ = m_t^+ - \theta_t$, which satisfies the stochastic differential equation

$$dp_t^+ = A_1 p_t^+ dt - \frac{1}{\sigma+k}\gamma_t^\bullet(A_1^+)^2 p_t^+ dt + \frac{1}{\sigma+k}\gamma_t^\bullet A_1^+ d\beta_t$$

$$+ \frac{1}{\sigma+k}\gamma_t^\bullet A_1^+\sigma^{1/2}d\tilde{w}_t - b_t dv_t, \quad p_0^+ = -\theta_0. \tag{25}$$

For a "negative" realization of $\eta^\bullet$, on the other hand,

$$x(t) = A_1^-[\theta_t - m_t^-], \quad A_1^-(t) = -A_1^+(t)$$

where m_t^- satisfies the SDE

$$dm_t^- = a(t)m(t)dt + \frac{1}{\sigma+k}\gamma_t^\bullet(A_1^-)^2[\theta_t - m_t^-]dt + \frac{1}{\sigma+k}\gamma_t^\bullet A_1^- d\beta_t$$

$$+ \frac{1}{\sigma+k}\gamma_t^\bullet A_1^-\sigma^{1/2}d\tilde{w}_t; \quad m_0^- = 0. \tag{26}$$

Letting $p_t^- = m_t^- - \theta_t$, we have

$$dp_t^- = a_t p_t^- dt - \frac{1}{\sigma+k}\gamma_t^\bullet(A_1^-)^2 p_t^- dt + \frac{1}{\sigma+k}\gamma_t^\bullet A_1^- d\beta_t$$

$$+ \frac{1}{\sigma+k}\gamma_t^\bullet A_1^-\sigma^{1/2}d\tilde{w}_t - b_t dv_t; \quad p_0^- = -\theta_0. \tag{27}$$

Now, in terms of p_t^+ and p_t^-, the distortion measure can be written as

$$\tilde{J}(\eta^*, \mu) = \frac{1}{2} E^\mu \{ \int_{t_0}^{t_f} [(p_t^+)^2 + (p_{t_f}^-)^2 dt\} \tag{28}$$

where μ is the probability measure yet to be determined by the jammer.

We now let $M_t^\pm = (p_t^\pm)^2$ and obtain the $M_t^\pm$ by direct application of Itô's formula:

$$dM_t^\pm = 2a_t M_t^\pm dt - \frac{2}{\sigma+k} \gamma_t^*(A_1^\pm)^2 M_t^\pm dt + \frac{2}{\sigma+k} p_t^\pm \gamma_t^* A_1^\pm d\beta_t$$

$$+ \frac{2}{\sigma+k} p_t^\pm \gamma_t^* A_1^\pm \sigma^{1/2} d\tilde{w}_t - 2p_t^\pm b_t dv_t + \frac{1}{(\sigma+k)^2} (\gamma_t^*)^2 (A_1^\pm)^2 (d\beta_t)^2 \tag{29}$$

$$+ \frac{\sigma}{(\sigma+k)^2} (\gamma_t^*)^2 (A_1^\pm)^2 (d\beta_t)^2 + b_t^2 dt, \quad M_0^\pm = \theta_0^2.$$

Finally, letting $\epsilon^\pm(t) = E^\mu[M_t^\pm]$, we obtain

$$\dot{\epsilon}^\pm = 2a_t \, \epsilon^\pm - \frac{2}{\sigma+k} \gamma_t^*(A_1^\pm)^2 \, \epsilon^\pm + \frac{2}{\sigma+k} \gamma_t^* A_1^\pm E^\mu[d\beta_t p_t^\pm]$$

$$\tag{30}$$

$$+ \frac{1}{(\sigma+k)^2} (\gamma_t^*)^2 (A_1^\pm)^2 E^\mu[d\beta_t]^2 + \frac{1}{(\sigma+k)^2} (\gamma_t^*)^2 (A_1^\pm)^2 \sigma + b_t^2, \quad \epsilon^\pm(o) = E[\theta_0^2] = \lambda.$$

Hence, $\epsilon = \frac{1}{2} \epsilon^+ + \frac{1}{2} \epsilon^-$ satisfies the differential equation:

$$\dot{\epsilon} = 2a_t \, \epsilon - \frac{2}{\sigma+k} \gamma_t^*(A_1^+)^2 \, \epsilon + \frac{1}{\sigma+k} \gamma_t^* A_1^+ E^\mu[d\beta_t(p_t^+ - p_t^-)]$$

$$\tag{31}$$

$$+ \frac{1}{(\sigma+k)^2} (\gamma_t^*)^2 (A_1^+)^2 E^\mu[d\beta_t]^2 + \frac{1}{(\sigma+k)^2} (\gamma_t^*)^2 (A_1^+)^2 \sigma + b_t^2, \quad \epsilon(o) = \lambda.$$

We now obtain an expression for $E^\mu[d\beta_t(p_t^+ - p_t^-)]$. Toward this end, let $\tilde{p}_t = p_t^+ - p_t^-$; then

$$d\tilde{p}_t = (a_t - \frac{P}{\sigma+k})\tilde{p}_t dt + \frac{2}{\sigma+k}\sqrt{P\gamma_t^*}\, d\beta_t + \frac{2}{\sigma+k}\sqrt{P\gamma_t^*}\, \sigma^{1\!/\!2} d\tilde{w}_t, \quad \tilde{p}_0 = 0 \tag{32}$$

leading to

$$\tilde{p}_t = \frac{2}{\sigma+k}\int_0^t \sqrt{P\gamma_s^*}\, e^{-\int_s^t (a_u - \frac{P}{\sigma+k})du}\, d\beta_s + \frac{2\sigma^{1\!/\!2}}{\sigma+k}\int_0^t \sqrt{P\gamma_s^*}\, e^{-\int_s^t (a_u - \frac{P}{\sigma+k})du}\, d\tilde{w}_s. \tag{33}$$

Hence, since β_t is independent of $\{\tilde{w}_t\}$,

$$E^\mu[d\beta_t \tilde{p}_t] = \frac{2}{\sigma+k}\int_0^t \sqrt{P\gamma_s^*}\, e^{-\int_s^t (a_u - \frac{P}{\sigma+k})du}\, E^\mu(d\beta_t d\beta_s). \tag{34}$$

Therefore, the maximization problem faced by the jammer is

$$\max_\mu E^\mu\{\frac{1}{2}\int_{t_0}^{t_f}[(p_t^+)^2 + \frac{1}{2}(p_t^-)^2]dt\} \equiv \max_\mu \int_{t_0}^{t_f} \epsilon(t)dt \tag{35}$$

where $\epsilon(t_f)$ is the solution of

$$\dot{\epsilon} = 2(a_t - \frac{P}{\sigma+k})\,\epsilon + \frac{2}{(\sigma+k)^2}\sqrt{P\gamma_t^*}\int_0^t \sqrt{P\gamma_s^*}\, e^{-\int_s^t (a_u - \frac{P}{\sigma+k})}\, duE^\mu(d\beta_t d\beta_s)$$

$$\tag{36}$$

$$+ \frac{P\gamma_t^*}{(\sigma+k)^2}E^\mu[d\beta_t]^2 + \frac{P\gamma_t^*}{(\sigma+k)^2}\sigma + b_t^2; \quad \epsilon(o) = \lambda.$$

Now, when the jammer's policy is restricted to the structure

$$\beta_t = \int_{t_0}^t B(s, \theta_s)dw_s$$

we have for $t > s$,

$$E^\mu[d\beta_t d\beta_s] = 0, \quad E^\mu[d\beta_t^2] = E^\mu[B^2(t, \theta_t)]dt.$$

Hence,

$$\epsilon = 2(a_1 - \frac{P}{\sigma+k}) \epsilon + \frac{1}{(\sigma+k)^2} P\gamma_t^*[\sigma + E^\mu[B^2(t, \theta_t)]] + b_t^2, \quad \epsilon(0) = \lambda. \tag{37}$$

Note that the jammer's maximization problem (35) depends only the second moment of B at each point in time, and thus could be chosen to be independent of θ_t. Letting

$$E^\mu[B^2(t, \theta_t)] \triangleq \tilde{B}(t)$$

we have

$$\epsilon(t) = \lambda e^{-2\int_0^t (a_u - \frac{P}{\sigma+k})du} + \frac{1}{(\sigma+k)^2} \int_0^t P\gamma_s^*[\sigma + \tilde{B}(s)] e^{-2\int_s^t (a_u - \frac{P}{\sigma+k})du} ds$$

$$\tag{38}$$

$$\leq \lambda e^{-2\int_0^t (a_u - \frac{P}{\sigma+k})du} + \frac{1}{(\sigma+k)^2} \int_0^t P\gamma_s^*[\sigma + \max_s \tilde{B}(s)] e^{-2\int_s^t (a_u - \frac{P}{\sigma+k})du} ds$$

and this upper bound can be attained by choosing

$$\tilde{B}(s) = k \quad \forall s \in [0, t].$$

Since this is also true for all $t \leq t_f$, a uniformly best policy for the jammer is

$$\beta_t = \sqrt{k} \, w_t.$$

Substitution of $B^2(t) = k$ in (37) leads to the differential equation (also noting that $\epsilon_t = \gamma_t^*$)

$$\dot{\gamma}_t^* = (2a_1 - \frac{P}{\sigma+k})\gamma_t^* + b_t^2, \quad \gamma_0^* = \lambda \tag{39}$$

which is identical to (13). This then completes the proof of the LHS inequality for Theorem 1.

Remark 1: When the message to be transmitted is a Gaussian random variable (i.e., a = 0, b = 0 in the formulation of the problem), the saddle point distortion level J^* admits a closed-form expression (from (38))

$$J^* = \int_{t_0}^{t_f} \gamma_t^* dt = \frac{\lambda(\sigma+k)}{P} [e^{-P\frac{t_0}{\sigma+k}} - e^{-\frac{Pt_f}{\sigma+k}}].$$

□

3. <u>Conclusion</u>: This paper has studied the problem of transmitting a Gaussian stochastic process $\{\theta_t, t \in [0, t_f]\}$ over a continuous-time channel which is subjected to an unknown jamming noise under a given power constraint. The noise process has been allowed to be correlated with the message process, and also a noiseless feedback has been allowed between the receiver and the transmitter. Under a quadratic distortion measure, we have obtained a saddle-point solution with probabilistic encoder structure, linear receiver structure, and a least favorable jamming policy given by a Wiener intergal.

Acknowledgements: Research of the first author was partially supported by the Illinois Institute of Technology Integrated Information and Telecommunications Systems Center, and research of the second author was supported in part by the Joint Services Electronics Program under Contract N00014-84-C-0149.

References

[1] T. Ü. Başar and T. Başar, "Optimum linear casual coding schemes for Gaussian stochastic processes in the presence of correlated jamming," *IEEE Transactions on Information Theory*, January 1989 (to appear).

[2] R. S. Liptser and A. N. Shiryayev, *Statistics of Random Processes II Applications*, Springer-Verlag, 1978.

EFFICIENT MSK SPREAD-SPECTRUM SIGNALING

J. S. Lehnert

Purdue University

1. Introduction. We are concerned with a direct-sequence spread-spectrum multiple-access (DS/SSMA) communication system that employs a minimum-shift keyed (MSK) signal format in the following. Minimum-shift keyed signaling may be preferred in DS/SSMA systems because it is a quaternary signaling scheme, it possesses desirable spectral characteristics, and it yields transmitted signals that have a constant envelope. In [1], such a system is examined as a quaternary system with two channels, a quadrature channel and an in-phase channel, i.e., it is examined as a parallel system. The signal-to-noise ratio (SNR) is obtained, and expressed as a function of certain parameters of the signature sequences. A good multiple-access capability in the system can be achieved by finding sequences with good correlation parameters.

In [2], a simplified signaling scheme for nonspread data transmission is shown to require fewer hardware components and to still achieve all the advantages of MSK systems implemented with a quadrature and an in-phase channel. The simplified scheme requires only one channel on which binary data transmission occurs at twice the rate of transmission on each of the two channels in the parallel implementation of MSK. This single-channel system, called serial MSK, produces exactly the same transmitted signal as the signal produced by the corresponding parallel system.

We are interested in serial MSK implementations of DS/SSMA systems. In particular, we are interested in the problem of choosing sequences for such systems to achieve a good multiple-access capability. Our approach is to first find sequences that produce a good multiple-access capability for a parallel MSK system. We then determine what sequences must be used by a serial MSK transmitter to produce the same transmitted signal that is produced by the parallel implementation of the MSK transmitter. This identifies the good sequences for the serial transmitter that correspond to the good sequences for the parallel transmitter.

We begin with a definition of the parallel MSK transmitter and receiver. We then describe the multiple-access channel and provide an alternative analysis to that in [1] that yields the SNR of both parallel and serial DS/SSMA systems. This expression for the SNR can be used to find good choices for the sequences in both parallel and serial MSK DS/SSMA systems. We demonstrate this by describing the serial MSK transmitter and determining the input that produces the same transmitted signal as the parallel system produces. When the transmitted signals are the same, the multiple-access capabilities of the systems are equivalent. Finally, we briefly describe potential serial receivers.

2. Parallel System Model. In the following we describe the model of the MSK DS/SSMA system for the parallel implementation of the transmitter and receiver. The corresponding serial transmitter is described in a later section.

2.1. Parallel Transmitter Model. In this section the transmitted MSK signal is defined as the sum of an in-phase component and a quadrature component. The k-th transmitter generates a pair of data signals $b_{2k-1}(t)$ and $b_{2k}(t)$ that are given by

$$b_n(t) = \sum_{\lambda=-\infty}^{\infty} b_\lambda^{(n)} p_T(t-\lambda T) \tag{1}$$

for $n=2k-1$ and $n=2k$, where the unit rectangular pulse $p_T(t)$ is given by $p_T(t) = 1$ for $0 \le t < T$, and $p_T(t) = 0$, otherwise. The data symbols $b_\lambda^{(n)}$ are independent and identically distributed random variables for which $\Pr\{b_\lambda^{(n)} = +1\} = \Pr\{b_\lambda^{(n)} = -1\} = 1/2$. An even integer n corresponds to an in-phase signal component, and an odd integer n corresponds to a quadrature signal component.

The k-th transmitter is assigned two binary sequences $(a^{(2k-1)})$ and $(a^{(2k)})$ for the quadrature channel and the in-phase channel, respectively. The sequence $(a^{(n)})$ is used to form the spectral spreading signal $a_n(t)$ that is given by

$$a_n(t) = \sum_{\lambda=-\infty}^{\infty} a_\lambda^{(n)} \psi(t-\lambda T_c), \tag{2}$$

where $\psi(t)$ is the common chip waveform for all signals. The chip waveform $\psi(t)$ is time limited to the interval $[0,T_c]$, where the chip duration T_c satisfies the relationship $T_c = T/N$, and N is the period of the signature sequences used in each of the two channels. For the MSK signals that we are considering, the chip waveform is the sine pulse given by $\psi(t) = \sqrt{2}\sin(\pi t/T_c) p_{T_c}(t)$.

The transmitted signals can be specified using the analytic signal representation described in [3]. The k-th transmitter sends $\text{Re}[z_k(t-T_k)]$, where

$$z_k(t) = \sigma_k(t)\exp(j\omega_0 t) \tag{3}$$

and

$$\sigma_k(t) = [b_{2k}(t-t_0)a_{2k}(t-t_0) - jb_{2k-1}(t)a_{2k-1}(t)]\exp(j\alpha_k). \tag{4}$$

In the above expressions, α_k is the carrier phase for the k-th transmitter, t_0 is the offset parameter needed for MSK, and T_k is a random variable that models the asynchronous system. We model each time delay T_k, for $k \neq i$, as a random variable that is uniformly distributed on the interval $[0,T)$ and independent of all other random variables that define the system. We model each carrier phase α_k, for $k \neq i$, as a random variable that is uniformly distributed on the interval $[0,2\pi)$ and independent of all other random variables that define the system. We choose the offset parameter $t_0 = T_c/2$, although other delays satisfying $t_0 = T_c(2\nu+1)/2$ also generate MSK signals. The transmitter is shown in Figure 1.

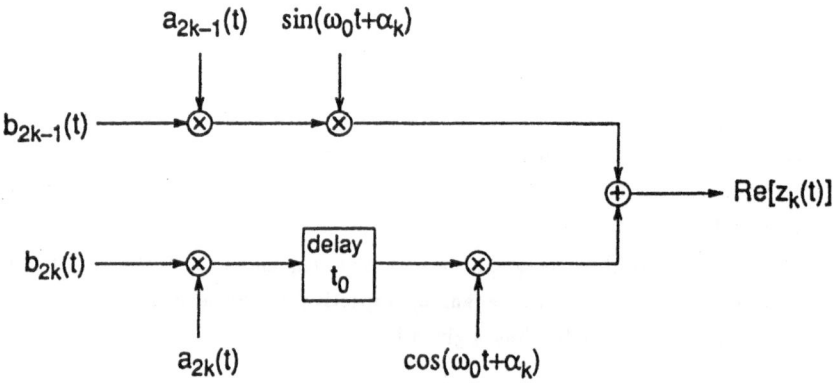

Figure 1. Parallel transmitter.

2.2. Receiver Model. The receiver consists of a quadrature section and an in-phase section. The parameter n specifies a particular section of the receiver. For the i-th receiver we are concerned with n=2i−1 and n=2i, and in general, $1 \leq n \leq 2K$. The impulse response of the matched filter in section n is given by $\text{Re}[h_n(t)]$, where

$$h_n(t) = 2 \left[\sum_{\lambda=0}^{N-1} a_\lambda^{(n)} \psi(T-t-\lambda T_c) \exp(j\omega_0 t) \right]. \tag{5}$$

For detection of an in-phase signal, corresponding to an even value of n, the output of the filter is multiplied by $2\cos(\omega_0 t + \phi_i)$, where $\phi_i = \alpha_i - \omega_0 T_i$, i.e., ϕ_i is a phase determined at the receiver to maintain phase coherence. For detection of a quadrature signal, corresponding to an odd value of n, the output of the filter is multiplied by $2\sin(\omega_0 t + \phi_i)$. In each case, the resulting product is passed through a low-pass filter, sampled, and compared with a threshold. A decision on the bit $b_\lambda^{(2i-1)}$ is made at $t=(\lambda+1)T+T_i$, and a decision on the bit $b_\lambda^{(2i)}$ is made at $t=(\lambda+1)T+t_0+T_i$. The structure of the receiver is shown in Figure 2 for the case in which $\alpha_i=T_i=\phi_i=0$.

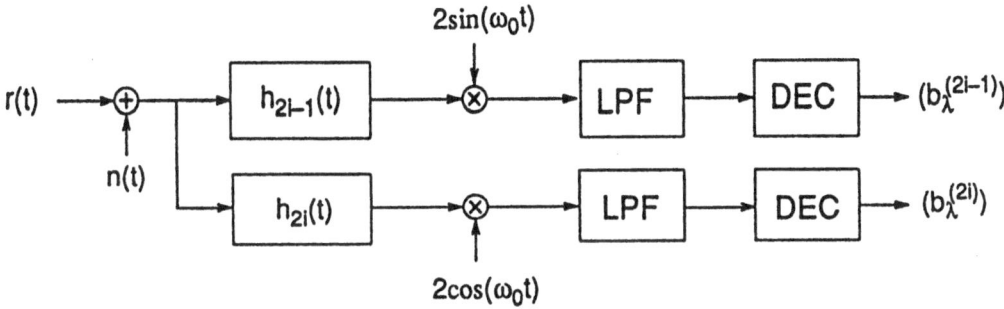

Figure 2. Parallel receiver.

2.3. Channel Model. In this section we describe the channel model for the asynchronous system. The k-th transmitted signal is delayed by τ_k and the phase of the k-th transmitted signal is shifted by θ_k. Hence, the received signal from the k-th transmitter is $Re[z_k(t-T_k-\tau_k)e^{j\theta_k}]$, and the total received signal is given by $r(t)=\sum_{k=1}^{K}Re[z_k(t-T_k-\tau_k)e^{j\theta_k}]$.

Because of the structure of the transmitted signals and because the thermal noise that is added at the receiver is stationary, the time delays $T_k+\tau_k$, $1\leq k\leq K$, are only important modulo T. We model each time delay T_k, for $k\neq i$, as a random variable that is uniformly distributed on $[0,T)$ and independent of all other random variables that define the system. Hence, the random variable $T_k+\tau_k$ considered modulo T is uniformly distributed on $[0,T)$, and we lose no generality by assuming that the channel delay τ_k has been incorporated into the delay T_k, i.e., by assuming that $\tau_k=0$. The received carrier phases $\alpha_k+\theta_k$, $1\leq k\leq K$, are only important modulo 2π. We model each phase α_k, for $k\neq i$, as a random variable that is uniformly distributed on $[0,2\pi)$ and independent of all other random variables that define the system. Hence the random variable $\alpha_k+\theta_k$ considered modulo 2π is uniformly distributed on $[0,2\pi)$, and we lose no generality by assuming that the channel phase shift θ_k has been incorporated into the phase α_k, i.e., by assuming that $\theta_k=0$. Since only relative time delays and phases are important and we are studying the i-th receiver, we set $T_i=0$ and $\alpha_i=0$. The multiple-access communication channel is depicted in Figure 3.

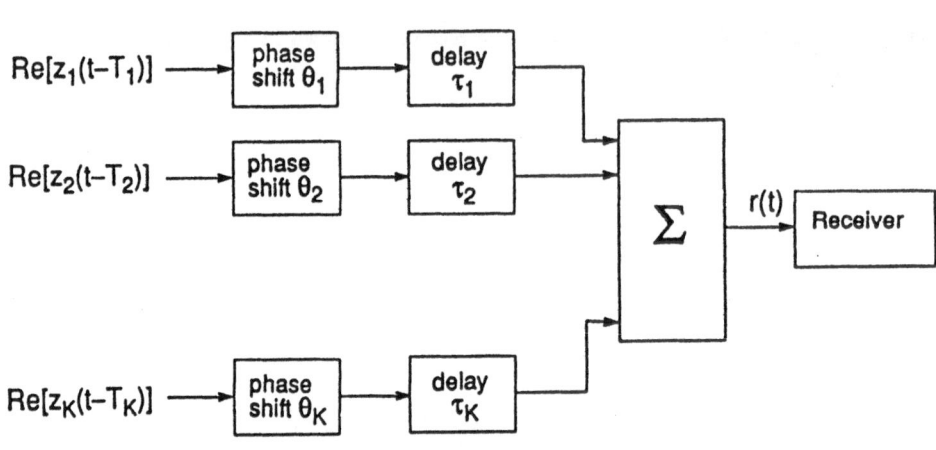

Figure 3. Multiple-access communication channel.

3. Multiple-Access Interference from the k-th Transmitter. In this section we compute the variance of the multiple-access interference (MAI) resulting from the k-th interfering transmitter by first computing the autocorrelation function of the wide-sense stationary random process modeling the received signal from the k-th transmitter. We then compute the autocorrelation function of the wide-sense stationary (WSS) process at the output of the matched filter and evaluate this function at zero.

3.1. Autocorrelation Function of Received Signal from k-th Transmitter. The autocorrelation function of the WSS random process modeling the received signal from the k-th transmitter is given by

$$\frac{1}{2}E\{z_k(t+\tau-T_k)z_k^*(t-T_k)\} = \frac{1}{2}E\{\sigma_k(t+\tau-T_k)\sigma_k^*(t-T_k)\}\exp(j\omega_0\tau) \tag{6}$$

$$= R_k(\tau)\exp(j\omega_0\tau),$$

where

$$R_k(\tau) = \frac{1}{2}E\{\sigma_k(t+\tau-T_k)\sigma_k^*(t-T_k)\}. \tag{7}$$

Evaluating this autocorrelation function yields

$$R_k(\tau) = \frac{1}{2T}[C(2k,2k)(\tau) + C(2k-1,2k-1)(\tau)], \tag{8}$$

where

$$C(m,m)(t) = \int_0^T a_m(x-t)p_T(x-t)a_m(x)dx, \tag{9}$$

i.e., $C(m,m)(t)$ is the continuous autocorrelation function of the spectral spreading waveform.

3.2. Autocorrelation Function of WSS Output Process. The autocorrelation function of the WSS random process modeling the signal at the output of the matched filter resulting from the k-th transmission can be found by performing two convolutions. Convolving the input autocorrelation function with the impulse response of the matched filter and then with the complex conjugate of the time-reverse of the impulse response yields the autocorrelation function of the output process after matched filtering $R_M(k,n;\tau)$, i.e.,

$$R_M(k,n;\tau) = \frac{1}{4}\int_{-\infty}^{\infty}\int_{-\infty}^{\infty}[R_k(s)\exp(j\omega_0 s)]h_n(u)h_n^*(s+u-\tau)dsdu. \tag{10}$$

3.3. Evaluation of the Variance of the MAI from the k-th Transmission. The variance of the multiple-access interference at the output of the matched filter resulting from the k-th transmission at the sampling time can be found by evaluating the autocorrelation function $R_M(k,n;\tau)$ at zero. Using (5), (9), and (10) yields an expression for the variance $N_M(k)$ given by

$$N_M(k) = R_M(k,n;0) = \int_{-T}^{T} R_k(x)C(n,n)(x)dx. \tag{11}$$

Substituting (8) in (11) yields

$$N_M(k) = T^2[\sigma_{2k,n}^2 + \sigma_{2k-1,n}^2], \tag{12}$$

where

$$\sigma^2_{m,n} = \tfrac{1}{2}T^{-3} \int\limits_{-\infty}^{\infty} C(m,m)(x)C(n,n)(x)dx \ . \tag{13}$$

However, since $C(m,m)(t)$ can be expressed as

$$C(m,m)(t) = T_c^{-1}C(m,m)(\lambda T_c)\hat{R}_\psi(s) + T_c^{-1}C(m,m)[(\lambda+1)T_c]R_\psi(s) \tag{14}$$

$$= C_{m,m}(\lambda)\hat{R}_\psi(s) + C_{m,m}[(\lambda+1)]R_\psi(s),$$

where $\lambda = \lfloor t/T_c \rfloor$, $s = t - \lambda T_c$, $R_\psi(s) = p_{T_c}(t)\int\limits_0^s \psi(t)\psi(t+T_c-s)dt$, $\hat{R}_\psi(s) = p_{T_c}(t)\int\limits_s^{T_c} \psi(t)\psi(t-s)dt$, and $C_{m,m}(i)$ is the discrete aperiodic autocorrelation function of the m-th signature sequence [1]. Hence, substituting (14) in (13) and evaluating yields

$$\sigma^2_{m,n} = T^{-3}[\mu_{m,n}(0)M_\psi + \mu_{m,n}(1)M'_\psi] \ , \tag{15}$$

where $M_\psi = \int\limits_0^{T_c} R_\psi^2(s)ds = T_c^3(15+2\pi^2)/(12\pi^2)$, $M'_\psi = \int\limits_0^{T_c} R_\psi(s)\hat{R}_\psi(s)ds = T_c^3(15-\pi^2)/(12\pi^2)$, and

$$\mu_{m,n}(L) = \sum_{\lambda-1-N}^{N-1} C_{m,m}(\lambda)C_{n,n}(\lambda+L). \tag{16}$$

Hence, from (12) and (15) the variance of the interference from the k-th transmitter can be expressed as

$$N_M(k) = \frac{1}{T}[\mu_{2k,n}(0)M_\psi + \mu_{2k,n}(1)M'_\psi + \mu_{2k-1,n}(0)M_\psi + \mu_{2k-1,n}(1)M'_\psi] \ . \tag{17}$$

4. Signal-to-Noise Ratio. Since the WSS random processes modeling the interferences from the various transmissions are independent, the total multiple-access interference is given by

$$N_M = \frac{1}{T}\sum_{\substack{k-1 \\ k \neq i}}^{K}[\mu_{2k,n}(0)M_\psi + \mu_{2k,n}(1)M'_\psi + \mu_{2k-1,n}(0)M_\psi + \mu_{2k-1,n}(1)M'_\psi] \ . \tag{18}$$

The signal contribution from transmitter i has a magnitude T. Hence, the SNR at the i-th receiver is given by

$$\text{SNR}_i = T/\sqrt{N_M + N_T}, \tag{19}$$

where N_T is the variance of the thermal noise. Using (18) yields as in [1]

$$\text{SNR}_i = \{T^{-3} \sum_{\substack{k=1 \\ k \neq i}}^{K} [\mu_{2k,n}(0)M_\psi + \mu_{2k,n}(1)M'_\psi + \mu_{2k-1,n}(0)M_\psi + \mu_{2k-1,n}(1)M'_\psi] + \frac{N_0}{2E_b}\}^{-1/2}, \tag{20}$$

where E_b is the energy transmitted per data bit.

5. Serial Implementation. In the serial implementation of the k-th transmitter, the data signal $c_k(t) = c_\lambda^{(k)} p_{T_0}(t - \lambda T_0)$ is multiplied by the signal $A\cos(2\pi f_1 t)$ and the resulting signal is applied to the input of a filter with impulse response $h(t) = \sin(2\pi f_2 t)$. If the parameters defining the serial transmitter are chosen appropriately, the generated signal is identical to that produced by the parallel transmitter. In particular, consider some fixed positive integer m, some constant A, and some odd N. If $T_0 = T_c/2$, $f_1 = m/(2T_0)$, $f_2 = (m+1)/(2T_0)$, and the sequence $(c_\lambda^{(k)})$ with period $M = 2N$ is chosen carefully, the same signal is generated by the serial transmitter as is generated by the parallel transmitter. The constant A can be adjusted to match the powers. The precise relationship between T_0 and the frequencies f_1 and f_2 can be relaxed for large m. For large m, $f_1 = f_0 - 1/(4T_0)$ and $f_2 = f_0 + 1/(4T_0)$, where $f_0 = (f_1 + f_2)/2$. The serial transmitter is shown in Figure 4.

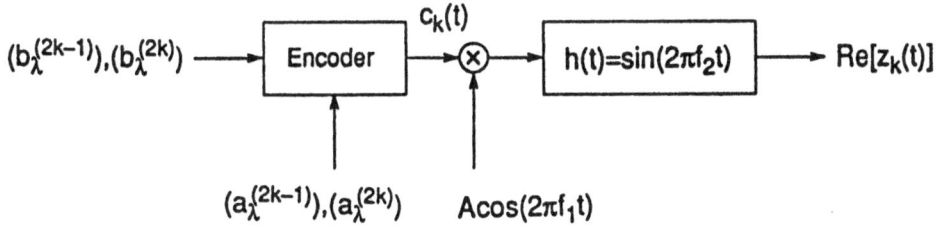

Figure 4. Serial MSK DS/SSMA transmitter.

Notice that an encoder is shown that has as inputs the data stream and the sequences from each of the two channels in the parallel system. This encoder produces the appropriate sequence $(c_\lambda^{(k)})$, and hence the appropriate signal $c_k(t)$. The encoder is defined by the equations

$$c^{(k)}_{\lambda+iM} = b_i^{(2k-1)}a_{\lambda/2}^{(2k-1)}, \quad c^{(k)}_{\lambda+1+iM} = -b_i^{(2k)}a_{\lambda/2}^{(2k)} \text{ for } \lambda=0,4,...,M-2 \tag{21}$$

$$c^{(k)}_{\lambda+iM} = -b_i^{(2k-1)}a_{\lambda/2}^{(2k-1)}, \quad c^{(k)}_{\lambda+1+iM} = b_i^{(2k)}a_{\lambda/2}^{(2k)} \text{ for } \lambda=2,6,...,M-4.$$

The signal may be detected as before with the parallel receiver shown in Figure 2. In that case the SNR is given by (20) as before. Alternatively, a single channel receiver can be implemented for the case in which the same data bit is sent on each channel during a symbol duration, or for the case in which the same sequence is used for transmission over the quadrature and in-phase channels. The analysis can be modified to accommodate these cases.

6. Conclusions. It is possible to implement an MSK DS/SSMA system in the parallel or serial mode. In the serial mode the hardware is simpler and a single channel system is possible. By precisely stating the relationship between the parallel and serial modes of transmission, results are obtained which provide the means to evaluate the effects of sequence selection on the multiple-access capability of the system.

7. References.

[1]   M. B. Pursley, F. D. Garber, and J. S. Lehnert, "Analysis of generalized quadriphase spread-spectrum communications," *Conference Record, IEEE International Conference on Communications*, June 1980, vol. 1, pp. 15.3.1-15.3.6.

[2]   Amoroso, F., and Kivett, J. A., "Simplified MSK signaling technique," *IEEE Transactions on Communications*, vol. COM-25, pp. 433-441, April 1977.

[3]   M. Schwartz, W. R. Bennett, and S. Stein, *Communication Systems and Techniques.* New York: McGraw-Hill, 1966.

The author is with the School of Electrical Engineering, Purdue University, West Lafayette, IN 47907.

ON PARAMETER ESTIMATION IN DS/SSMA FORMATS

H. Vincent Poor
University of Illinois

1 Introduction

During the past few years there has been a growing interest in the problem of optimum demodulation in multiple user communications. This interest is motivated largely by the work of Verdú [1], in which it is shown that the performance to be gained by optimum, as opposed to conventional, demodulation in multiuser environments can be remarkable. This is particularly evident in coherent formats, such as Direct-Sequence Spread-Spectrum Multiple-Access (DS/SSMA), in which such well-known limitations as the near-far problem are seen to be eliminated.

In addition to the usual requirements of coherent demodulation (i.e., carrier phaselock and symbol synchronization), the use of optimum demodulation requires knowledge of the relative amplitudes at which the various users in the channel arrive at the receiver. Thus, a problem that must be addressed before optimum demodulators such as those proposed in [1] can become practical, is how one would perform this additional required amplitude estimation. The purpose of this paper is to formulate the general problem of multiuser amplitude estimation in coherent formats, and to suggest some promising solution techniques that can be brought to bear on this problem. In addition to the potential utility of such solutions in the optimum multiuser demodulation problem, this formulation serves as a paradigm for the general problem of multiuser parameter estimation. This latter problem is an interesting and promising research topic in its own right, in view of the important gains that have been demonstrated in the multiuser detection problem in [1]. Moreover, nonlinear processing techniques for direct-sequence spread-spectrum demodulators, such as the nonlinear correlation receiver proposed by Aazhang and Poor [2], and the adaptive nonlinear interference suppression filters of Poor and Vijayan [3], require knowledge of (or adaptivity to) the signal amplitude for optimum operation. Thus, the techniques described here may be of interest in a number of other single-user and multiuser communications applications as well.

This paper is organized as follows. In Section 2, we formulate the coherent multiuser communication model of interest, and review the problem of optimum demodulation within this context. In Section 3, we briefly review amplitude-free demodulation techniques, referring the reader to [4] in this same volume for a more thorough discussion of this variant on multiuser demodulation. The proposed multiuser estimation techniques can be categorized as being based either on maximum-likelihood data recovery or on minimum-error-probability (i.e., Bayesian) data recovery. Amplitude estimation techniques based on these two approaches are developed in Sections 4 and 5, respectively. Some conclusions and interesting problems for further study are discussed in Section 6.

2 Optimum Multiuser Demodulation

In this section, we describe the multiuser communications model of interest, and discuss briefly the structure of optimum demodulators for this problem as developed by Verdú [1].

Consider the following model for a received waveform consisting of K superimposed modulated data signals observed in additive white Gaussian noise:

$$r(t) = S_t(\mathbf{b}) + \sigma N_t, \quad -\infty < t < \infty \tag{2.1}$$

where

$$S_i(\mathbf{b}) = \sum_{k=1}^{K} A_k \sum_{i=-M}^{M} b_k(i) s_k(t - iT - \tau_k), \tag{2.2}$$

and where the parameters and other quantities are defined as follows:

K = # users
A_k = received amplitude of the kth user
$(2M + 1)$ = frame length
$b_k(i)$ = i^{th} symbol of the kth user (binary , ± 1)
b = $K \times (2M + 1)$ matrix consisting of all symbols of all users
$s_k(t)$ = waveform of the kth user ($s_k(t) = 0, t \notin [0, T]$; $\int_0^T s_k^2(t) dt = 1$)
T = symbol interval
τ_k = relative delay of the kth user.

The data signals may be asynchronous, in which case the relative delays with which the various data signals arrive at the receiver are different from one another. It is assumed throughout that these delays, and the normalized signaling waveforms are known to the receiver.

Note that the principal thing that distinguishes multiuser formats of the type described in (2.1) from one another is the choice of the set of signaling waveforms (i.e., the signal constellation). One of the most important formats of this type is the Direct-Sequence Spread-Spectrum Multiple-Access (DS/SSMA) format, which corresponds to a set of signaling waveforms of the form

$$s_k(t) = \begin{cases} 2^{\frac{1}{2}} a_k(t) \sin(\omega_c t + \phi_k) & , \quad t \in [0, T] \\ 0 & , \quad t \notin [0, T] \end{cases} \tag{2.3}$$

where ω_c is a common carrier frequency, ϕ_k is the phase of the kth user relative to some reference, and where the spreading waveforms $a_k(t)$ are of the form:

$$a_k(t) = \sum_{\ell=0}^{N-1} a_\ell^k \psi(t - \ell T_c). \tag{2.4}$$

Here, a_ℓ^k is a signature sequence of +1's and -1's assigned to the kth user, and ψ is a normalized "chip" waveform of duration T_c (where $NT_c = T$). The signature sequences are typically chosen to be shift-register sequences with autocorrelation and cross-correlation properties that reduce multipath, multiple-access interference, and unintended detectability. Since these requirements usually call for flat spectral properties, such signature sequences are often called *pseudonoise* (PN) sequences and the bandwidth of the underlying data signal is spread by a factor of N .

The conventional approach to the demodulation of multiuser formats of the form (2.1) - (2.2) is to demodulate each user as if the other users were not in the channel. This approach leads to a demodulator that estimates the ith bit of the kth user as:

$$\hat{b}_k(i) = sgn\{y_k(i)\} = \begin{cases} +1, & \text{if} \quad y_k(i) \geq 0 \\ -1, & \text{if} \quad y_k(i) < 0 \end{cases} \tag{2.5}$$

where

$$y_k(i) \equiv \int_{\tau_k + iT}^{\tau_k + (i+1)T} s_k(t - iT - \tau_k) r(t) dt, \quad -M \leq i \leq +M \tag{2.6}$$

is the output of a filter matched to the kth user's signaling waveform and sampled at the end of the ith symbol interval of the kth user. A realization of this receiver for the particular case of DS/SSMA signals (2.3) - (2.4) is shown in Fig. 2.1.

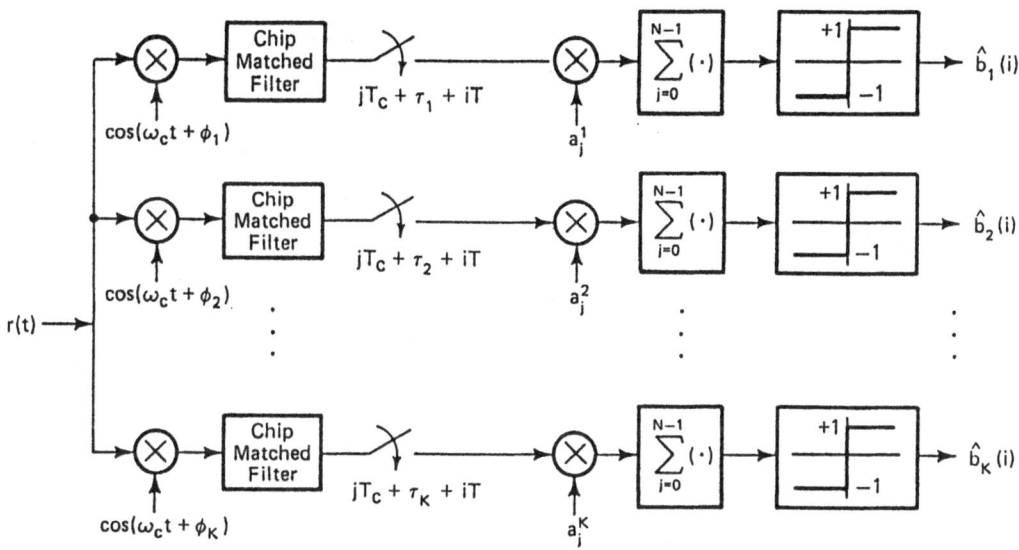

Figure 2.1. Conventional System for Simultaneous Demodulation
of K Asynchronous DS/SSMA Users

Note that this conventional demodulator would be optimum against the white Gaussian noise
background (see, e.g., Poor [5]) but is not optimum against the highly structured and nonGaussian
multiple-access interference caused by the presence of the other $K-1$ users in the channel. Nevertheless, this system works well under most circumstances provided that the signal constellation is
well-designed. It does, however, suffer from some well-known and serious disadvantages, including:
the *near-far problem*, which refers to the phenomenon of high-power interferers completely destroy-
ing communications from lower power transmitters; and the fact that it can become multiuser noise
limited when the signal-to-background-noise ratio is large. Moreover, even for spread-spectrum
systems not suffering from either of these two problems, the complexity required of the signal constellation in order to achieve good performance is such that a loss in bandwidth efficiency by a
factor of about ten is experienced relative to traditional methods of multiplexing (such as FDMA).

These disadvantages of conventional multiuser demodulation raise the question of whether significant performance improvement might be possible by employing demodulation schemes that are
optimum against both the white Gaussian background noise and the multiple-access interference.
The general structure of such schemes can be inferred by examining the likelihood function of the
observed waveform (2.1), conditioned on the knowledge of all data symbols. On assuming that
the signal amplitudes $A_1, A_2, \ldots, A_K$ are known, this likelihood function can be written via the
Cameron-Martin formula [5] as

$$\ell(\{r(t); -\infty < t < \infty\}) = C \exp\{\Omega(\mathbf{b})/2\sigma^2\} \qquad (2.7)$$

where

$$\Omega(b) = 2 \int_{-\infty}^{\infty} S_t(b)r(t)dt - \int_{-\infty}^{\infty} S_t^2(b)dt \tag{2.8}$$

where, as before, b is the $K \times (2M+1)$ matrix of data symbols. Note that the part of (2.8) that depends on the received waveform can be written as

$$\int_{-\infty}^{\infty} S_t(b)r(t)dt = \sum_{k=1}^{K} A_k \sum_{i=-M}^{M} b_k(i)y_k(i) \tag{2.9}$$

from which it follows that the matrix of matched filter outputs $\{y_k(i); i = -M,\ldots,M; k = 1,\ldots,K\}$, forms as sufficient statistic for the matrix b of data symbols (and for the vector $\underline{A} = (A_1, A_2, \ldots, A_k)^T$ of received amplitudes as well). Thus, the general structure of optimum procedures for determining the data symbols from the received waveform consists of an analog front-end that extracts the matched filter outputs, followed by a decision algorithm that infers optimum decisions from the collection of these outputs. A realization of this structure for the particular case of DS/SSMA signaling is illustrated in Fig. 2.2.

The nature of the decision algorithm in this process depends on the optimality criterion that one wishes to apply to the decision. If one adopts a maximum-likelihood (ML) philosophy, in which the optimum decision solves

$$\hat{b} = \arg \left\{ \max_{b \in \{-1,1\}^{K(2M+1)}} \Omega(b) \right\} \tag{2.10}$$

then, as shown by Verdú [1], the optimum decision algorithm can be implemented as a dynamic program of the forward type having time complexity per binary decision that is $O(2^K)$. Alternatively, one could assume a statistical model for the data streams, and adopt a minimum-error-probability (MEP) criterion:

$$\hat{b}_k(i) = \arg\{\min P\left(\hat{b}_k(i) \neq b_k(i)\right)\}. \tag{2.11}$$

Under the common assumption that the data streams of the users are independent sequences of independent and identically distributed (i.i.d.) binary random variables with $P(b_k(i) = +1) = P(b_k(i) = -1) = 1/2$, (2.11) can be solved with a backward-forward type of dynamic program of complexity comparable to that of the ML algorithm (Verdú [1], Verdú and Poor [6]). It should be noted that either of these decision algorithms requires knowledge of the received amplitudes for implementation, a fact which is more or less obvious upon examination of (2.9).

As shown by Verdú in [1] and [7], the use of the optimum demodulators described above alleviates the performance disadvantages exhibited by the conventional demodulator for coherent multiuser formats. Moreover, for reasonably high bit-energy-to-background-noise ratios, performance very near that of single-user communications (and hence close to traditional multiplexing schemes) is possible with the optimum multiuser demodulator. Thus, the disadvantages that were once thought to be inherent limitations of coherent multiuser formats such as DS/SSMA, are now known to be limitations only of the conventional demodulation technique.

The much improved performance afforded by the optimum multiuser algorithms comes at the expense of three possible disadvantages: centralized implementation; higher complexity; and the need to know the relative amplitudes of the various users' signals at the receiver. The first two of these are basically implementation considerations. Centralized processing is often used even with the conventional type of demodulator, particularly in satellite links and in master nodes of communication networks. (Optimum *decentralized* multiuser detectors are derived by Poor and Verdú in [8]). Similarly, since the front-end complexity of the optimum demodulator is identical to that of the conventional demodulator, the additional complexity is in the software stage and thus may not be overly burdensome. (Lower complexity centralized demodulators have been studied

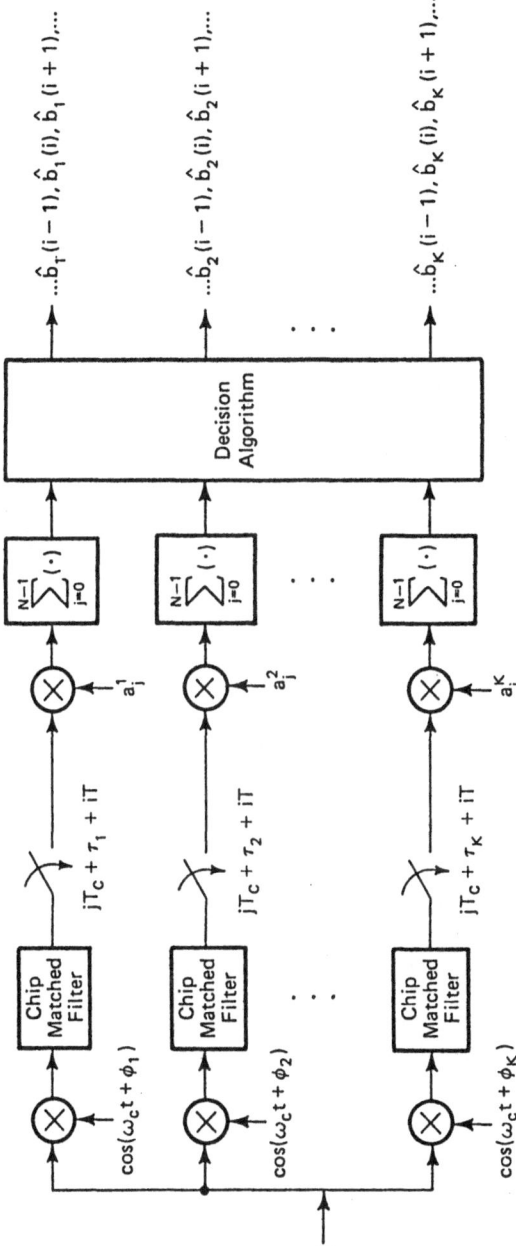

Figure 2.2. Optimum System for Simultaneous Demodulation of K Asynchronous DS/SSMA Users

by Lupas and Verdú in [9] and [10], and by Varanasi and Aazhang in [11] and [12].) The third possible disadvantage (i.e., the need to know amplitudes) raises the interesting question of multiuser amplitude estimation that we address in the remainder of this paper.

3 Amplitude-Free Multiuser Demodulation

Before discussing the problem of multiuser amplitude estimation as motivated in the preceding section, it is of interest to note that there are alternatives to optimum demodulation that retain many of the desirable properties of the optimum algorithms without requiring knowledge of the amplitudes for their implementation. Since these techniques are reviewed elsewhere in this volume [4], the basic idea will be outlined here only very briefly.

Observe that the linear modulation of (2.2), the additivity of the users' signals in the channel, and the linearity of the matched filter bank producing the sufficient-statistic matrix (e.g., Fig. 2.2) constitute a linear mapping on the vector bit streams:

$$\begin{pmatrix} A_1 b_1(i) \\ A_2 b_2(i) \\ \vdots \\ A_K b_K(i) \end{pmatrix}, \quad -M \leq i \leq M. \tag{3.1}$$

That is, we can write the observables as

$$\begin{pmatrix} y_1(i) \\ y_2(i) \\ \vdots \\ y_K(i) \end{pmatrix} = \mathbf{H}(z) \begin{pmatrix} A_1 b_1(i) \\ A_2 b_2(i) \\ \vdots \\ A_k b_K(i) \end{pmatrix} + \begin{pmatrix} n_1(i) \\ n_2(i) \\ \vdots \\ n_K(i) \end{pmatrix} \tag{3.2}$$

where the sequence of noise vectors $\underline{n}(i), -M \leq i \leq M$, represents the noise produced in the matched-filter outputs by the additive white Gaussian background noise, and where the filter $\mathbf{H}(z)$ is the linear mapping described above. In the synchronous case ($\tau_1 = \tau_2 = \ldots = \tau_K$) this filter is a $K \times K$ memoryless system, and in the asynchronous case it has a memory of one time lag.

In the absence of the additive Gaussian noise, the data can be recovered perfectly from the observations (3.2) simply by (componentwise) hardlimiting the outputs of an inverse filter driven by the outputs of the matched-filter bank. This also works well when the bit-energy-to-background-noise ratio is high (i.e., σ small), and in fact has some asymptotic optimality properties in this regime. For further details, the reader is referred to [4] and the references contained therein. The connection between this approach and multiuser demodulation based on amplitude estimation will be discussed in the following section.

4 Multiuser Amplitude Estimators Based on Maximum-Likelihood Data Recovery

We now turn to the problem of estimating the amplitudes $\underline{A} = (A_1, A_2, \ldots, A_K)^T$, of the various users from observation of the received signal $\{r(t); -\infty < t < \infty\}$. We adopt a maximum-likelihood estimation criterion with respect to these amplitudes. Since the data symbols are a fortiori unknown, we must also adopt a decision criterion for them. As discussed in Section 2, there are (at least) two ways in which we can proceed in this regard: maximum-likelihood symbol decisions; and minimum-error-probability symbol decisions. Here, we consider these two philosophies separately, with maximum-likelihood data recovery being treated in this section and minimum-error-probability data recovery being treated in the following section.

We begin by observing from (2.9) that the matrix y of matched filter outputs $\{y_k(i); -M \leq i \leq M, 1 \leq k \leq K\}$ is sufficient for $\underline{A}$. Thus, we may, without loss of inferential information, reduce the received waveform to this matrix. So the (RF) hardware requirements of multiuser amplitude estimation are the same as for the conventional decentralized multiuser demodulator. To investigate the estimator structure further we treat first the case of estimating the amplitude of a single user in the channel alone (i.e., $K = 1$).

In the single-use case the observables can be written as

$$y_1(i) = A_1 b_1(i) + n_1(i), \quad -M \leq i \leq M \tag{4.1}$$

where $\{n_1(i); -M \leq i \leq M\}$ is a sequence of i.i.d. Gaussian random variables with zero means and variances σ^2. From this it follows that the maximum-likelihood choices for A_1 and $\underline{b}_1 = [b_1(-M), \ldots, b_1(M)]$ are described by:

$$\hat{A}_1, \hat{\underline{b}}_1 = \arg \left\{ \max_{A \in \mathcal{R}, \underline{b} \in \{-1,1\}^{2M+1}} \ell(\underline{y}_1 | A, \underline{b}) \right\} \tag{4.2}$$

where $\underline{y}_1$ denotes the set of observables described in (4.1), and the likelihood function is given by:

$$\ell(\underline{y}_1 | A, \underline{b}) = \exp\left\{ \frac{1}{2\sigma^2} [2A \sum_{i=-M}^{M} b(i) y_1(i) - (2M + 1) A^2] \right\}. \tag{4.3}$$

Note that, with $\underline{b}$ fixed, the log-likelihood is quadratic in A. Thus, on holding $\underline{b}$ fixed and maximizing with respect to A, we see that $\hat{A}_1$ and $\hat{\underline{b}}_1$ are related via

$$\hat{A}_1 = \frac{1}{2M + 1} \sum_{i=-M}^{M} \hat{b}_1(i) y_1(i). \tag{4.4}$$

Substituting this relationship into (4.3), we see after some manipulation that

$$\hat{\underline{b}}_1 = \arg \left\{ \max_{\underline{b}} \left(\sum_{i=-M}^{M} b(i) y_1(i) \right)^2 \right\}, \tag{4.5}$$

from which

$$\hat{b}_1(i) = sgn\{y_1(i)\} \tag{4.6}$$

and thus,

$$\hat{A}_1 = \frac{1}{2M + 1} \sum_{i=-M}^{M} |y_1(i)|. \tag{4.7}$$

Equation (4.6) is, of course, the conventional demodulator of (2.5), which is well-known to be optimum in this single-user case. (The amplitude estimator of (4.7) is the same as that employed by Poor and Vijayan [3] for use in an adaptive nonlinear interference suppression filter for direct-sequence spread-spectrum signals, except that the "symbols" in that application are spreading code "chips" rather than data bits.) Note that there is a sign ambiguity in the above optimization. Here we have arbitrarily taken the amplitude to be positive; if we do not know this to be the case, then there is a one-bit ambiguity in the demodulated $(2M + 1)$-long data sequence. Of course, such ambiguity will always arise for BPSK without the aid of a test pattern since phaselocking can only be accomplished modulo π for this case.

We now turn to the multiuser case, in which $K > 1$. Here, we can write the observables as in (3.2). The sequence of noise vectors is the result of driving the linear front-end matched-filter bank with white Gaussian noise, and thus is a zero-mean stationary Gaussian sequence with (matrix) correlation sequence determined straightforwardly by the crosscorrelation properties of the signal constellation and by the relative delays of the users. So, the likelihood function for fixed A and b can be written from such description. However, it is simpler to write the likelihood function directly from (2.7) - (2.9) as

$$\ell(\mathbf{y}|A, \mathbf{b}) = \exp\left\{\frac{1}{2\sigma^2}[2A^T\alpha(\mathbf{b}) - A^T\beta(\mathbf{b})A]\right\} \tag{4.8}$$

where α is a K-vector whose k^{th} component is given by

$$\alpha_k(\mathbf{b}) = \sum_{i=-M}^{M} b_k(i)y_k(i), \quad k = 1,\ldots,K, . \tag{4.9}$$

and where β is a symmetric $K \times K$ matrix whose $k - \ell^{th}$ entry is given by

$$(\beta(\mathbf{b}))_{k,\ell} = \sum_{i=-M}^{M}\sum_{j=-M}^{M} b_k(i)b_\ell(j)H_{k,\ell}(i-j) \tag{4.10}$$

$$H_{k,\ell}(i-j) = \int_{-\infty}^{\infty} s_k(t)s_\ell(t+(i-j)T+\tau_k-\tau_\ell)dt. \tag{4.11}$$

Note that $H_{k,\ell}(i-j)$ has the property that $H_{k,\ell}(i-j) = H_{\ell,k}(j-i)$, and is nonzero only when either $i = j$ or $(i-j) = sgn(\tau_\ell - \tau_k)$. Further note that $H_{k,k}(0) = 1$ by virtue of the fact that the signals s_k are normalized, and that $|H_{k,\ell}(-1)| + |H_{k,\ell}(0)| + |H_{k,\ell}(+1)|$ is bounded above by unity for any k and ℓ via the Schwarz Inequality. These properties imply that

$$(\beta(\mathbf{b}))_{k,\ell} = \begin{cases} (2M+1) & \text{if } k = \ell \\ \\ \sum_{i=-M}^{M} b_k(i)[b_\ell(i)H_{k,\ell}(0) + b_\ell(i-d_{k,\ell})H_{k,\ell}(d_{k,\ell})], & \text{if } k \neq \ell \end{cases}$$

where $d_{k,\ell} = sgn(\tau_\ell - \tau_k)$. Note that the off-diagonal elements of β are bounded in magnitude by $(2M+1)$, and this bound could be achieved only for very impractical signal sets and for specific bit sequences.

Proceeding as in the single-user case, we note that, with b fixed, the maximization of (4.8) with respect to A is a least-squares problem. Its solution is found by solving the equation

$$\beta(\mathbf{b})A(\mathbf{b}) = \alpha(\mathbf{b}) \tag{4.12}$$

(Note: the invertibility of β will be assured if the sum of the absolute values of the elements in each row is smaller than 2, a condition which can be attained with proper signal design.) We then may substitute this solution into (4.8) to solve for $\hat{\mathbf{b}}$:

$$\hat{\mathbf{b}} = \arg\left\{\max_{\mathbf{b}} \ell(\mathbf{y}|A(\mathbf{b}), \mathbf{b})\right\}$$

$$\tag{4.13}$$

$$= \arg\left\{\max_{\mathbf{b}}[A^T(\mathbf{b})\alpha(\mathbf{b}) - \frac{1}{2}A^T(\mathbf{b})\beta(\mathbf{b})A(\mathbf{b})]\right\}.$$

It is instructive to illustrate this procedure explicitly for the two-user case ($K = 2$).

Assume $K = 2$, and without loss of generality, that $0 = \tau_1 \leq \tau_2 \leq T$. After some manipulation, we then have that the solution to (4.13) is given by

$$\left(\begin{array}{c} \hat{A}_1(\mathbf{b}) \\ \hat{A}_2(\mathbf{b}) \end{array} \right) = \frac{1}{(2M+1)(1-\hat{\rho}_{12}^2)} \left(\begin{array}{c} \alpha_1(\mathbf{b}) - \hat{\rho}_{12}\alpha_2(\mathbf{b}) \\ \alpha_2(\mathbf{b}) - \hat{\rho}_{12}\alpha_2(\mathbf{b}) \end{array} \right) \tag{4.14}$$

where

$$\hat{\rho}_{12} = \frac{1}{2M+1}(\beta(\mathbf{b}))_{12} = \frac{1}{2M+1} \left[H_{1,2}(0) \sum_{i=-M}^{M} b_1(i)b_2(i) + H_{1,2}(+1) \sum_{i=-M}^{M} b_1(i)b_2(i-1) \right] \tag{4.15}$$

with

$$H_{1,2}(0) = \int_{\tau_2}^{T} s_1(t)s_2(t-\tau_2)dt \text{ and } H_{1,2}(+1) = \int_{0}^{\tau_2} s_1(t)s_2(t-\tau_2+T)dt. \tag{4.16}$$

Inserting this solution into (4.14) yields

$$\max_{\underline{A}} \left[\underline{A}^T \underline{\alpha}(\mathbf{b}) - \frac{1}{2}\underline{A}^T \beta(\mathbf{b})\underline{A} \right] = \frac{1}{2} \frac{\alpha_1^2(\mathbf{b}) + \alpha_2^2(\mathbf{b}) - 2\hat{\rho}_{12}\alpha_1(\mathbf{b})\alpha_2(\mathbf{b})}{1-\hat{\rho}_{12}^2} \tag{4.17}$$

which must be maximized to find the overall maximum-likelihood data streams.

It should be noted that the maximization of (4.18) is a highly complex nonlinear integer program, which, unlike its known-amplitude counterpart, does not appear to be configurable as a dynamic program. However, it can be shown that, in the noiseless case, this maximization yields the correct data streams. This suggests that this procedure will be consistent and thus (in view of its maximum-likelihood nature) possibly efficient as the noise level vanishes. As a practical matter, the complexity of maximizing (4.18) may be alleviated by noting that with $\underline{A}$ fixed the maxization over b has a time complexity per binary decision of $O(2^K)$, and with b fixed the maximization over $\underline{A}$ is at most $O(K^3)$. Thus, each iteration of a Gauss-Seidel algorithm [13] which alternately maximizes over $\underline{A}$ and b with the other fixed will still be only $O(2^K)$. As a final comment here, we note that the assumption that the amplitudes are constant throughout the frame (i.e., they do not depend on the symbol time i), complicates the above analysis. If these amplitudes were allowed instead to vary arbitrarily from symbol to symbol, application of the above optimization procedure would result in the channel inversion demodulator discussed in Section 3. Thus, although both the channel inverter and the above procedure are perfect in the noiseless case, one would expect the maximization of (4.18) to yield better performance within the scenario in which the amplitudes do not vary substantially over the frame.

5 Multiuser Amplitude Estimators Based on Minimum-Error-Probability Data Recovery

We now turn to an alternative formulation of the multiuser amplitude-estimation problem in which the data streams are given a statistical model, and the nominal data recovery objective is to minimize the probability of error. In particular, we assume that the data sequences of the users are independent sequences of i.i.d. random variables each taking the values +1 and -1 with equal probability $\frac{1}{2}$.

As in the preceding section, it is useful to treat the single-user case first. In this situation, the observables are described by (4.1). The likelihood function of these observations with A_1 fixed is simply the likelihood function with A_1 and $\underline{b}_1$ fixed (i.e., (4.3)) averaged over the statistics of $\underline{b}_1$. Under the assumed statistical model for $\underline{b}_1$, this implies that the ML estimate of A_1 is given by

$$\text{MLE of } A_1 = \arg\left\{\max_{A_1} \prod_{i=-M}^{M} e^{-A_1^2/2\sigma^2} \cosh(A_1 y_1(i)/\sigma^2)\right\}. \tag{5.1}$$

The maximization problem of (5.1) is a complicated problem which requires numerical solution. Note that this problem would be much easier if we could observe $\underline{b}_1$ in addition to $\underline{y}_1$, since then the ML estimate of A_1 would be given (as in (4.4)) by

$$\hat{A}_1 = \frac{1}{2M+1} \sum_{i=-M}^{M} b_1(i)y_1(i). \tag{5.2}$$

Of course, since we do not know $\underline{b}_1$, we cannot do this. However, we can exploit this simplicity via the EM algorithm, which was proposed by Baum, et al. [14] for estimation problems in which the observables are probabilistically related to an unobserved Markov chain (in this case the data stream), and which was expanded upon by Dempster, et al. in [15].

The EM algorithm involves iteration between two steps: a conditional expectation (E) step and a maximization (M) step. In going from the estimate at iteration n to the estimate at iteration $n+1$, these two steps are given by:

$$E - Step : Q(A_1'; \hat{A}_1(n)) = E\left\{\log \ell(\underline{x}|A_1)|\underline{y}, \hat{A}_1(n)\right\} \tag{5.3}$$

$$M - Step : \hat{A}_1(n+1) = \arg\left\{\max_{A_1'} Q(A_1'; \hat{A}_1(n))\right\} \tag{5.4}$$

where $\underline{x}$ denotes the pair $(\underline{b}_1, \underline{y}_1)$. Thus, the algorithm is very similar to the Gauss-Seidel iteration described in Section 4, except that maximization of log-likelihood over $\underline{b}$ with A fixed is replaced by conditional expectation of log-likelihood over $\underline{b}$ with A fixed.

This iteration takes a particularly simple form for the model of interest, since we have

$$\log \ell(\underline{x}|A_1') = \frac{A_1'}{\sigma^2} \sum_{i=-M}^{M} b_1(i)y_1(i) - \frac{(A_1')^2}{2\sigma^2}(2M+1) + \text{const}. \tag{5.5}$$

from which the E-step is given by

$$E\{\log \ell(\underline{x}|A_1')|\underline{y}, \hat{A}_1(n)\} = \frac{A_1'}{\sigma^2} \sum_{i=-M}^{M} y_1(i) \tanh(\hat{A}_1(n)y_1(i)/\sigma^2) - \frac{(A_1')^2}{2\sigma^2}(2M+1) + \text{const}. \tag{5.6}$$

Since (5.6) is quadratic in A_1', the M-step is quite straightforward, yielding the iteration equation

$$\hat{A}_1(n+1) = \frac{1}{2M+1} \sum_{i=-M}^{M} y_1(i) \tanh(\hat{A}_1(n)y_1(i)/\sigma^2). \tag{5.7}$$

This algorithm has some nice properties. In particular, it can be shown that the likelihood function in (5.1) increases on each iteration of the algorithm. Moreover, under mild conditions, the EM iteration will converge to a stationary point in likelihood.

The term $\tanh(\hat{A}_1(n)y_1(i)/\sigma^2)$ appearing in (5.7) is the conditional expectation of $b_1(i)$ given $\underline{y}_1$ and $\hat{A}_1(n)$. This quantity is the difference between the posterior probability that $b_1(i) = +1$ and the posterior probability that $b_1(i) = -1$. In the multiuser case ($K > 1$), the counterpart of (5.5) is (4.8). By inspection, we see that the E-step will similarly involve the computation of posterior probabilities (univariate and bivariate) of bit values, while the M-step will involve the minimization of a quadratic (as in (5.6)). The M-step is thus a straightforward $O(K^3)$ procedure, while the M-step is more complex. However, since the posterior probabilities are the metric of interest in minimum-error-probability bit decisions, exponential complexity in K should be possible here through the use of dynamic programming as discussed in Section 2 .

6   Conclusion

In this paper, we have discussed the problem of estimating the amplitudes of a set of asynchronous modulated data signals superimposed in an additive white Gaussian noise channel. The motivation for this work is twofold: to suggest implementations for the optimum multiuser detectors developed by Verdú [1]; and to consider amplitude estimation as a paradigm for the interesting theoretical problem of multiuser parameter estimation. Two estimation philosophies have been proposed, and these are based respectively on maximum-likelihood (ML) data recovery and minimum-error-probability (MEP) data recovery. In each case, the most promising algorithm is iterative, with Gauss-Seidel iteration and the EM algorithm being proposed for the ML and MEP approaches respectively. The convergence analysis and estimation performance of these algorithms are interesting topics for further study, as is the comparison of these algorithms with others based on channel inversion (as discussed in Section 3).

The techniques discussed here can be readily extended to the estimation of other signal parameters, although the sufficient statistic may be different in other cases. For example, to estimate the carrier phases in a DS/SSMA signal, one could apply similar methods to complex observables consisting of in-phase and quadrature components. Other parameters of interest, such as the relative delays, might require infinite-dimensional statistics, unless the parameter space could be assumed to be finite (e.g., in DS/SSMA, it might be possible to assume chip synchrony, but not data synchrony).

7   Acknowledgement

This work was supported by the U. S. Army Research Office under Contract DAAL03-87-K-0062.

References

[1] S. Verdú, "Minimum Probability of Error for Asynchronous Gaussian Multiple-Access Channels," *IEEE Trans. Inform. Theory*, vol. IT-32, pp. 85-96, January 1986.

[2] B. Aazhang and H. V. Poor, "An Analysis of Nonlinear Direct-sequence Correlators," *IEEE Trans. Commun.*, vol. COM-37, 1989 (to appear).

[3] H. V. Poor and R. Vijayan, "Analysis of a Class of Adaptive Nonlinear Predictors," this volume.

[4] S. Verdú, "Recent Progress in Multiuser Detection," this volume.

[5] H. V. Poor, *An Introduction to Signal Detection and Estimation.* (Springer-Verlag: New York, 1988)

[6] S. Verdú and H. V. Poor, "Abstract Dynamic Programming Models under Commutativity Conditions," *SIAM J. Control and Optimization*, vol. 25, pp. 990-1006, July 1987.

[7] S. Verdú, "Optimum Multiuser Asymptotic Efficiency," *IEEE Trans. Commun.* vol. COM -34, 890-897, Sept. 1986.

[8] H. V. Poor and S. Verdú, "Single-user Detectors for Multiuser Channels," *IEEE Trans. Commun.*, vol. COM-36, pp. 50-60, January 1988.

[9] R. Lupas and S. Verdú, "Linear Multiuser Detectors for Synchronous Code-Division Multiple-Access Channels," *IEEE Trans. Inform. Theory*, vol. IT-34, 1988 (to appear).

[10] R. Lupas and S. Verdú, "Near-far Resistance of Linear Detectors for Code-Division Multiple-access Channels," *IEEE Trans. Commun.*, vol. COM-37, 1989 (to appear).

[11] M. K. Varanasi and B. Aazhang, "Iterative Multiuser Detection for Asynchronous Code-division Multiple-access Systems," *IEEE Trans. Commun.*, vol. COM-37, 1989 (to appear).

[12] M. K. Varanasi and B. Aazhang, "Probability of Error Comparison of Linear and Iterative Multiuser Detectors," this volume.

[13] J. M. Ortega and W. C. Rheinbolt, *Iterative Solution of Nonlinear Equations in Several Variables.* (Academic Press: New York, 1970).

[14] L. E. Baum, T. Petrie, G. Soules, and N. Weiss, "A Maximization Technique in Statistical Estimation for Probabilistic Functions of Markov Chains," *Ann. Math. Stat.*, vol. 41, pp. 164-171, 1970.

[15] A. P. Dempster, N. M. Laird, and D. B. Rubin, "Maximum Likelihood from Incomplete Data via the EM Algorithm," *J. Royal Statist. Soc.*, Ser. B, vol. 37, pp. 1-38, 1977.

ORTHOGONAL SIGNALLING AND DIVERSITY IN PARTIAL-BAND INTERFERENCE

Manjunath V. Hegde   Wayne Stark
Louisiana State University   The University of Michigan

1. Introduction: We investigate the performance of simple signalling and demodulation schemes over the partial-band jammed channel. When communicating over the added white Gaussian noise channel, orthogonal signalling suffices asymptotically to achieve capacity, i.e. by choosing M large enough the error probability can be made arbitrarily small for all rates less than capacity or equivalently, provided that the ratio of the energy transmitted per information bit and the one-sided power spectral-density, $\gamma_0 = E_b/N_0$ is greater than ln 2. Conversely no other signals can achieve arbitrarily small error probability when $\gamma_0 < \ln 2$. It is also known [Stark 85a],[Stark 85b], that provided codes of small enough rate are used the capacity of a partial-band jammed channel is the same as that of a white Gaussian noise channel. In the light of this it seems plausible that for the partial-band jammed (PBJ) channel, orthogonal signals with correlation detection which suffice in the white Gaussian case, could be used as a simple scheme to form a reliable communication system. Unfortunately this turns out not to be true as has been shown in [Hegd 88]. The result which is described there for the case with partial-band jamming but no background noise turns out to be true even in the more general case when such background noise is incorporated. In section 2 we demonstrate this by considering the limiting value of the error probability of orthogonal signals in worst case two-level partial-band jamming with background noise. For any $\overline{\gamma}_J = E_b/(N_0 + N_J)$ (where N_J is the equivalent one-sided power spectral density of the jamming noise) the asymptotic probability of error is not zero. The analysis shows that for large $\overline{\gamma}_J$ the worst-case jammer ρ (ρ is a constant indicating the fraction of band that has interference) is very small. Hence simple redundancy in the form of diversity is next analyzed with majority logic decoding as well as linear combining at the receiving end. Both schemes turn out to be insufficient. The linear combining case indicates that when the outputs of the L diversity transmissions are summed up, small values of ρ_L (the jammer chooses ρ as a function of L) play a significant part in degrading the performance when using the sum statistic.

This naturally suggests clipped linear combining, wherein we clip the output of each diversity transmission, as a more effective combining technique. The rationale for such a scheme is the expectation that our probability of error will decrease because now the infrequent transmissions with the large noise components will affect our sum much less. To allow for greater generality we allow the clipping level to be a function of L too. Our analysis indicates that for this case we recapture the same threshold phenomenon with orthogonal signals that we had for the AWGN channels. We do this by use of certain Central Limit Theorem approximations. Since such a threshold phenomenon is very sensitive to the kind of approximation used, we need to use a powerful non-uniform "Berry-Esseen" bound due to Michel and a less well-known form of the Central Limit Theorem due to Sirazhdinov and Mamatov. Our analysis shows that provided a certain relation is satisfied by the clipping level, the number of diversity transmissions and the number of signals, then, asymptotically, orthogonal signalling with diversity and clipped linear combining suffices to achieve capacity over the partial-band jammed channel.

This paper is organized as follows: in section 2 we show why orthogonal signalling does not suffice in channels with background noise and a partial-band jammer. In section 3 we consider orthogonal signalling with diversity and examine two receivers; one employing majority logic decoding and the other employing linear combining. In section 4 we analyze orthogonal signalling with diversity and clipped linear combining at the receiver. This scheme is sufficient to recover the threshold behavior of the probability of error as in the case of orthogonal signalling over the AWGN channel. In section 5 we summarize our results and state our conclusions.

2. Orthogonal Signaling over the AWGN Channel: If we use orthogonal signalling with M orthogonal signals each with energy E over the AWGN channel with one-sided noise power spectral density N_0 watts/Hz then with coherent correlation detection the error probability P_e is known to be.

$$P_e(\gamma_0, M) = 1 - \frac{1}{\sqrt{2\pi}} \int_{-\infty}^{\infty} (\Phi(u))^{M-1} \exp\left\{-\frac{1}{2}(u - \sqrt{2\gamma_0 \log_2 M})^2\right\} du$$

where $\Phi(u)$ is the distribution function of a standard normal random variable. For noncoherent detection the error probability is

$$P_e^{nc}(\gamma_0, M) = 1 - \int_0^{\infty} u \exp\left\{-\frac{1}{2}(u^2 + 2\gamma_0 \log_2 M)\right\} I_0(\sqrt{2\gamma_0 \log_2 M} u)(1 - \exp(-\frac{u^2}{2}))^{M-1} du$$

where $I_0(\cdot)$ is the modified Bessel function of the zero-th order. In what follows we use $P_e(\gamma_0, M)$ to refer to the probability of error in both the coherent and noncoherent cases.

It is well known that [Vite 66]

$$\lim_{M \to \infty} P_e(\gamma_0, M) = \begin{cases} 1, & \gamma_0 < \ln 2 \\ 0, & \gamma_0 > \ln 2 \end{cases}$$

in either the coherent or the noncoherent case. However as we show in the following section when we have partial-band jamming, orthogonal signals perform much more poorly.

A. Orthogonal Signaling in Partial-band Jamming and Background Noise: We show in this section that the following partial-band jamming strategy causes the error probability to be non-zero asymptotically for any value of $\overline{\gamma}_J$. In this case where there is both background noise as well as partial-band jamming we model the effect of partial-band interference as adding additional Gaussian noise in each of the M dimensions with probability ρ and no additional noise with probability $1-\rho$. We see that if the signal i has been transmitted then the decision statistics at the modulator can be written as

$$y_i = Z_i n_i + N_i$$

for $i \neq j$ and

$$y_j = \sqrt{E} + Z_j n_j + N_j$$

for $i=j$ where N_i, $i=1,...,M$ are i.i.d. Gaussian random variables with mean 0 and variance $N_0/2$ and the n_i, $i=1,...,M$ are i.i.d. Gaussian random variables with mean 0 and variance $N_J/2$ and Z_i, $i=1, \cdots, M$ are i.i.d. random variables with distribution

$$\Pr(Z_i = 0) = 1-\rho, \quad \Pr(Z_i = \rho^{-\frac{1}{2}}) = \rho.$$

Denoting the probability of error in this case by $P_e(\rho, E_b, N_0, N_J, M)$ we see that by conditioning on Z_i and then averaging the error probability may be written as

$$P_e(\rho, E_b, N_0, N_J, M) = (1-\rho)P_e(\gamma_0, M) + \rho P_e(\gamma_J(\rho), M)$$

where $\gamma_J(\rho) = E_b/(N_0 + N_J/\rho)$. For the worst-case partial-band jammer the error probability is

$$P_e(E_b, N_0, N_J, M) = \sup_{0 \le \rho \le 1} P_e(\rho, E_b, N_0, N_J, M)$$

For $\rho=0$, $P_e(\rho, E_b, N_0, N_J, M) = P_e(\gamma_0, M)$.

Now we show that for any signal-to-noise ratio the worst case jammer can ensure that the asymptotic probability of error does not go to zero. This is stated precisely in the following Theorem.

Theorem 1:
i) For $\overline{\gamma}_J < \ln 2$

$$\lim_{M \to \infty} \max_{0 \le \rho \le 1} \rho P_e(\gamma_J(\rho), M) + (1-\rho)P_e(\gamma_0, M) = 1$$

ii) For $\overline{\gamma}_J > \ln 2$

$$\lim_{M \to \infty} \max_{0 \le \rho \le 1} \rho P_e(\gamma_J(\rho), M) + (1-\rho)P_e(\gamma_0, M) = \overline{\rho}$$

where $\overline{\rho}$ is the solution of the equation $\gamma_J(\rho) = \ln 2$, i.e. $\overline{\rho} = \dfrac{\ln 2}{\gamma_J - \dfrac{N_0}{N_J}\ln 2}$ where $\gamma_J = E_b/N_J$. Note that

$$\frac{\ln 2}{\gamma_J} \le \overline{\rho} = \frac{\ln 2}{\gamma_J - \dfrac{N_0}{N_J}\ln 2} \le \frac{\ln 2}{\overline{\gamma}_J}$$

Proof of Theorem 1: Consider first the case $\overline{\gamma}_J < \ln 2$. Then we have

$$\lim_{M \to \infty} \max_{0 \le \rho \le 1} \rho P_e(\gamma_J(\rho), M) + (1-\rho)P_e(\gamma_0, M) \ge \lim_{M \to \infty} \max_{0 \le \rho \le 1} \rho P_e(\gamma_J(\rho), M)$$

$$\ge \lim_{M \to \infty} \max_{0 \le \rho \le 1} \rho P_e(\overline{\gamma}_J, M)$$

$$\ge \lim_{M \to \infty} P_e\left(\frac{E_b}{N_0 + N_J}, M\right) = 1 \tag{1}$$

since $P_e(x, M)$ is a decreasing function of x (see Appendix A).

For $\overline{\gamma}_J > \ln 2$ for any $\varepsilon > 0$, let $\rho_1 = \overline{\rho} - \varepsilon$. Then

$$\frac{E_b}{\dfrac{N_J}{\rho_1} + N_0} < \ln 2.$$

Hence

$$\liminf_{M \to \infty} \max_{0 \le \rho \le 1} \rho P_e(\gamma_J(\rho), M) + (1-\rho)P_e(\gamma_0, M) \ge \liminf_{M \to \infty} \max_{0 \le \rho \le 1} \rho P_e(\gamma_J(\rho), M)$$

$$\ge \liminf_{M \to \infty} \rho_1 P_e(\gamma_J(\rho_1), M)$$

$$= \rho_1. \tag{2}$$

Since this is true for any $\varepsilon > 0$ we can infer that

$$\liminf_{M \to \infty} \max_{0 \le \rho \le 1} \rho P_e(\gamma_J(\rho), M) + (1-\rho)P_e(\gamma_0, M) \ge \bar{\rho}.$$

Now let ρ_M be the value of ρ which achieves the maximum in

$$\max_{0 \le \rho \le 1} \rho P_e(\gamma_J(\rho), M).$$

Then for sufficiently large M, $\rho_M \le \bar{\rho}$. Else for every M there would be a $M_1 > M$ and $\rho_{M_1} > \bar{\rho}$ such that

$$\rho_{M_1} P_e(\gamma_J(\rho_{M_1}), M_1) \ge \rho P_e(\gamma_J(\rho), M) \tag{3}$$

for all ρ, $0 \le \rho \le 1$. Now for any $\varepsilon > 0$ for sufficiently large M ($M > M_2$ say) the right-hand side of (3) can be made greater than $\bar{\rho} - \varepsilon$. Since the left-hand side of (3) is obviously approaching 0 for $\rho_M > \bar{\rho}$, we have a contradiction. Therefore

$$\limsup_{M \to \infty} \max_{0 \le \rho \le 1} \rho P_e(\gamma_J(\rho), M) + (1-\rho)P_e(\gamma_0, M) \le \bar{\rho}.$$

From (1) and (2) it follows that

$$\lim_{M \to \infty} \max_{0 \le \rho \le 1} \rho P_e(\gamma_J(\rho), M) + (1-\rho)P_e(\gamma_0, M) = \begin{cases} 1, & \bar{\gamma}_J < \ln 2 \\ \bar{\rho}, & \bar{\gamma}_J > \ln 2. \end{cases}$$

The jammer thus is able to thwart reliable communication at any signal-to-noise ratio by choosing ρ small enough (but not too small). Since the choice of a small ρ by the jammer allows a number of transmissions to pass through unscathed but corrupts the rest substantially it seems likely that simple coding such as through diversity may be able to recover some of this loss in rate due to partial-band jamming. We pursue such an investigation in the following section.

3. **Orthogonal Signalling with Diversity:** In this section we try to use time diversity to overcome the partial-band jammer, i.e. make the worst-case partial-band jammer no more deleterious than a broad-band jammer of equivalent power. With diversity L, the energy per bit, E_b, is $LE/\log_2 M$ and we shall use E_b' to denote $E/\log_2 M$, γ_0' to denote E_b'/N_0, $\bar{\gamma}_J'$ to denote $E_b'/(N_0+N_J)$ and $\gamma_J'(\rho)$ to denote $E_b'/(N_J/\rho+N_0)$.

Assume that symbol j was sent, i.e. $s_j(t)$ was transmitted. Then using the earlier demodulator we have L M-component vector decision variables:

$$y_l = (y_{l,1}, ..., y_{l,M}) \qquad l = 1, 2, ..., L$$

where for $i \ne j$

$$y_{l,i} = Z_{l,i} n_{l,i} + N_{l,i}$$

and for $i=j$

$$y_{l,j} = \sqrt{E} + Z_{l,j} n_{l,j} + N_{l,j}$$

where $n_{l,j}$ is a $N(0, N_J/2)$ Gaussian random variable, $Z_{l,j}$ is 0 with probability $1-\rho$ and $\sqrt{1/\rho}$, with probability ρ, $N_{l,j}$ is a $N(0, N_0/2)$ Gaussian random variable and $Z_{l,i}$, $n_{l,i}$ and $N_{l,i}$, $i=1,...,M$ are mutually independent. Also, $y_1,..., y_L$ are i.i.d. random vectors.

A. **Majority Logic Combining:** We first investigate the following majority logic decoding strategy. The receiver observes the output during each interval of duration T and picks i if the output of the ith correlator is maximum. At the end of L intervals the output of the decoder is the symbol which has been picked a maximum number of times. The jammer is assumed to know both L and M and chooses ρ based on this knowledge. We do the asymptotic analysis assuming the diversity L to be an increasing function of M for sufficiently large M. Now in each time interval of length T the probability of error (i.e. the probability that j will not be picked) is

$$\kappa = \rho P_e(\gamma_J'(\rho), M) + (1-\rho)P_e(\gamma_0', M)$$

where $P_e(x,y)$ is the probability of error for y orthogonal signals with bit energy to noise ratio x. Thus the probability of j being picked in each interval is $1-\kappa$ and the probability of $i \neq j$ being picked in each interval is $\kappa/(M-1)$. Using the above scheme we denote the probability of error as $P_e^L(\rho_{L,M}, E_b', N_J, N_0, M)$ where $\rho_{L,M}$ is used to denote the jammer's choice of ρ as a function of L and M. Since the channel is independent between repetitions and the same input is applied to the channel during each of the L repetitions, the outputs of the channel during the L diversity transmissions are i.i.d. random variables with finite mean. Thus we can utilize the Weak Law of Large Numbers as $L \to \infty$, i.e. the probability that the proportion of times we choose j out of the L repetitions differs from $1 - \kappa_{L,M}$ (where $\kappa_{L,M} = \rho_{L,M} P_e(\gamma_J'(\rho_{L,M}), M) + (1-\rho_{L,M})P_e(\gamma_0', M)$) by $\varepsilon > 0$ goes to zero:

$$\lim_{L \to \infty} \text{Prob}\left\{ \left| \frac{\text{number of times } j \text{ is picked}}{L} - (1-\kappa_{L,M}) \right| > \varepsilon \right\} \to 0$$

$$\lim_{L \to \infty} \text{Prob}\left\{ \left| \frac{\text{number of times } i \ (i \neq j) \text{ is picked}}{L} - \frac{\kappa_{L,M}}{(M-1)} \right| > \varepsilon \right\} \to 0.$$

Thus we see that using the above decoding strategy the limiting probability of error will be zero or one according as

$$1 - \kappa_{L,M} > \frac{\kappa_{L,M}}{M-1}$$

or

$$\frac{\kappa_{L,M}}{M-1} > 1 - \kappa_{L,M}$$

and we examine when this is true. As shown in [Hegd 87], it can now be shown that for $\overline{\gamma}_J' > \ln 2$

$$\lim_{M,L \to \infty} \sup_{0 \leq \rho_{L,M} \leq 1} P_e^L(\rho_{L,M}, E_b', N_J, N_0, M) = 0 \qquad (4)$$

and for $\overline{\gamma}_J' < \ln 2$

$$\lim_{M,L\to\infty} \sup_{0\le\rho_{L,M}\le1} P_e^L(\rho_{L,M}, E_b', N_J, N_0, M) = 1. \tag{5}$$

Note however that in our diversity signalling scheme the actual bit energy to noise ratio is not $\overline{\gamma}_J'$ but $\overline{\gamma}_J = L\overline{\gamma}_J'$ and thus in both (4) and (5) we have allowed very large bit energy to noise ratios. Therefore we can say that when we use orthogonal signalling with diversity against an intelligent jammer our scheme allows reliable communication when the bit energy to noise ratio $(\overline{\gamma}_J)$ increases fast enough with diversity i.e. $\overline{\gamma}_J > L\ln 2$ for sufficiently large L. If $\overline{\gamma}_J < L\ln 2$ for sufficiently large L then the worst-case jammer can frustrate even such high energy to noise ratios.

B. Linear Combining: Another commonly used method of diversity combining is linear combining. Here we process the output to get the following decision variables:

$$D_i=\sum_{l=1}^{L} y_{l,i} \quad i=1,...,M$$

When symbol j is sent then for $i\ne j$

$$D_j=\sum_{l=1}^{L} (Z_{l,j} n_{l,j} + N_{l,j})$$

and for $i=j$

$$D_j=\sum_{l=1}^{L} (\sqrt{E} + Z_{l,j} n_{l,j} + N_{l,j}).$$

Again we use $P_e^L(\rho_{L,M}, E_b', N_J, N_0, M)$ to denote the probability of error. By conditioning on the number of diversities jammed we can write

$$P_e^L(\rho_{L,M}, E_b', N_J, N_0, M) = \sum_{k=0}^{L} \binom{L}{k} \rho_{L,M}^k (1-\rho_{L,M})^{L-k} P_e\left(\frac{LE_b'}{\frac{kN_J}{\rho_{L,M}} + LN_0}, M\right).$$

Again it can be shown (see [Hegd 87]) that for $\overline{\gamma}_J' > \ln 2$

$$\lim_{M,L\to\infty} \sup_{0\le\rho_{L,M}\le1} P_e^L(\rho_{L,M}, E_b', N_J, N_0, M) \ge 1-e^{-\overline{\rho}} \tag{6}$$

and for $\overline{\gamma}_J' < \ln 2$

$$\lim_{M,L\to\infty} \sup_{0\le\rho_{L,M}\le1} P_e^L(\rho_{L,M}, E_b', N_J, N_0, M) = 1 \tag{7}$$

where $\overline{\rho}$ is as defined in section 2. Thus (6) indicates that the worst-case jammer (when $\overline{\gamma}_J' > \ln 2$) jams such that $\rho_{L,M}\to0$ and (7) shows that if $\overline{\gamma}_J' < \ln 2$ then the worst-case jammer can choose $\rho_{L,M} = 1$. In either case the jammer can frustrate very high bit energy to noise ratios $(\overline{\gamma}_J = \frac{LE_b'}{N_0+N_J})$.

We note that although for $\overline{\gamma}_J' > \ln 2$ the worst-case jammer chooses $\rho_{L,M}$ such that $\rho_{L,M}\to0$, $\rho_{L,M}$ cannot approach 0 too fast for the following reason. Let the jammer choose $\rho_{L,M} = \frac{1}{L^\alpha}(\alpha > 1)$. Now

$$P_e^L(\rho_{L,M}, E_b', N_J, N_0, M) = (1-\rho_{L,M})^L P_e\left(\frac{E_b'}{N_0}, M\right) + \sum_{k=1}^{L} \binom{L}{k}\rho^k(1-\rho_{L,M})^{L-k}P_e\left(\frac{\rho_{L,M}LE_b'}{kN_J+\rho_{L,M}LN_0}, M\right).$$

By the jammer's choice of $\rho_{L,M}$

$$P_e^L(\frac{1}{L^\alpha}, E_b', N_J, N_0, M) = (1-1/L^\alpha)_L P_e(E_b'/N_0, M) + \sum_{k=1}^{L} \binom{L}{k}(1/L^\alpha)^k(1-1/L^\alpha)^{L-k} P_e(L^{\alpha-1}E_b'/(kN_J+L^{\alpha-1}N_0), M)$$

We point out that the first term goes to 0 with L, M. The second term is

$$\sum_{k=1}^{L} \binom{L}{k}\rho_{L,M}^k(1-\rho_{L,M})^{L-k} P_e(E_b'/(L^{\alpha-1}kN_J'+N_0), M) \qquad (8)$$

Every term $P_e(E_b'/(L^{\alpha-1}kN_J + N_0), M)$ in the summation goes to 1 as L increases and so as $L\to\infty$, (8) becomes .

$$\lim_{L\to\infty} (1-(1-\rho_{L,M})^L).$$

Next we show that $\lim_{L\to\infty} (1-\rho_{L,M})^L = 1$ and consequently $\lim_{M,L\to\infty} P_e^L(\rho_{L,M}, E_b', N_J, 'N_0, M) = 0$. Now

$$(1-\rho_{L,M})^L = (1-1/L^\alpha)^L = (1-(1/L^{\alpha-1})(1/L))^L.$$

We know $\lim_{L\to\infty} (1-t/L)^L = e^{-t}$ and so for all t in $(0,1)$ there exists and L_0 such that for all $L \geq L_0$

$$(1-t/L)^L \leq (1-(1/L^{\alpha-1})(1/L))^L$$

and so

$$\lim_{L\to\infty} (1-(1/L^{\alpha-1})(1/L))^L \geq \lim_{L\to\infty} (1-\frac{t}{L})^L = e^{-t}.$$

Since t is arbitrary we have

$$\lim_{L\to\infty} (1-\rho_{L,M})^L = 1.$$

In linear combining we see that since the output of the L diversity transmissions is summed up, small values of $\rho_{L,M}$, while making it less likely that a diversity transmission is jammed, make the probability of error on such a jammed transmission very high because of the low bit energy to noise ratio on such a transmission. The jammer's strategy of choosing $\rho_{L,M}$ to be inversely proportional with diversity level L is intuitively explicable. Since L outputs are added he jams such that if he hits one transmission there is enough jamming power to corrupt the sum statistic. On the other hand, trying to put too much jamming power in a single jammed transmission turns out not to be too effective because then the number of good transmissions increases sufficiently enough to overcome the jamming noise. We thus see that in linear combining the few jammed transmissions have a significant effect on the probability of error. This suggests that if we use a form of clipped linear combining wherein we clip the output of each diversity transmission our probability will improve because the infrequent transmissions with the high P_e values will affect our sum much less. Possibly the clipping level can be chosen as a function of L. In the next section we conduct the analysis using this idea.

4. **Clipped Linear Combining:** The analysis in the previous sections suggests that the jammer contributes infrequently to the decision statistics but when he does so his contribution is large. This suggests that some form of limiting the jammer's contribution to the decision statistics may be effective. Here we first clip each of the diversity transmission outputs by a symmetric limiter and then combine linearly.

Thus the decision variables we use are

$$D_j' = \sum_{l=1}^{L} C_L(Z_{l,j} n_{l,j} + N_{l,j} + \sqrt{E}) \text{ for } i = j \text{ and}$$

$$D_i' = \sum_{l=1}^{L} C_L(Z_{l,i} n_{l,i} + N_{l,i}) \quad \text{for } i \neq j$$

where

$$C_L(x) = \begin{cases} -\alpha_{L,M}, & x < -\alpha_{L,M} \\ x, & -\alpha_{L,M} < = x < = \alpha_{L,M} \\ \alpha_{L,M}, & x > \alpha_{L,M}. \end{cases}$$

The decision rule is to decide that $\hat{i}$ was sent where $D_{\hat{i}}' = \max_j D_j'$. Using this decision rule we calculate the probabilities of being correct.

$$Pr\,(\text{correct} \mid j \text{ is sent}) = Pr\,(\hat{i} = j \mid j \text{ is sent}) = Pr\,(D_i' < D_j', \ i \neq j \mid j \text{ is sent})$$

Again we use $P_e^L(\rho_{L,M}, E_b', N_J, N_0, M)$ to denote the probability of error. In this case however we restrict ourselves to only the coherent detection case as our analysis requires the specific structure of the probability of error. Now let

$$D_j = \frac{D_j'}{\sqrt{L}\,\sigma_L'} = \frac{1}{\sqrt{L}} \sum_{l=1}^{L} \frac{C_L(Z_l n_{j,l} + N_{j,l} + \sqrt{E})}{\sigma_L'}$$

$$= \frac{1}{\sqrt{L}} \sum_{l=1}^{L} \frac{Y_{L,l}}{\sigma_L'}$$

where $Y_{L,l} = C_L(Z_l n_{l,j} + N_{l,j} + \sqrt{E})$ (i.e. the unnormalized decision variable containing the signal) $\sigma_L' = \sqrt{\mathrm{Var}D_j'}$ and let

$$D_i = \frac{D_i'}{\sqrt{L}\,\sigma_L'} = \frac{1}{\sqrt{L}} \sum_{l=1}^{L} C_l \frac{(Z_l n_{l,i} + N_{l,i})}{\sigma_L'}$$

$$= \frac{1}{\sqrt{L}} \sum_{l=1}^{L} \frac{X_{L,l}}{\sigma_L'}$$

where $X_{L,l} = C_L(Z_l n_{l,j} + N_{l,j})$ (i.e. the unnormalized decision variables with only noise). Note that $Pr(D_i' < D_j', i \neq j \mid j \text{ is sent}) = Pr(D_i < D_j, i \neq j \mid j \text{ is sent})$. Using the D_i's as decision variables we show that we can recapture the asymptotic performance of orthogonal signals over the AWGN channel.

Specifically if the number of orthogonal signals, M, the clipping level and the diversity level L increase in a certain relation to each other(such that $M \alpha_{L,M}^3 / \sqrt{L} \to 0$) then the probability of error with clipped linear combining exhibits the same threshold behavior in worst-case partial-band jamming that orthogonal signalling in AWGN achieves, i.e.

Theorem 2:

i) For $\bar{\gamma}_J < \ln 2$

$$\lim_{M,L \to \infty} \sup_{0 \leq \rho \leq 1} P_e^L(\rho_{L,M}, E_b', N_J, N_0, M) = 1.$$

ii) For $\overline{\gamma}_J > \ln 2$ if $M\alpha_{L,M}^3/\sqrt{L} \to 0$ then

$$\lim_{M,L \to \infty} \sup_{0 \le p \le 1} P_e^L(\rho_{L,M}, E_b', N_J, N_0, M) = 0.$$

Proof of Theorem 2: See [Hegd 87]. The proof referred to uses the following steps. First we show that the above threshold phenomenon is true if our decision variables were all Gaussian with parameters chosen in a certain way. The corresponding probability of correct decoding we denote as $P_{c,a}$. Then by use of some fairly powerful Central Limit approximations we show that the difference between $P_{c,a}$ and the actual probability of error goes to zero establishing the Theorem.

5. Conclusions: We have investigated the asymptotic performance of orthogonal signals over channels with both thermal noise as well as unknown partial-band interference. Knowing that such signalling suffices asymptotically to communicate over the AWGN at the limits prescribed by the channel capacity theorem we tried to recover from the effects of the worst-case unknown partial-band interference on the performance of such signalling. The worst-case partial-band jammer does degrade the asymptotic performance severely but he needs to optimize his strategy for each value of the bit energy to noise ratio $(E_b/(N_J+N_0))$ chosen by the communicator. Our analysis reveals that for bit energy to noise ratios over a constant (ln 2) the jammer can be most effective if he jams only a fraction (ρ) of the band. This is because the probability of error near the values of $E_b/(N_J+N_0)$ around ln 2 rises dramatically with a small decrease in $E_b/(N_J+N_0)$. The fraction ρ jammed gets smaller as $E_b/(N_J+N_0)$ gets larger (observe however that it does not get too small). This indicates that the jammer is wilfully reducing the probability of affecting a transmission in order that he may cause more serious damage when he does affect a transmission. This observation suggests that simple coding such as diversity in such a case may be very effective.

Diversity over the partial-band interference channel was next investigated using, at first, majority logic decoding. In this scheme, since the jammer is willing to accept a small probability of affecting transmissions, we expect that the majority of the received diversity transmissions would be received error free. However, the worst-case jammer optimizes his ρ with respect to the diversity level and he is able to ensure that even for very large $E_b/(N_J+N_0)$ the asymptotic symbol error probability is 1. Reliable communication in this case is possible only if $E_b/(N_J+N_0)$ increases with diversity level faster than a certain rate.

Our next diversity scheme was linear combining wherein we simply added the outputs of each diversity transmission. The hope is that the few diversity transmissions that are jammed are nullified in the sum statistic by the many good receptions. In this case the jammer can, by an appropriate choice of $\rho_{L,M}$ ensure that for any $E_b/(N_J+N_0)$ the asymptotic error probability is non-zero. The jammer's choice of $\rho_{L,M}$ is inversely proportional to the diversity level and thus for large L the jammer is jamming a very small fraction but with a large power. The effect hoped for i.e. the swamping out of the few bad receptions by the many good ones does take place but only if the jammer's choice of $\rho_{L,M}$ goes to zero fast enough.

All this suggests that in our diversity combining we must find a way of limiting the contribution of any individual diversity transmission to the overall decision statistic. We therefore proceed with clipped linear combining wherein we first clip the output of each diversity transmission and then simply add. The output statistics are thus the sums of many i.i.d. random variables which suggests the use of some form of Central Limit Theorem Approximations. To ensure the threshold behavior that we are looking for in the error probability we need to use powerful versions of the Central Limit Theorem for which we need the clipping level, the diversity level and the number of signals to satisfy a certain relation. Doing so we can show that the worst-case jammer can be neutralized asymptotically, i.e. he is seen to be no more detrimental to reliable communication than the AWGN channel of equivalent noise spectral density.

Appendix A:

We show that $P_e(x, M)$ is a monotone decreasing function of x.

i) Coherent detection:

As x increases, $Q\left[z + \sqrt{2x \log M}\right]$ decreases. Therefore

$$K = \frac{1}{\sqrt{2\pi}} \int_{-\infty}^{\infty} e^{-\frac{z^2}{2}} \left[1 - Q(z + \sqrt{2x \log_2 M})\right]^{M-1} dz$$

increases. Therefore $P_e(x,M) = 1 - K$ decreases.

ii) Non-coherent detection:

As x increases, $I_0(\sqrt{x}z)e^{-\frac{x^2}{2}}$ decreases. Therefore

$$P_e(x,M) = \int_0^{\infty} z I_0(xz) \exp\left[-\frac{z^2 + x^2}{2}\right] \left[1 - \prod_{i \neq M}(1 - \exp\left[-\frac{z^2}{2}\right])\right] dz$$

decreases.

References

[Arau80]
A.Araujo, E.Gine, *The Central Limit Theorem for Real and Banach Valued Random Variables*, Wiley, 1980.

[Bhat76]
R.N.Bhattacharya, R.Ranga Rao, *Normal Approximation and Asymptotic Expansions*, Wiley, 1976.

[Crep87]
P.N. Crepeau, "Performance of coded FH/MFSK with a quantizer-limiter in a worst-case partial-band gaussian interference Channel", *Proceedings of the Military Communications Conference, 1987*, 12.2.1-12.2.6, 1987.

[Fell68]
W. Feller, "On the Berry-Esscen theorem", *Zeitschrift fur Wahrscheinlichkeitstheorie*, no.10, pp.261-268, 1968.

[Frie86]

    K.J. Friedrichs, "Error analysis for noncoherent M-ary orthogonal communication systems in the presence of arbitrary Gaussian interference", *IEEE Transactions on Communications*, vol. 34, pp. 817-821, Aug. 1986.

[Gned54]

    B.V. Gnedenko, A.N. Kolmogorov, *Limit Distributions for Sums of Independent Random Variables*, Addison-Wesley, 1954.

[Hegd87],

    M.V. Hegde, *Performance analysis of coded frequency-hopped spread-spectrum systems with unknown interference*, Ph.D Thesis, University of Michigan at Ann Arbor, August 1987.

[Hegd88]

    M.V. Hegde, W.E. Stark, "Asymptotic performance of M-ary orthogonal signals in worst-case partial-band interference and rayleigh fading", *IEEE Transactions on Communications*, vol. 36, no. 8 , Aug 1988, pp 789-792 .

[Kell85]

    C.M. Keller, *Diversity combining for frequency-hop spread-spectrum communications with partial-band interference and fading*, Ph.D. Thesis, University of Illinois at Urbana-Champaign, Sept. 1985.

[Mich81]

    R. Michel, "On the constant in the nonuniform version of the Berry-Esseen Theorem", *Zeitschrift fur Wahrscheinlichkeitstheorie*, no. 55, 109-117, 1981.

[Petr75]

    V.V. Petrov, *Sums of Independent Random Variables*, Springer-Verlag, 1975.

[Sira62]

    S.K. Sirazhdinov, M. Mamatov, "On convergence in the mean for densities", *Theory of Probability and its Applications*, no. 4, pp. 425-428, 1962.

[Star82]

    W.E. Stark, *Coding for frequency-hopped spread-spectrum channels with partial-band interference*, Ph.D. Thesis, University of Illinois at Urbana-Champaign, 1982.

[Star85a]

    W.E. Stark, "Coding for frequency-hopped spread-spectrum communication with partial-Band interference-Part 1: capacity and cutoff rate", *IEEE Transactions on Communications*, vol. 33, no. 10, Oct. 1986.

[Star85b]

W.E. Stark, "Coding for frequency-hopped spread-spectrum communication with partial-band interference-Part 2: coded performance", *IEEE Transactions on Communications*, vol. 33, no. 10, Oct. 1986.

[Vite64]

A.J. Viterbi, "Phase coherent communication over the continuous Gaussian Channel" in *Digital Communications with Space Applications*, edited by S. Golomb.

[Vite66]

A.J. Viterbi, *Principles of Coherent Communication*, McGraw-Hill, 1966.

APPLICATIONS OF ROBUSTNESS MEASURES IN SIGNAL DETECTION

Michael W. Thompson
Colorado State University

The Robustness Issue:

In recent years the robustness issue in signal detection has received increased attention. The central goal of much of the work in this area has focused on the attempt to arrive at robust signal detection schemes which perform reasonably well, given the fact that there may exist an imperfect knowledge of the underlying statistical model for the process under consideration. For our purposes, we link the robustness property associated with a given detector to the *stability* of performance over the possible perturbations from the nominal assumptions.

The traditional approach toward the robustness problem has been to employ saddlepoint based techniques such as described by Huber in [1], A variety of work in the literature, some of which has been summarized in a survey article by Kassam and Poor [2], has verified that such an approach can lead to tractable results. A fundamental question which arises when considering robust detection schemes is in regard to the extent of protection which is provided. It may be argued that the degree of robustness obtained form the classical saddlepoint methods owes much to the types of noise models admitted by the method, and in practice, it may not be easy to verify that the types of models which are appropriate for saddlepoint techniques sufficiently represent the full extent of variation from the nominal model used in the detector design. On the other hand, some argue that the classical techniques may, in some sense, be overly cautious as dictated by a minimax approach. Furthermore, saddlepoint based techniques are inherently nonquantitative approaches toward imparting robustness, whereas intuitively, we suspect that robustness is obtained at the expense of detection performance. If this intuition is valid, then it is obviously desirable to have the means to judiciously tradeoff between performance and robustness.

With these comments in mind, it seems clear that there exists the need to obtain a quantitative measure on the degree of robustness which can be associated with a given detector. Such a measure would allow a comparison of the robustness properties of various detectors and also could motivate a new fidelity criterion which consists of some combination of performance and robustness. In the previous work of [3]-[4] we have presented a new approach toward the robustness question which considers the problem from a differential geometric perspective.

Since this is a new approach, we devote a section of this paper to review the development of our geometrically based robustness measures. The focus of this paper is to discuss the evaluation of the local, general measure ϕ for more realistic sample sizes. In our earlier work [3], we had demonstrated that the application of ϕ was quite straightforward for smaller sample sizes of say $n = 2, 3$ and 4. In this paper it is shown that the evaluation of the robustness measures can also be quite straightforward for larger sample sizes and in many cases, methods which are valid for evaluating the detection probability β and the false alarm probability α also apply in the evaluation of ϕ.

The Robustness Measures:

In [3] we have developed quantitative measures of robustness by employing ideas which have differential geometric origins. In this work we started by first considering an admissible class of distribution functions which is characterized by m parameters, with the idea being that there is some choice of the m parameters which affords an accurate model of the underlying process. Two measures have resulted from this approach; one based on the l_2 norm, and the other based on l_1. It is then possible for the independent sample case to extend the l_1 based robustness measure to account for much more general perturbations than is allowed by the robustness measure for the parameterized noise family. A more detailed summary of the results in [3] is presented in the subsequent paragraphs.

Consider the discrete-time detection of a signal in the midst of corrupting noise, where based on n-observations, the detector determines whether or not the signal is present. Let $\mathcal{D}_n$ denote the class of n-dimensional distribution functions which are parameterized by m parameters. From this point of view, the performance of the detector is thus expressed by considering the performance functional $h : \mathcal{D}_n \to R$. We then wish to choose the detector so that h has an acceptable nominal value and does not greatly deviate from this value, which is obtained by evaluating h under the nominal distribution function of $\mathcal{D}_n$. Viewing h as a height function over $\mathcal{D}_n$, we could say that a desirable robust detector would yield a "surface" which above the nominal element has an acceptable value and is not strongly sloped.

We can apply these observations to the robust detection problem. Note that the canonical detector structure, which consists of a nonlinearity, followed by an accumulator, followed by a threshold comparator, can be reformulated into a setting where one compares an n-dimensional observation vector to an appropriately defined decision region $\mathcal{B}_n$. If the observation vector is in $\mathcal{B}_n$ we would announce H_1 whereas if the observation vector is not in $\mathcal{B}_n$ we would announce H_0. If we consider the common case where randomization of the test is not necessary, then the decision region $\mathcal{B}_n$ for the canonical detector structure is specified via

$$\mathcal{B}_n = \{(x_1, x_2, \ldots, x_n) \in R^n : \sum_{i=1}^{n} g_i(x_i) > T\}$$

where g_i is the detector nonlinearity for the ith sample, and T is a threshold chosen to achieve a prescribed false alarm probability α. By taking this approach, a popular choice of h would be of the form

$$h = \int_{\mathcal{B}_n} dF_{\overline{X}}(y_1, y_2, \ldots, y_n | H_i) \quad ,$$

where $F_{\overline{X}}$ corresponds to the n-dimensional distribution function of the data and we condition on hypothesis H_i. Note that h corresponds to the false alarm probability α if $i = 0$, whereas h

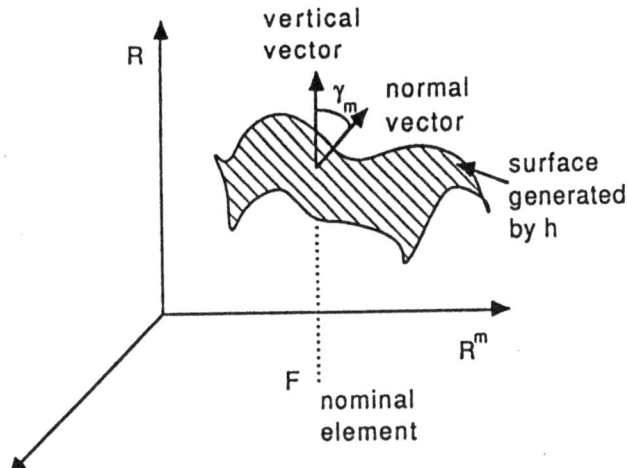

Figure 1: Illustration of the robustness measure $\phi_{2,m} = \cos\gamma_m$

corresponds to the detection probability β when $i = 1$.

Now that we have illustrated how one may specify the performance functional h, we are then interested in determining the "slope" of the surface generated by h at the nominal parameter values. Associating the slope of the surface with the cosine of the angle γ_m, which is the angle between the normal and vertical as illustrated in Figure 1 (where $\mathcal{D}_n = R^m$), we have proposed to use $\phi_{2,m} = \cos(\gamma_m)$ as a measure of local "first order" robustness. The interpretation for this measure is that larger values of $\phi_{2,m}$ suggest a more level surface and thus h should be more stable near the nominal. By an elementary, although somewhat lengthy, differential geometric argument which appears in [5], it is shown that $\phi_{2,m}$ is given by

$$\phi_{2,m} = \frac{1}{\sqrt{1 + \sum_{i=1}^{m}\left(\frac{\partial h}{\partial a_i}\right)^2}} \ . \tag{1}$$

Suppose that we form the vector $(1, \frac{\partial h}{\partial a_1}, \ldots, \frac{\partial h}{\partial a_m})$. We then observe that equation (1) can be expressed in terms of the l_2 norm of this vector. This motivates another natural choice for a robustness measure which applies to a noise class which is parameterized by m parameters. By basing the robustness measure on l_1 instead of l_2, we obtain the robustness measure $\phi_{1,m}$ which is given by

$$\phi_{1,m} = \frac{1}{1 + \sum_{i=1}^{m}\left|\frac{\partial h}{\partial a_i}\right|} \ . \tag{2}$$

One natural way to evaluate the measures $\phi_{1,m}$ and $\phi_{2,m}$ is at the nominal parameter values; that is, those parameter values which one perceives to be "most likely". After considering the nature of expressions (1) and (2), we see that these robustness measures are "local" measures which are meaningful in a sufficiently small neighborhood about the nominal. In [5] we show that it is possible to develop "nonlocal" robustness measures by finding the "worst-case" robustness over an appropriately defined parameter space.

Recall that the robustness measures $\phi_{1,m}$ and $\phi_{2,m}$ were obtained for the special case of an admissible noise class which is parameterized by m parameters. For example, these measures would apply to a situation where we have n-observations and we wish to detect the presence of a constant signal in Gaussian noise. If the signal strength and the elements of the covariance and mean matrices are imprecisely known, then we could view these as the parameters which form a basis for the admissible noise class. Note that as this example suggests, the measures of robustness for the parameterized case can be applied to situations where there exists dependency, and furthermore, it is possible to apply these measures in situations where the noise is nonstationary.

In many practical situations we would wish to employ a robustness measure which will allow for more arbitrary variations about the assumed noise model. As an illustration, suppose that we suspect for a particular application that the noise is Gaussian, but concede that perturbations from this assumption beyond that which can be represented by variations in the mean and covariance matrices are possible. This and similar scenarios emphasize the need for a more general measure of robustness than that which is given by $\phi_{1,m}$ and $\phi_{2,m}$. Ideally, we would wish to place virtually no constraints on the perturbations from the nominal model, and such a measure is feasible for the independent (but not necessarily identically distributed) case.

Suppose that $X_1, X_2, \ldots, X_n$ are independent random variables which represent the sampled data under a given hypothesis and that $F_1, F_2, \ldots, F_n$ are the corresponding distribution functions for these random variables. Since the false alarm probability α and the detection probability β are expressed via an integral over a Borel set B_n with respect to the appropriate n–dimensional distribution under hypotheses H_0 and H_1 respectively, we can, without loss of generality, note the independence of the observations and investigate perturbations in α and β by limiting consideration to the class of those univariate distributions given by step functions. That is, those distribution functios of the form

$$F_i(\cdot) = \sum_{j=1}^{m_i} a_{ij} I_{A_{ij}}(\cdot) \quad,$$

where the m_i are natural numbers, the a_{ij} are real numbers between zero and unity, the A_{ij} are bounded subsets of the reals which generate the partition $\mathcal{P}$, and $I_A(\cdot)$ denotes the indicator function of the set A. Letting m denote the total number of bounded partitioning subsets of R^n, we note that the a_{ij} parameterize $F_1, F_2, \ldots, F_n$, and we thus define the general robustness measure ϕ as

$$\phi = \lim_{|\mathcal{P}| \to 0} \phi_{1,m} \quad,$$

where (as is the custom for extended integrals) the existence of the limit corresponds to the same value being obtained regardless of how the union of the bounded partitioning sets approaches R^n. Thus we see that this general measure of robustness is obtained by extending the $\phi_{1,m}$ measure. For this class of step functions we have (defining $a_{i,m_{i+1}} = 1$ for $i = 1, \ldots, n$) that h is given by

$$h = \sum_{j_1=1}^{m_1} \cdots \sum_{j_n=1}^{m_n} \prod_{i=1}^{n} (a_{i,j_i+1} - a_{i,j_i}) I_{B_n}(x_{1,j_1}, \ldots, x_{n,j_n}) \quad ,$$

where $x_{i,j_i} = \sup A_{i,j_i}$ for $i = 1, \ldots, n$ and $j_i = 1, \ldots, m_i$. Taking the appropriate partial derivatives we obtain for $k = 1, \ldots, n$ and $l = 2, \ldots, m_k$

$$\frac{\partial h}{\partial a_{k,l}} = \sum_{j_1=1}^{m_1} \cdots \sum_{j_{k-1}=1}^{m_{k-1}} \sum_{j_{k+1}=1}^{m_{k+1}} \cdots \sum_{j_n=1}^{m_n} \prod_{\substack{i=1 \\ i \neq k}}^{n} (a_{i,j_i+1} - a_{i,j_i})$$

$$\cdot [I_{B_n}(x_{1,j_1}, \ldots, x_{k-1,j_{k-1}}, x_{k,l-1}, x_{k+1,j_{k+1}}, \ldots, x_{n,j_n})$$
$$- I_{B_n}(x_{1,j_1}, \ldots, x_{k-1,j_{k-1}}, x_{k,l}, x_{k+1,j_{k+1}}, \ldots, x_{n,j_n})] \quad .$$

(3)

Noting the difference of indicator functions which appears in (3), we are motivated to define the sets $\partial_i^+ B_n$ and $\partial_i^- B_n$ as follows:

$$\begin{aligned} \partial_i^+ B_n = \ & \{(x_1, x_2, \ldots, x_n) : \text{ there exists } \epsilon > 0 \text{ such that} \\ & (x_1, \cdots, x_{i-1}, y, x_{i+1}, \cdots, x_n) \in \overline{B_n} \text{ for } y \in (x_i - \epsilon, x_i) \text{ and} \\ & (x_1, \cdots, x_{i-1}, z, x_{i+1}, \cdots, x_n) \in B_n \text{ for } z \in (x_i, x_i + \epsilon)\} \quad , \end{aligned}$$

and

$$\begin{aligned} \partial_i^- B_n = \ & \{(x_1, x_2, \cdots, x_n) : \text{ there exists } \epsilon > 0 \text{ such that} \\ & (x_1, \cdots, x_{i-1}, y, x_{i+1}, \cdots, x_n) \in B_n \text{ for } y \in (x_i - \epsilon, x_i) \text{ and} \\ & (x_1, \cdots, x_{i-1}, z, x_{i+1}, \cdots, x_n) \in \overline{B_n} \text{ for } z \in (x_i, x_i + \epsilon)\} \quad . \end{aligned}$$

A careful inspection of (3) reveals that as the norm of the partition approaches zero, it is the $\partial_i^+ B_n$ and the $\partial_i^- B_n$ sets which contribute to the robustness measure. From an intuitive perspective, $\partial_i^+ B_n$ consists of those elements of the boundary of B_n which are intersected by rays parallel to the i axis and moving in the positive direction from the exterior of B_n to its interior. The set $\partial_i^- B_n$ is formed in analogous manner with the rays moving in the negative direction from the exterior of B_n to its interior.

Now suppose that if $x \notin \partial_1^+ B_n \cap \cdots \cap \partial_n^+ B_n$ then $x \in \text{int}(B_n) \cup \text{int}(\overline{B_n})$. For the common situation where the sets $\partial_i^- B_n$ are empty for $i = 1, \ldots, n$, then we have

$$\phi = \frac{1}{1 + \sum_{k=1}^{n} |\int_{\partial_k^+ B_n} dF_1(x_1) \cdots dF_{k-1}(x_{k-1}) dF_{k+1}(x_{k+1}) \cdots dF_n(x_n)|} \quad , \quad (4)$$

where again, F_k is the nominal distribution of X_k. The situation where all of the sets $\partial_i^- B_n$ are empty is often encountered in practical applications. For example, detectors which have the canonical structure with continuous, nondecreasing nonlinearities will have the property that the $\partial_i^- B_n$ sets are empty. A result similar to (4) holds for the situation where all of the $\partial_i^+ B_n$ sets are empty, but this case is not often encountered in practice. The case where there exists both nonempty $\partial_i^+ B_n$ sets and nonempty $\partial_i^- B_n$ sets is more complex and is considered in [6].

Evaluating ϕ

In much of our previous work, the evaluation of ϕ was restricted to small sample sizes. This restriction was due in part to the difficulty in evaluating the multi-dimensional integrals which appear in (4). Hence, the major focus of this paper is to consider the evaluation of ϕ for larger sample sizes. We show here that in some situations a method for evaluating ϕ can be quite similar to evaluating the false alarm probability α or the detection probability β.

For these discussions, we shall employ an identically distributed assumption for the element of $\mathcal{D}_n$ which is used to evaluate the robustness measure ϕ or, for that matter, the detector performance as indicated by α and β. We still, however, admit nonstationary perturbations from the selected element. Note that the results for ϕ do not require evaluation at the nominal and thus a "least favorable" element (such as discussed by Huber [1]) could also be employed.

Careful consideration of the expression for ϕ motivates the vertical censoring of the detector nonlinearity for imparting robustness. Letting $v(\cdot)$ denote a nondecreasing, piecewise continuous function, we define a nonlinearity $g(\cdot)$ via

$$g(x) = \begin{cases} k & \text{if } v(x) > k \\ v(x) & \text{if } -k \leq v(x) \leq k \\ -k & \text{if } v(x) < -k \end{cases} ,$$

where $k \in R$ is the censoring height. Thus, the decision region $\mathcal{B}_n$ is expressed as

$$\mathcal{B}_n = \{(x_1, \ldots, x_n) : \sum_{i=1}^{n} g(x_i) > T\} ,$$

where T is the detector threshold. We now note that the decision region $\mathcal{B}_n$ is convex and one consequence of this convexity property is that, when the detection probability is used as a performance criterion, then the sets $\partial_i^- \mathcal{B}_n$, $i = 1, \ldots, n$ are empty. Furthermore, by exploiting certain symmetry properties of $\mathcal{B}_n$, we may write ϕ as

$$\phi = \frac{1}{1 + n \int_{\partial_n^+ \mathcal{B}_n} dF(x_1) \cdots dF(x_{n-1})} .$$

Consider the integral which appears in the previous expression. The fact that we have i.) independent observations, ii.) a convex decision region, and iii.) an interpretation of the $\partial_n^+ \mathcal{B}_n$ set as the collection of elements obtained by intersecting the boundary of $\mathcal{B}_n$ with a ray parallel to the n axis moving in a positive direction from the exterior of $\mathcal{B}_n$ to its interior, allows us to write

$$\int_{\partial_n^+ \mathcal{B}_n} dF(x_1) \cdots dF(x_{n-1}) = \int_{\Gamma_{n-1}} dF(x_1) \cdots dF(x_{n-1}) ,$$

where Γ_{n-1} is defined as

$$\Gamma_{n-1} = \{(x_1, x_2, \ldots, x_{n-1}) : \text{there exists } z \in R : \sum_{i=1}^{n-1} g(x_i) + g(z) > T\} .$$

The intuitive interpretation is that, due to independence, we may evaluate the integral by "projecting" the $\partial_n^+ \mathcal{B}_n$ set onto the previous dimension. Also, due to the convexity of $\mathcal{B}_n$, we have that

$$\Gamma_{n-1} = \widehat{\Gamma}_{n-1} = \{(x_1, x_2, \ldots, x_{n-1}) : \sum_{i=1}^{n-1} g(x_i) > T - k\} ,$$

since for all $z \in R$, $g(z) \leq k$. We then note that the form of Γ_{n-1} is very similar to that of $\mathcal{B}_{n-1}$ with the only difference being that the threshold T is replaced with $T - k$. Thus letting

Table 1: β, ϕ_α and ϕ_β for $s = 1.0$ and $\alpha = 0.05$.			
$n=5$	β	ϕ_α	ϕ_β
$k=0.25$	0.5156	0.6420	0.2385
$k=0.5$	0.5938	0.6156	0.2189
$k=1.0$	0.6229	0.5916	0.2103
$n=10$	β	ϕ_α	ϕ_β
$k=0.25$	0.7659	0.5155	0.1080
$k=0.5$	0.8518	0.4862	0.0995
$k=1.0$	0.8698	0.4717	0.0979

$\alpha_n(T_n)$ and $\beta_n(T_n)$ (the subscript n is added to denoting the dependency on the sample size) denote the false alarm and detection probabilities, respectively, for an n sample detector which employs threshold T_n, we have that

$$\phi_\alpha = \frac{1}{1 + n\alpha_{n-1}(T_n - k)} \quad \text{and} \quad \phi_\beta = \frac{1}{1 + n\beta_{n-1}(T_n - k)} \ .$$

We thus conclude that methods for evaluating the false alarm probability α and the detection probability β can also be employed to evaluate ϕ_α and ϕ_β.

Example:

We illustrate the application of the aforementioned results by considering the detection of a constant signal in additive, nominally Laplace noise, where we employ the Neyman-Pearson nonlinearity and censor at the vertical heights of $-k$ and k. A lengthy extension of the work in [7] yields the distribution function of the canonical test statistic under both H_0 and H_1. The resulting expressions for the distribution of the test statistic are very lengthy and we omit these here due to space considerations. Although the results quoted in this example are obtained under a nominal assumption, the distribution of the test statistic under a "least favorable" assumption is also available and can be used when evaluating ϕ.

In Table 1 we have given a sampling of the results which can be computed for this example. In each case we have constrained the false alarm probability to be 0.05. The general trends are that robustness increases for heavier censoring and that robustness decreases as the number of samples becomes larger. robustness increases for heavier censoring and that robustness decreases as the number of samples becomes larger.

Conclusions:

In this paper we have reviewed a new approach toward robust signal detection which is based on geometric concepts. By providing a quantitative measure of robustness, this approach can be an attractive alternative to employing classical saddlepoint techniques since important limitations associated with such techniques can be circumvented.

Presented also is a result that, for many cases of interest, can be used to evaluate the robustness measure ϕ for larger sample sizes. It is seen that the technique for evaluating ϕ in these situations is essentially equivalent to evaluating the detector performance in terms of the false alarm probability α or the detection probability β under the nominal conditions.

Acknowledgement

The author is with the Electrical Engineering Department at Colorado State University, Fort Collins, Colorado 80533.
e-mail: mikethom@elbert.lance.ColoState.EDU

References

[1] P. J. Huber, *Robust Statistics*. New York: Wiley. 1981.

[2] S. A. Kassam and H. V. Poor, "Robust techniques for signal processing: A Survey," *Proceedings of the IEEE*, Vol. 73, No. 3, pp. 433-480. March 1985.

[3] M. W. Thompson and D. R. Halverson, "A differential geometric approach toward robust signal detection in nonstationary and/or dependent noise," Proceedings of the Twenty-Fifth Annual Allerton Conference on Communications, Control and Computing, Monticello, Illinois, September 30- October 2, 1987. pp. 163-170.

[4] M. W. Thompson and D. R. Halverson, "A novel differential geometric approach toward robust signal detection," *Proc. of the 1986 Conference on Information Sciences and Systems*, Princeton, New Jersey, March 19-21, 1986, pp. 11-16.

[5] M. W. Thompson, "Applications of Geometric Concepts toward Robust Signal Detection", Ph.D. thesis. Texas A&M University, College Station, Texas, 1987.

[6] M. W. Thompson and D. R. Halverson, "Geometric measures of robustness in signal processing," to appear in the proceedings of ComCon 88: Advances in Communications and Control Systems, Baton Rouge, Louisiana, October 19-21, 1988.

[7] D. R. Halverson, G. L. Wise, and M. W. Thompson, "Robust detection of a constant signal in Laplace noise," *Proceedings of the Seventeenth Annual Conf. on Information Sciences and Systems*, Baltimore, Maryland, March 23-25, pp. 228-232.

ROBUST MULTIPLE HYPOTHESIS TESTS

Patrick A. Kelly and Xiaoke Duan
University of Massachusetts

1. Introduction. There has long been interest in problems of deciding between two
hypotheses when there is uncertainty in the statistical models for the observations.
Huber (1965) showed that for some finite-sample problems, tests can be found that
are minimax-robust in terms of risk. Specifically, Huber considered problems in
which the observation under each of two hypotheses consists of a sequence of
independent random variables whose densities are known only to lie in uncertainty
classes. He showed that, in some cases, there exist least-favorable densities in
the classes, having the property that likelihood ratio tests based on those
densities minimize the maximum risk as the true observation densities vary. Huber's
results have been extended and modified for different situations by many subsequent
researchers. However, there are not many results concerning robust multiple-
hypothesis tests. Problems of deciding among more than two hypotheses occur in many
applications, including image processing, and there is often uncertainty in the
observation models. In this paper, we consider finite-sample multiple-hypothesis
tests, such as those that arise in image segmentation problems. Our goal is to see
to what extent methods used to define robust two-hypothesis tests can be used to
define robust or approximately robust multiple-hypothesis tests.

Suppose that we have a problem of deciding among M hypotheses. Most multiple-
hypothesis tests are Bayes tests, and in general Bayes tests can be put in the form
of comparing likelihood ratios for pairs of hypotheses to thresholds. That is, if
the observation $\underline{Z} = \{Z_1, \ldots, Z_N\}$ has density $f^{(i)}(\underline{z})$ under hypothesis H_i, we have the
decision rule:

Choose H_j if $\ln[f^{(j)}(\underline{z})/f^{(k)}(\underline{z})] \geq \tau_{jk}$ for all $k = 1, \ldots, M$ (1.1)

where the τ_{jk}'s are determined by costs and the prior probablities of the
hypotheses. This suggests that one approach to making the rule (1.1) robust if
there is uncertainty in the densitites would be to use the least-favorable densities
found for two-hypothesis tests. We could replace each likelihood ratio $f^{(j)}/f^{(k)}$
with the ratio $f'^{(j)}/f'^{(k)}$, where $f'^{(j)}$ and $f'^{(k)}$ are the least favorable densities
associated with the two-hypothesis test of H_j versus H_k. In section 2 we show that
this approach leads to minimax-robust rules when the risk is the probability of
error and when, under each hypothesis, the observations are sample-to-sample
independent, with each Z_ℓ nominally having one of only two densities (such a
situation occurs, for example, when the observations are binary signals in noise).

We call this the case of binary marginal densities. When each Z_ℓ may nominally have one of more than two densities, using least-favorable pairs of densities found for two-hypothesis tests does not lead to a minimax-robust multiple-hypothesis test. Results from Kuznetsov (1982) can be used to find least-favorable sets of densities that define robust one-sample tests in this situation; however, these densities do not generally lead to robust tests when the observations contain more than one sample, which is the usual case of interest. We discuss this difficulty in section 3. We then show that a reasonable test can be defined for this problem that is based on robust two-hypothesis tests and is approximately minimax-robust in terms of probability of error when the observation consists of one or more samples.

2. <u>Robust tests for the Binary Marginal Density Case</u>. A problem that frequently occurs in applications such as image segmentation is the following: Suppose that we have a random signal sequence $\underline{S} = \{S_\ell, \ \ell=1, \ \ldots, \ N\}$ such that each S_ℓ takes the value 0 or 1. Given a value $\underline{s}$ of the signal, the observed sequence $\underline{Z} = \{Z_\ell, \ \ell=1, \ \ldots, \ N\}$ is such that the observations are sample-to-sample independent, with Z_ℓ nominally having density $f_\ell(z)$ if $s_\ell = 0$ and $g_\ell(z)$ if $s_\ell = 1$. There are 2^N hypotheses, one corresponding to each possible value of $\underline{S}$. Under a given hypothesis, the nominal joint density of $\underline{Z}$ is

$$\prod_{\{\ell:s_\ell=0\}} f_\ell(z_\ell) \cdot \prod_{\{\ell:s_\ell=1\}} g_\ell(z_\ell) \tag{2.1}$$

where $\underline{s}$ is the signal value corresponding to the hypothesis. We assume that prior probabilities of the different hypotheses are known, and that we want to decide which hypothesis is true. A common approach is to find the MAP estimate of $\underline{S}$, given $\underline{Z}$; that is, to decide on a value $\underline{s}$ so as to minimize the probability of error. (A special case of this problem, with $Z_\ell = S_\ell + W_\ell$ where $\{W_\ell\}$ is a sequence of independent random variables, was discussed by Kelly and Duan (1988).)

Let uncertainty be introduced into this model as follows: assume that the actual density of Z_ℓ, given s_ℓ, is a member of the class

$$\Phi_\ell = \{f_\ell'': f_\ell'' = (1-\epsilon_0) f_\ell + \epsilon_0 q_0, \ q_0 \in Q\} \tag{2.2}$$

if $s_\ell = 0$, or the class

$$\Gamma_\ell = \{g_\ell'': g_\ell'' = (1-\epsilon_1)g_\ell + \epsilon_1 q_1, \ q_1 \in Q\} \tag{2.3}$$

if $s_\ell = 1$, where Q is the set of all densities. It follows from results in Huber

(1965) that for ε_0 and ε_1 sufficiently small there exist densities $f'_\ell \in \Phi_\ell$ and $g'_\ell \in$

Γ_ℓ such that, if $L_\ell(z)$ denotes $\ln[g'_\ell(z)/f'_\ell(z)]$, we have

$$F''_\ell\{L_\ell(Z_\ell) \geq t\} \leq F'_\ell\{L_\ell(Z_\ell) \geq t\} \tag{2.4}$$

$$G''_\ell\{L_\ell(Z_\ell) \geq t\} \geq G'_\ell\{L_\ell(Z_\ell) \geq t\} \tag{2.5}$$

for all t, where F'_ℓ, G'_ℓ denote the probability measures corresponding to f'_ℓ, g'_ℓ and

F''_ℓ, G''_ℓ denote measures corresponding to any other densities $f''_\ell \in \Phi_\ell$ and $g''_\ell \in \Gamma_\ell$.

Then (2.4) and (2.5) imply that the "least-favorable" pair (f'_ℓ, g'_ℓ) leads to a

saddlepoint solution of the problem of minimizing the worst-case probability of

error for the one-sample test of $s_n = 0$ versus $s_n = 1$: the error probability using

a test based on $L_\ell(z)$ is maximized when the densities are (f'_ℓ, g'_ℓ), while under

those densities a test based on $L_\ell(z)$ has minimum error probability. As shown in

Huber (1965), f'_ℓ and g'_ℓ are such that L_ℓ is a censored version of the nominal log-

likelihood ratio $\ln[g_\ell/f_\ell]$.

Of course, in most problems of interest the observation consists of more than

one sample point. Huber made use of Lemma 1 on p. 73 of Lehmann (1959) to define

robust N-sample, two-hypothesis tests based on the one-sample results (2.4) and

(2.5). To define a robust N-sample, 2^N-hypothesis test we need to generalize

Lehmann's result as follows:

Lemma 1. Let $\{X_\ell\}$ and $\{Y_\ell\}$ be two sequences of independent random variables such

that for each ℓ, $\text{Prob}\{X_\ell \geq t\} \leq \text{Prob}\{Y_\ell \geq t\}$ for all t. Let $\{r_{m,\ell}: R \rightarrow R; m = 1,$

$..., M; \ell = 1, ..., N\}$ be a set of monotone nondecreasing functions. Then for any

real numbers $t_1, ..., t_M$,

$$\text{Prob}\{ \sum_{\ell=1}^{N} r_{1,\ell}(X_\ell) \geq t_1, ..., \sum_{\ell=1}^{N} r_{M,\ell}(X_\ell) \geq t_M \}$$

$$\leq \text{Prob}\{ \sum_{\ell=1}^{N} r_{1,\ell}(Y_\ell) \geq t_1, ..., \sum_{\ell=1}^{N} r_{M,\ell}(Y_\ell) \geq t_M \}. \tag{2.6}$$

A proof for this lemma is outlined in Kelly and Duan (1988). Now, suppose that

$\underline{S}$ has the value $\underline{s}^{(j)}$. Let $\{X_\ell\}$ be a sequence of independent random variables such

that if $s_\ell^{(j)} = 0$, then X_ℓ is distributed as $-L_\ell(Z_\ell)$ with Z_ℓ having density f'_ℓ, while if $s_\ell^{(j)} = 1$, then X_ℓ is distributed as $L_\ell(Z_\ell)$ with Z_ℓ having density g_ℓ'. Let $\{Y_\ell\}$ be a second sequence of independent random variables defined similarly to $\{X_\ell\}$, but with Z_ℓ having a density equal either to $f_\ell'' \in \Phi_\ell$ if $s_\ell = 0$ or $g_\ell'' \in \Gamma_\ell$ if $s_\ell = 1$. Let $f'^{(j)}(\underline{z})$ denote the joint density of $\underline{Z}$, given $\underline{s}^{(j)}$, when the Z_ℓ's have the appropriate densities f_ℓ' or g_ℓ', and let $f''^{(j)}(\underline{z})$ denote the density of $\underline{Z}$, given $\underline{s}^{(j)}$, when the Z_ℓ's have densities f_ℓ'' or g_ℓ''. Let $F'^{(j)}$ denote the probability measure corresponding to $f'^{(j)}$ and $F''^{(j)}$ denote the measure corresponding to $f''^{(j)}$. Finally, define $r_{k,\ell}(t) = t$ if $s_\ell^{(k)} \neq s_\ell^{(j)}$ and $r_{k,\ell}(t) = 0$ otherwise. By Lemma 1 and eqs. (2.4) and (2.5),

$$F'^{(j)}\{\{z_1,\ldots,z_n\}: \ln[f'^{(j)}(\underline{z})/f'^{(k)}(\underline{z})] \geq \tau_{jk} \text{ for all } k = 1,\ldots,M\}$$

$$= \text{Prob}\{\sum_{\ell=1}^{N} r_{1,\ell}(X_\ell) \geq \tau_{j1}, \ldots, \sum_{\ell=1}^{N} r_{M,\ell}(X_\ell) \geq \tau_{jM}\}$$

$$\leq \text{Prob}\{\sum_{\ell=1}^{N} r_{1,\ell}(Y_\ell) \geq \tau_{j1}, \ldots, \sum_{\ell=1}^{N} r_{M,\ell}(Y_\ell) \geq \tau_{jM}\}$$

$$= F''^{(j)}\{\{z_1,\ldots,z_N\}: \ln[f'^{(j)}(\underline{z})/f'^{(k)}(\underline{z})] \geq \tau_{jk} \text{ for all } k = 1,\ldots,M\} \qquad (2.7)$$

where τ_{jk}'s are the appropriate thresholds for MAP tests (i.e., $\tau_{jk} = \ln[p_k/p_j]$, where p_m is the prior probability of the signal $\underline{s}^{(m)}$). Note that, by (1.1), the first term in (2.7) is the probability of selecting the signal $\underline{s}^{(j)}$ when $\underline{s}^{(j)}$ is the true signal, using the MAP test for the least-favorable densities when those are the true densities. The last term is the probability of selecting $\underline{s}^{(j)}$ given that $\underline{s}^{(j)}$ is the true signal using the MAP test for the least-favorable densities, but when the true densities are some other elements of the uncertainty classes. Equation (2.7) holds for each $j = 1, \ldots, 2^N$. Since the MAP test for any set of densities is the test having minimum probability of error under those densities, we have shown the following result.

Theorem 1. Let $f'^{(m)}(\underline{z}) = \prod_{(\ell:s_\ell^{(m)}=0)} f_\ell'(z_\ell) \cdot \prod_{(\ell:s_\ell^{(m)}=1)} g_\ell'(z_\ell)$. Consider the decision rule:

Choose $\underline{s}^{(j)}$ if $\ln[f'^{(j)}(\underline{z})/f'^{(k)}(\underline{z})] \geq \tau_{jk}$ for all $k = 1,\ldots,M$ $\qquad (2.8)$

where $\tau_{jk} = \ln[p_k/p_j]$ and p_m is the prior probability that $\underline{S} = \underline{s}^{(m)}$. The rule (2.8) is minimax-robust in terms of probability of error; it is the rule having minimum probability of error when $f'^{(m)}$ is the true observation density for each m, while the probability of error using the rule is maximized over allowable densities by the densities $f'^{(m)}$.

Hence, to obtain a rule for deciding the value of the signal that is minimax-robust in terms of probability of error, we simply replace the nominal densities in a MAP rule of the form (1.1) with the least-favorable densities found for two-hypothesis tests.

3. Robust Tests for Non-Binary Marginal Densities. Now suppose that the random signal sequence $\underline{S}$ is such that each S_ℓ takes one of the three values $\{0,1,2\}$. Given a value $\underline{s}$ of $\underline{S}$, the observed sequence $\underline{Z}$ is again sample-to-sample independent, but now Z_ℓ has the nominal marginal densities $f_\ell(z)$ if $s_\ell = 0$, $g_\ell(z)$ if $s_\ell = 1$, and $h_\ell(z)$ if $s_\ell = 2$. Again, there is one hypothesis (and one nominal joint density for $\underline{Z}$) corresponding to each possible realization of $\underline{S}$; assuming that we know the prior probabilities of the realizations, we want to find the MAP estimate of $\underline{S}$, given $\underline{Z}$. As in section 2, uncertainty can be introduced into the model by assuming that the true densities for Z_ℓ, given s_ℓ, are known only to be in the uncertainty classes Φ_ℓ, Γ_ℓ, or

$$H_\ell = \{h''_\ell: h''_\ell = (1-\epsilon_2)h_\ell + \epsilon_2 q_2, \quad q_2 \in Q\}. \tag{3.1}$$

For this case, it is clear that it is not possible, in general, to get a minimax-robust rule by simply replacing the nominal likelihood ratios in (1.1) with likelihood ratios defined by least-favorable pairs for comparing two hypotheses at a time. The reason is that, unlike the binary marginal density case, the resulting rule is not optimal for any fixed density, because the least-favorable density corresponding to any one hypothesis in the likelihood ratios differs depending on which other hypothesis it is being compared to. (For example, the least-favorable element of Φ_ℓ for testing $s_\ell = 0$ against $s_\ell = 1$ is different from the least-favorable element for testing $s_\ell = 0$ against $s_\ell = 2$.)

It is possible to use results in Kuznetsov (1982) to find minimax-robust tests for this problem when the observation consists of only one sample point. Let $p_{\ell,i}$ be the probability that $S_\ell = i$, $i = 0,1,2$. Then the following lemma, which is a slight generalization of Lemma 2 in Kelly and Duan (1988), is a consequence of Theorem 1 of Kuznetsov.

<u>Lemma 2</u>. The rule for choosing the value of S_ℓ, given Z_ℓ, that has the minimum worst-case probability of error for densities in the classes Φ_ℓ, Γ_ℓ and H_ℓ, has the following form: If

$$\max_i \{p_{\ell,i}\} \geq \int \max\{(1-\epsilon_0)\ p_{\ell,0}f_\ell(z);\ (1-\epsilon_1)p_{\ell,1}g_\ell(z);\ (1-\epsilon_2)p_{\ell,2}h_\ell(z)\}dz \qquad (3.2)$$

then choose $s_\ell = j$, where $j = \arg_i \max\{p_{\ell,i}\}$; otherwise choose $s_\ell = m$, where

$m = \arg \max_{0,1,2} \{p_{\ell,0}f_\ell(z);\ p_{\ell,1}g_\ell(z);\ p_{\ell,2}h_\ell(z)\}$ (the nominal MAP rule).

Note that the robust rule defined in Lemma 2 is similar to the robust one-sample two-hypothesis rule, in that it reduces to the nominal rule for a range of prior probabilities (corresponding to low thresholds for the rule written in the form (1.1)). When one probability gets too large, the rule says to discard the observation and always choose the signal value corresponding to the largest probability. Unfortunately, it also follows from (3.2) that the value that the largest probability must take in order for the robust rule to differ from the nominal rule is a function, not just of the sizes of the uncertainty classes (as it is in the binary case), but also of the other prior probabilities. This causes problems in trying to implement a robust MAP test using least-favorable densities. As is discussed in Kelly and Duan (1988), one can find densities in the uncertainty classes for which the one-sample robust rule defined in Lemma 2 is the optimal (MAP) rule, and under which the probability of error using the rule is maximized over all possible densities. However, these one-sample least-favorable densities depend on the prior probabilities, and lead to robust rules only for specific values of the prior probabilities. We are not able, in general, to get an ordering of distributions over their entire support, as we had in the binary case with (2.4) and (2.5). This in turn implies that we do not have a result like Lemma 1 that enables us to use the one-sample densities to define an N-sample robust test.

To try to get around this difficulty, we instead define a test that intuitively seems robust, and see if there are conditions under which it truly is either exactly or approximately robust. It is easiest to consider a special case, with $Z_\ell = S'_\ell + W_\ell$ for each ℓ, where $\{W_\ell\}$ is a sequence of iid, mean-zero nominally Gaussian random variables, and where $S'_\ell = \mu_i$ if $S_\ell = i$, where $\mu_0 < \mu_1 < \mu_2$. Without loss of generality, we can assume that $\mu_1 = 0$. For this special case, we use the notation $f_0(z)$ in place of $f_\ell(z)$, $f_1(z)$ in place of $g_\ell(z)$, and $f_2(z)$ in place of $h_\ell(z)$. Assume that the contamination levels ϵ_0 in (2.2) and ϵ_2 in (3.1) are fixed small numbers, but let ϵ_1 in (2.3) be a variable. Define

$$f_0'(z) = (1-\epsilon_0)f_0(z) \qquad \text{if } f_2(z)/f_0(z) < 1/c_0 \qquad (3.3)$$
$$\qquad\;\; = (1-\epsilon_0)c_0 f_2(z) \qquad \text{otherwise}$$

$$f_1'(z) = (1-\epsilon_1)c_{10}f_0(z) \qquad \text{if } f_1(z)/f_0(z) \leq c_{10}$$
$$\qquad\;\; = (1-\epsilon_1)f_1(z) \qquad\quad \text{if } f_1(z)/f_0(z) > c_{10} \text{ and } f_2(z)/f_1(z) < 1/c_{12} \qquad (3.4)$$
$$\qquad\;\; = (1-\epsilon_1)c_{12}f_2(z) \qquad \text{otherwise}$$

$$f_2'(z) = (1-\epsilon_2)c_2 f_0(z) \qquad \text{if } f_2(z)/f_0(z) < c_2$$
$$\qquad\;\; = (1-\epsilon_2)f_2(z) \qquad\quad \text{otherwise.} \qquad (3.5)$$

To find the unknowns c_0, c_{10}, c_{12}, c_2 and ϵ_1, we first solve the equations $\int f_i'(z)dz$ = 1, i = 0,2, which give c_0 and c_2. Next, let z_L be the value of z at which $f_2(z)/f_0(z) = c_2$, and let z_u be the value at which $f_2(z)/f_0(z) = 1/c_0$. We then define $c_{10} = f_1(z_L)/f_0(z_L)$ and $1/c_{12} = f_2(z_u)/f_1(z_u)$. Finally, ϵ_1 is determined by setting $\int f_1'(z)dz$ = 1. It can be shown that, if ϵ_0 and ϵ_2 are small enough, then z_L < z_u and the densities are well-defined. Let $L_{ij}(z)$ denote $\ln[f_i'(z)/f_j'(z)]$. Note that f_0' and f_2' are the least-favorable pair for testing $S_\ell = \mu_0$ against $S_\ell = \mu_2$, and that we have defined f_1' so that the censored log-likelihood ratios L_{10}, L_{21}, and L_{20} all cutoff at the same values of z.

Now, suppose that the ϵ_1 determined above is the contamination level in (2.3). Then $f_0' \epsilon \Phi$, $f_1' \epsilon \Gamma$, and $f_2' \epsilon H$ (where we have dropped the subscript ℓ for the uncertainty classes). Let F_i' denote the measure corresponding to f_i', and let F_i'' denote the measure corresponding to any other element of the appropriate uncertainty class. We have the following result:

Lemma 3. (1) For any real numbers t_0 and t_1,

$$F_0'\{L_{10}(Z_\ell) \leq t_0, L_{20}(Z_\ell) \leq t_1\} \geq F_0''\{L_{10}(Z_\ell) \leq t_0, L_{20}(Z_\ell) \leq t_1\} \qquad (3.5)$$

$$F_2''\{L_{20}(Z_\ell) \geq t_0, L_{21}(Z_\ell) \geq t_1\} \geq F_2'\{L_{20}(Z_\ell) \geq t_0, L_{21}(Z_\ell) \geq t_1\}. \qquad (3.6)$$

(ii) For any (t_0, t_1) such that either $t_0 > L_{10}(z_L)$ and $t_1 < L_{21}(z_u)$, or $t_0 \leq L_{10}(z_L)$ and $t_1 \geq L_{21}(z_u)$,

$$F_1''\{L_{10}(Z_\ell) \geq t_o,\ L_{21}(Z_\ell) \leq t_1\} \geq F_1'\{L_{10}(Z_\ell) \geq t_o,\ L_{21}(Z_\ell) \leq t_1\}. \tag{3.7}$$

Proof. (i) Note that $L_{10}(z)$ and $L_{21}(z)$ are both monotone increasing functions of

$L_{20}(z)$. Since (f_o', f_2') is the least-favorable pair for testing $S_\ell' = \mu_o$ against $S_\ell' = \mu_2$, (3.5) and (3.6) follow from (2.4), (2.5), and Lemma 1.

(ii) If $t_o \leq L_{10}(z_L)$ and $t_1 \geq L_{21}(z_u)$, then the set $\{L_{10}(Z_\ell) \geq t_o,\ L_{21}(Z_\ell) \leq t_1\}$ has probability equal to one under any density for Z_ℓ. If $t_o > L_{10}(z_L)$ and $t_1 < L_{21}(z_u)$, then

$$F_1'\{L_{10}(Z_\ell) \geq t_o,\ L_{21}(Z_\ell) \leq t_1\} = (1-\epsilon_1)F_1\{L_{10}(Z_\ell) \geq t_o,\ L_{21}(Z_\ell) \leq t_1\}$$

where F_1 is the measure corresponding to f_1. Then (3.7) follows from the definition of the uncertainty class Γ. .//.

Let $\tau_{ij} = \ln[p_{\ell,j}/p_{\ell,i}]$, where $p_{\ell,m}$ is the prior probability that $S_\ell' = \mu_m$. Then we have the result:

Theorem 2. If $p_{\ell,0}$, $p_{\ell,1}$ and $p_{\ell,2}$ are such that either $\tau_{10} > L_{10}(z_L)$ and $\tau_{21} < L_{21}(z_u)$, or $\tau_{10} \leq L_{10}(z_L)$ and $\tau_{21} \geq L_{21}(z_u)$, then the following one-sample decision rule for choosing the value of S_ℓ', given Z_ℓ:

Choose μ_i if $\ln[f_i'(Z_\ell)/f_j'(Z_\ell)] \geq \tau_{ij}$ for each $j = 0,1,2$ \hfill (3.8)

is minimax-robust in terms of probability of error for the uncertainty classes Φ, Γ and H.

Proof. First, the rule (3.8) is the MAP rule for the densities (f_o', f_1', f_2'), and hence has minimum probability of error. With the appropriate thresholds, the set in (3.5) is the set for which μ_o is chosen, given that μ_o is the true value, using the rule (3.8). Thus, (3.5) implies that the probability of correctly choosing μ_o using the rule is minimized over densities in Φ by f_o'. Similarly, (3.6) and (3.7) imply that the probability of correctly choosing μ_1 or μ_2 using (3.8) is minimized over allowable densities by f_1' or f_2', respectively. .//.

By allowing ϵ_1 to depend on ϵ_o and ϵ_2, we have thus been able to define densities (f_o', f_1', f_2') so that the optimal one-sample test under those densities is robust for a wide range of prior probabilities; roughly, the test is minimax-robust provided that we don't have either $p_{\ell,0} \cong p_{\ell,1} \gg p_{\ell,2}$ or $p_{\ell,2} \cong p_{\ell,1} \gg p_{\ell,0}$. Now, we want to see if we can use the densities to define a robust N-sample test. First, note that (3.5) and (3.6) are analogous to (2.4) and (2.5), in that there is an ordering of distributions over their entire support. This is the property needed in order to use Lemma 1. However, we do not have this property for the middle distribution (i.e., the one corresponding to $S_\ell' = \mu_1$). Hence, the best we can do is to argue that the densities (f_0', f_1', f_2') lead to an N-sample test that is approximately robust in some sense. The simplest case of interest is when $\underline{S}$ takes only the value $\underline{s}^{(0)}$, or $\underline{s}^{(1)}$, or $\underline{s}^{(2)}$, where $s_\ell^{(i)} = \mu_i$ for all $\ell = 1, \ldots, N$. Let $f'^{(i)}(\underline{z}) = \prod_\ell f_i'(z_\ell)$, $i = 0, 1, 2$. Then the MAP rule for choosing the value of $\underline{S}$, given $\underline{Z}$, under the joint densities $f'^{(i)}$ is:

Choose $\underline{s}^{(j)}$ if $\ln[f'^{(j)}(\underline{z})/f'^{(k)}(\underline{z})] \geq \tau_{jk}$, $k = 0, 1, 2$ (3.9)

where $\tau_{jk} = \ln[p_k/p_j]$ and p_m is the prior probability that $\underline{S} = \underline{s}^{(m)}$. Let $F'^{(i)}$ be the probability measure corresponding to $f'^{(i)}$, and let $F''^{(i)}$ be the measure corresponding to $f''^{(i)}(\underline{z}) = \prod_\ell f_{i,\ell}''(z_\ell)$, where $f_{i,\ell}''$ is any element of the uncertainty class for f_i (not necessarily the same element for each ℓ). Then it follows from (3.5), (3.6) and Lemma 1 that

$$F''(0)\left\{ \sum_{\ell=1}^N L_{10}(Z_\ell) \leq \tau_{10}, \ \sum_{\ell=1}^N L_{20}(Z_\ell) \leq \tau_{20} \right\} \geq F'(0)\left\{ \sum_{\ell=1}^N L_{10}(Z_\ell) \leq \tau_{10}, \right.$$

$$\left. \sum_{\ell=1}^N L_{20}(Z_\ell) \leq \tau_{20} \right\} \tag{3.10}$$

$$F''(2)\left\{ \sum_{\ell=1}^N L_{20}(Z_\ell) \geq \tau_{20}, \ \sum_{\ell=1}^N L_{21}(Z_\ell) \geq \tau_{21} \right\} \geq F'(2)\left\{ \sum_{\ell=1}^N L_{20}(Z_\ell) \geq \tau_{20}, \right.$$

$$\left. \sum_{\ell=1}^N L_{21}(Z_\ell) \geq \tau_{21} \right\}. \tag{3.11}$$

The set in (3.10) is the set of $\underline{Z}$'s for which we choose $\underline{s}^{(0)}$ when $\underline{s}^{(0)}$ is the true signal, using the rule (3.9). Similarly, the set in (3.11) is the set for which we choose $\underline{s}^{(2)}$ when $\underline{s}^{(2)}$ is correct, using (3.9). Hence the probability of making an incorrect decision using (3.9) when either $\underline{s}^{(0)}$ or $\underline{s}^{(2)}$ is the true signal is maximum when the observation density is either $f'^{(0)}$ or $f'^{(2)}$, respectively. To complete the proof that $(f'^{(0)}, f'^{(1)}, f'^{(2)})$ is least-favorable, we would also like to say that

$$F''^{(1)} \{ \sum_{\ell=1}^{N} L_{10}(Z_\ell) \geq \tau_{10}, \sum_{\ell=1}^{N} L_{21}(Z_\ell) \leq \tau_{21} \} \geq F'^{(1)} \{ \sum_{\ell=1}^{N} L_{10}(Z_\ell) \geq \tau_{10},$$

$$\sum_{\ell=1}^{N} L_{21}(Z_\ell) \leq \tau_{21} \}. \tag{3.12}$$

However, (3.12) is not necessarily true; it is true for some $F''^{(1)}$'s and some thresholds, but not all. To see this, note that it follows from the noise being nominally Gaussian that if σ^2 is the nominal noise variance, then

$$L_{10}(z) = - \frac{\mu_0}{\mu_2 - \mu_0} L_{20}(z) - \frac{\mu_0 \mu_2}{2\sigma^2} + \frac{\mu_0}{\mu_2 - \mu_0} \ln(\frac{1-\epsilon_2}{1-\epsilon_0}) + \ln(\frac{1-\epsilon_1}{1-\epsilon_0}) \tag{3.13}$$

$$L_{21}(z) = \frac{\mu_2}{\mu_2 - \mu_0} L_{20}(z) + \frac{\mu_2 \mu_0}{2\sigma^2} - \frac{\mu_2}{\mu_2 - \mu_0} \ln(\frac{1-\epsilon_2}{1-\epsilon_0}) + \ln(\frac{1-\epsilon_2}{1-\epsilon_1}) \tag{3.14}$$

Thus, we want

$$F''^{(1)} \{ a \leq \sum_{\ell=1}^{N} L_{20}(Z_\ell) \leq b \} \geq F'^{(1)} \{ a \leq \sum_{\ell=1}^{N} L_{20}(Z_\ell) \leq b \} \tag{3.15}$$

where a and b are given by

$$a = -\{ \tau_{10} - N[- \frac{\mu_0 \mu_2}{2\sigma^2} + \frac{\mu_0}{\mu_2 - \mu_0} \ln(\frac{1-\epsilon_2}{1-\epsilon_0}) + \ln(\frac{1-\epsilon_1}{1-\epsilon_0})]\} \cdot \{ \frac{\mu_2 - \mu_0}{\mu_0} \} \tag{3.16}$$

$$b = \{ \tau_{21} + N[- \frac{\mu_0 \mu_2}{2\sigma^2} + \frac{\mu_2}{\mu_2 - \mu_0} \ln(\frac{1-\epsilon_2}{1-\epsilon_0}) - \ln(\frac{1-\epsilon_2}{1-\epsilon_1})]\} \cdot \{ \frac{\mu_2 - \mu_0}{\mu_2} \} \tag{3.17}$$

Clearly, (3.15) is not true for every $F''^{(1)}$ and every pair of thresholds. However, it can be shown that for N sufficiently large, (3.15) is true for all $F''^{(1)}$'s corresponding to densities $f''_{1,\ell} \in \tilde{\Gamma} \subset \Gamma$, where $\tilde{\Gamma}$ is some set that includes f_1 and f'_1. The size of the set $\tilde{\Gamma}$ generally depends on N, (τ_{10}, τ_{21}), (μ_0, μ_2) and (ϵ_0, ϵ_2). For example, if $\mu_0 = -\mu_2$ and $\epsilon_0 = \epsilon_2$, then there is some small δ such that when N is large enough that $a < -\delta$ and $b > \delta$, then (3.15) is true for $F''^{(1)}$ equal to the nominal measure $F^{(1)}$; furthermore, as N increases the set $\tilde{\Gamma}$ grows to include all symmetric ϵ_1-contaminations of f_1 (as well as other elements of Γ).

In more general cases the set $\tilde{\Gamma}$ is harder to specify, but there is always such a set for N large enough. (This follows from the facts that the variance of $L_{20}(Z_\ell)$ is smaller under f_1 than it is under f'_1, that a and b eventually lie on opposite sides of the mean of $L_{20}(Z_\ell)$ under both f_1 and f'_1, and that the difference between either a or b and the mean of $L_{20}(Z_\ell)$ increases to infinity as N increases.) Thus, we have the following:

Theorem 3. For N suficiently large, there is a subset $\bar{\Gamma}$ of Γ such that $f_1 \in \bar{\Gamma}$, $f_1' \in \bar{\Gamma}$, and the decision rule (3.9) is minimax-robust in terms of probability of error for marginal densities in the uncertainty classes Φ, $\bar{\Gamma}$, and H .

Theorem 3 is, of course, less than completely satisfactory, but it does at least indicate that the rule (3.9) is robust in some sense. The same ideas can be used to argue that the densities (f_0', f_1', f_2') lead to an approximately robust rule for deciding the value of $\underline{S}$, given $\underline{Z}$, when $\underline{S}$ can take any of the 3^N possible realizations that result when each S_ℓ takes one of the three values 0,1, or 2, which is the typical problem of interest in image segmentation. The argument is significantly more complicated, however, and involves more approximation. Also, similar methods can be used to develop approximately robust rules when each S_ℓ can take any of the values $\{0,1,\ldots,K\}$ for K > 2 by fixing ϵ_0 and ϵ_K, and letting $\epsilon_1,\ldots,\epsilon_{K-1}$ vary. However, even when K = 2, ϵ_0 and ϵ_2 must be small to get a reasonable value for ϵ_1. For K > 2, ϵ_0 and ϵ_K must be so small to get reasonable values for $\epsilon_1,\ldots,\epsilon_{K-1}$ that the usefulness of the result is questionable.

4. Conclusion. In this paper we have considered some approaches to defining robust multiple hypothesis tests when there is uncertainty in the observation distribution under each hypothesis. We have seen that, when the observation is sample-to-sample independent under each hypothesis and when at each sample time the observation has one of only two nominal densities, N-sample decision rules can be found that are minimax-robust in terms of probability of error. The robust rules are MAP rules for least-favorable densities. When the observation at each sample has one of more than two-nominal densities, results in Kuznetsov (1982) can be used to find robust one-sample MAP rules; however, these do not immediately lead to robust N-sample rules. For the case where the observation is one of three constant signals in independent noise, we have seen that N-sample rules that are approximately minimax-robust in terms of probability of error can be found if the uncertainty level in the distribution of the observation for middle signal is some function of the uncertainty levels corresponding to the other two signals.

As we have noted in the paper, these results can be generalized somewhat; in particular, the result for three constant N-sample signals can be extended to give an approximately robust rule for deciding among the 3^N possible time-varying signals that at each sample point take one of three values. One generalization that would be desirable would be to have robust Bayes rules for risks other than probability of error. While one would guess that the least-favorable densities that lead to rules that are robust in terms of probability of error should also lead to robust rules for other risks, our results do not prove that. It follows, for example, from (2.7) or from (3.10) - (3.12) that Bayes tests based on the least-favorable densities that we have described will have minimum worst-case probability of error, since different cost functions will just lead to different thresholds in the decision rules. However, this is not enough to imply that the tests have minimum worst-case risk for risks other than probability of error.

ACKNOWLEDGEMENT

This research was supported by the National Science Foundation under Grant No. MIP-8710871.

REFERENCES

Huber, P.J. (1965), "A Robust Version of the Probability Ratio test", Ann. Math. Stat., vol. 36, pp. 1753-1758.

Kelly, P.A. and Duan, X. (1988), "Least Favorable Densities for Some Robust Map Tests", Proc. 1988 Conf. on Info. Sciences and Systems, Princeton Univ., pp. 596-598.

Kuznetsov, V.P. (1982), "Stable Rules for Discrimination of Hypotheses", Prob. Info. Transmission, vol. 18, pp. 41-51.

Lehmann, E.L. (1959), Testing Statistical Hypotheses, Wiley, New York.

DETECTING A TRANSIENT SIGNAL BY BISPECTRAL ANALYSIS

Melvin J. Hinich

The University of Texas at Austin

Introduction. This paper presents a method for detecting an unknown transient signal $a(t)$ of duration T in broadband noise. All that is known about the signal is that its frequency band is in the interval (f_a, f_b), and that the value of T is within some error band. Thus, matched filtering detection is not appropriate for this problem since the signal waveform is unknown.

The method uses a newly discovered mathematical property of the bispectrum of a bandlimited continuous time random process. The bispectrum will be defined and the detection method explained after discussion of the signal and noise models is completed.

To simplify exposition, shift the frequency band to the band $(0, f_o)$ where $f_o = f_b - f_a$. It will be shown that the detectability of the signal can be large if the time-bandwidth product Tf_o is large, even if the energy signal-to-noise ratio is small. This result implies that the bispectrum based test may be used to detect a weak signal of unknown form when standard energy detection methods will have small detection probabilities.

Assume that the received signal is passed through an anti-aliasing lowpass filter whose cutoff frequency is f_o, and the output of the filter is sampled at the rate $2f_o$. The sampling interval is then $\tau = 1/2f_o$. Define $N = T/\tau$ and assume for convenience that N is odd and thus $K = (N-1)/2$ is an integer. Then the signal has the finite Fourier sum representation

$$a(t) = \sum_{k=-K}^{K} A(f_k) \exp(i2\pi f_k t) \tag{1.1}$$

for the Fourier frequencies $f_k = k/T$. The complex amplitude $A(f)$ is related to the signal by the Fourier transform

$$A(f) = T^{-1} \int_0^T a(t) \exp(-2\pi i f t) dt \quad . \tag{1.2}$$

By Parcevel's Theorem (Tretter, 1976: Theorem 10.6.1)

$$\frac{1}{T} \int_0^T a^2(t)dt = 2 \sum_{k=1}^{K} |A(f_k)|^2 + |A(0)|^2 \quad . \tag{1.3}$$

The mean of the transient is nearly zero for most applications, so let's simplify the model by setting $A(0)=0$.

A standard method for scanning for signals is to use a sliding time window on the incoming data stream. A buffer of $N=T/\tau$ numbers is stored for each window, and this buffer is emptied into the signal processing input file. Each data window is analyzed for the presence of the transient signal.

This paper uses the framework of classical statistical hypothesis testing to develop the trade-off between a false alarm probability α (the Type I error) and the probability of detection which is called power in statistical theory (Deutsch, 1965). The null hypothesis is only noise present in the window. The test's power is a function of the signal-to-noise ratio (SNR).

Let $\{y(0),y(\tau),\ldots,y((N-1)\tau)\}$ denote the sampled observation of the received signal in a given time window. Use the standard convention of setting the time origin at the time of the first observation in the sampling window of $N=2f_oT$ discrete observations. Also, in keeping with detection theory conventions, suppose that the signal is completely in the window if the null hypothesis of noise alone is not true. In other words, ignore the decision problem of determining the time of arrival for the signal when the transient slides through the window as it is moved forward in time. The null and alternative hypotheses are as follows for $n=0,\ldots,N-1$:

H_o: Noise Alone $\qquad\qquad\qquad y(n\tau) = e(n\tau)$.

H_a: Signal plus Noise $\qquad\qquad y(n\tau) = a(n\tau) + e(n\tau)$.

The deterministic signal waveform is unknown, and thus both the null and alternative hypotheses are not simple.

Assume that the signal's complex amplitude $A(f)$ is slowly varying over the bandwidth of order $O(1/\tau\sqrt{N})$ that is used to smooth the sample bispectral estimator which is used in the test statistic. This assumption is a weak restriction on the signal waveform when Tf_o is large.

Assume that the noise, denoted {e(t)}, is a strictly stationary zero mean random process with bounded moment functions of all order. The stationary assumption is needed for the asymptotic sampling distribution of standard spectral and bispectral estimators. The validity of the use of asymptotic sampling results basically requires that the noise remain stationary during the sampling period, and that the span of dependence between e(t) and e(t+τ) be less than T. The central limit theorem that is used to prove that the estimator is asymptotically Gaussian requires a generalized finite memory condition that is called mixing in the probability theory literature (Billingsley, 1979). Note that the noise is not assumed to be Gaussian. This assumption is not needed.

Suppose that the noise can be observed for relatively long periods when there is no signal. Then standard spectral estimators will yield consistent and asymptotically unbiased estimates of the noise spectrum. Assume then that the noise can be prewhitened with great precision, so that the problem can be simplified by assuming that the noise is white. Denote its standard deviation by σ, a known parameter.

The next section presents a review of the bispectrum of a stationary random process. The distinction between the bispectrum of a continuous and discrete-time process is important to the understanding of the key result in this paper. A brief review of bispectral estimation is given in Section II.

Symmetries of the Bispectrum. A thorough review of the bispectrum is given by Brillinger and Rosenblatt (1967a and b). For refinements and advances in the statistical methods involving bispectral analysis, see Lii and Rosenblatt (1982), Subba Rao and Gabr (1980), Hinich (1982), Subba Rao (1983), and Hinich and Wolinsky (1988).

It will now be shown that the set of positive support of the continuous-time bispectrum is a proper subset of the principal domain of the bispectrum of the sampled process. The bispectrum must be zero in that extra triangle if there is no aliasing.

To review the mathematics of bispectra, let {x(t)} denote a real zero mean stationary continuous-time stochastic process. Assume all expected values, sums, and integrals used below hold. The bicovariance function of the process is $c_x(u,v)=Ex(t)x(t+u)x(t+v)$, which does not depend on t since the process is stationary. Its Fourier transform

$$B_x(f,g) = \int_{-\infty}^{\infty} \int_{-\infty}^{\infty} c_x(u,v)\exp[-i2\pi(fu+gv)]dudv \qquad (2.1)$$

is called its _bispectrum_. Although this two frequency index notation is standard, it hides the three frequency interaction that is so important for applications of bispectral estimation.

To help the exposition, we will use the three frequency notation $B_x(f,g,-f-g)$, and the Cramer representation of $x(t)$ used by Brillinger and Rosenblatt (1967). This representation is

$$x(t) = \int_{-\infty}^{\infty} \exp(i2\pi ft)dA_x(f) \quad , \qquad (2.2)$$

where $\{dA_x(f)\}$ is a stochastic orthogonal increments process. Since $x(t)$ is real, $dA_x(-f)$ is the complex conjugate of $dA_x(f)$. The spectral density at f of $\{x(t)\}$ is $S_x(f)df=E[dA_x(f)dA_x(-f)]$, and the bispectral density for $h=-f-g$ is given by

$$B_x(f,g,h)dfdg = E[dA_x(f)dA_x(g)dA_x(h)] \quad . \qquad (2.3)$$

Note that the right hand side of (2.3) is invariant to permutations of the frequency indices f,g, and $h=-f-g$. Thus the bispectrum's symmetry lines are $f=g$, $f=h$, $(2f=-g)$, and $g=h$ $(2g=-f)$. Since $dA_x(-f)=dA_x{}^*(f)$, $B_x(-f,-g,-h)=B_x^*(f,g,h)$. This skew symmetry yields another three symmetry lines: $f=-g$, $f=-h(g=0)$, and $g=-h(f=-0)$ (Fig. 1). Thus the pointed cone $C=\{f,g: 0\le f, g\le f\}$ is a principal domain of this continuous time bispectrum in the (f,g) plane.

Suppose that the process is bandlimited at frequency f_0. Then there is no variance in the process for frequencies beyond f_0 and thus the bispectrum cuts off at $f=\pm f_0$, $g=\pm f_0$, and $f+g=\pm f_0$. Then the continuous-time support set of the absolute value of B_x is the isosceles triangle $IT=\{f,g: 0\le f\le f_0, g\le f, f+g=f_0\}$. But the discrete-time principal domain is a larger triangle if the process is sampled with the frequency $2f_0$.

The principal domain of the discrete-time bispectrum will be presented. Consider the sequence $\{x(n\tau)\}$ where $1/\tau=2f_0$ is the sampling rate. then

$$x(t) = \int_{-f_0}^{f_0} \exp(i2\pi ft)dA_{x,\tau}(f) \qquad (2.4)$$

where $dA_{x,\tau}(f) = \Sigma_n dA_x(f+n/\tau)$ is the sampled data orthogonal increments process (for $n=0,\pm1,\pm2,\ldots$). It follows from Eq. (2.4) that the discrete-time bispectral density $B_{x,\tau}(f,g,h)dfdg$ is

$$B_{x,\tau}(f,g,h)dfdg = E[dA_{x,\tau}(f)dA_{x,\tau}(g)dA_{x,\tau}(h)]$$

(2.5)

$$= \Sigma_k\Sigma_n\Sigma_m B_x(f+k/\tau,g+m/\tau,h+n/\tau)dfdg$$

for $f+g+h+(k+m+n)/\tau=0$ where the signed integers are restricted to keep the indices in B's principal domain (see Brillinger and Rosenblatt, 1967, p. 190). For example if $h=-f-g$, then $n=-k-m$ where $m\leq k$. If $h=(1/\tau)-f-g$, then $n=-1-k-m$. But B_x is bandlimited at f_o and so the sum is restricted to the k,m, and n such that $0\leq f+k/\tau\leq f_o$, $0\leq g+m/\tau\leq f_o$, and $0\leq h+n/\tau-(k+m)/\tau-f-g\leq f_o$.

Sampling introduces an infinite set of parallel symmetry lines $2f+g=n/\tau$ and $f+2g=n/\tau$. The cone C is first cut by the symmetry line $2f+g=1/\tau$. The line segment of $f+2g=1/\tau$ in C lies above and to the right of $2f+g=1/\tau$ in C. Thus the principal domain of $B_{x,\tau}$ is the triangle $\{f,g: 0\leq f\leq 1/2\tau, g\leq f, 2f+g=1/\tau\}$ in C.

We will now show that $B_{x,\tau}=0$ in the odd triangle of area $1/48\tau^2$,

$$OT = \{f,g:g\leq f,1/2\tau\leq f+g\leq 1/\tau-f\}$$

(2.6)

adjoining the isosceles triangle $IT=\{f,g: g\leq f, 0\leq f+g\leq 1/2\tau\}$ (Fig. 2). If f and g are in OT, then $f+g>1/2\tau-f_o$, $f<f_o$, and $g<f_o$. Thus the only term in the sum to consider is $B_x(f,g,h-1/\tau)$ where $h=-f-g+1/\tau$, but this is $B_x(f,g,-f-g)$ which is zero since $f+g>f_o$. All the other terms are zero for a similar reason.

An Hypothesis Test Approach to Detection. This result that the bispectrum of a non-aliased sampled random process is zero in OT will now be applied to the problem of detecting a(t) in the window of observations of y(t). This application will use a hypothesis test statistic that is a function of an asymptotically unbiased and consistent estimator of the bispectrum of the {y(t)} that is a bifrequency analog to the smoothed periodogram spectrum estimator that is discussed in detail by Brillinger

(1975), Priestley (1981), Rosenblatt (1985), and many others. I am generalizing a result in Hinich (1982).

Select a bifrequency smoothing kernel $W(f,g)$ with the following properties:

1. W is a bounded periodic and continuous function with period $1/\tau$ in each variable.

2. W has finite support, i.e., $W(f,g)=0$ for (f,g) outside some subset in the principle domain.

3. W has unit volume over its support.

4. W shares the symmetries of the bispectrum.

Conditions 1 and 2 are slightly different from a number of equivalent conditions on the smoothing kernel that are often assumed in the literature on the asymptotic properties of spectrum estimators. A review of smoothing windows for spectral and bispectral analysis is given by Subba Rao and Gabr (1980).

Let $Y(f_k)=\sum_{n=0}^{N-1}y(n\tau)\exp(-i2\pi kn/N)$ denote the discrete Fourier transform ordinate of the data window at the frequency $f_k=k/T$. If there is no signal in the window (H_o is true) then $Y(f_k)=U(f_k)$ where $U(f_k)=\sum_{n=0}^{N-1}e(n\tau)\exp(-i2\pi kn/N)$ is the discrete Fourier transform of the noise.

The unsmoothed bispectral estimator is

$$B_y^{(N)}(j,k) = Y(f_j)Y(f_k)Y(-f_{j+k})/N. \tag{3.1}$$

Smoothing this complex array by convolving it with the weights $\{W_N(j,k)\}$ where Δ_N is a bandwidth of order $O(1/\tau\sqrt{N})$ and

$$W_N(j,k) = (N\Delta_N)^{-2}W(j\Delta_N,k\Delta_N) \tag{3.2}$$

yields the following estimator of the bispectrum at the bifrequency $(j\Delta_N,k\Delta_N)$:

$$\hat{B}_y(j,k) = \sum_m\sum_n W_N(m,n)B_y^{(N)}(j-m,k-n) \quad . \tag{3.3}$$

The discrete convolution in Eq. (3.3) for each variable is a finite weighted average of approximately $M = INT(N\Delta_N\tau)$ terms, where $\Delta_N\tau$ is $0(1/\sqrt{N})$ and thus $M = 0(\sqrt{N})$.

Suppose that $W(f,g) = 1$ on the square $\{f,g: -1/2\tau \leq f,g \leq 1/2\tau\}$ and $W(f,g) = 0$ outside the square. Then $W_N(m,n) = 1/M^2$ for $-M/2 \leq m,n \leq M/2$ and $W_N(m,n) = 0$ otherwise. The $B_y^{(N)}$ values are thus averaged over a square of M^2 adjacent bifrequencies. This is the smoothing that is used in Hinich (1982) and Brockett, Hinich, and Wilson (1987).

The distribution of this estimator is approximately complex Gaussian for large N with a variance $c\sigma^6/N\Delta_N^2$ where $c = \int |W(f,g)|^2 dfdg$ is the kernel's scaling constant. The estimators for different discrete bifrequencies on the coarser grid $\{j\Delta_N, k\Delta_N\}$ are asymptotically independent (see Brillinger and Rosenblatt, 1967a for exact statements and a derivation of the asymptotic properties alluded to here). These large sample properties can be heuristically derived by using the key result that the joint distribution of $\{U(k_1\Delta_N)/\sigma\sqrt{N}, \ldots, U(k_L\Delta_N)/\sigma\sqrt{N}\}$ is approximately a complex Gaussian multivariate distribution whose mean is zero and whose covariance matrix is the identity for any integer L.

Since the noise is not aliased, the bispectrum of the observations is zero in OT under the null hypothesis. Thus when H_0 is true, then the expected value of $\hat{B}_y$ is zero in the triangle OT even if the noise is not Gaussian. If the noise is Gaussian, then the bispectrum is zero for all bifrequencies. The importance of restricting attention to the OT triangle is that we can then detect transient signals in a noise background where there are nonlinear random sources that make the noise's bispectrum non-zero in the IT triangle.

Suppose that H_0 is true. Since $\hat{B}_y$ is complex Gaussian with mean zero, the distribution of the statistic

$$CHI(j,k) = 2N\Delta_N^2 |\hat{B}_y(j,k)|^2 / c\sigma^6 \tag{3.4}$$

is approximately a central chi-square with two degrees of freedom for bifrequencies in OT. Thus the sum of the CHI(j,k) for $(j\Delta_N, k\Delta_N)$ in OT, denoted CHISUM, is approximately a central chi square statistic with 2P degrees of freedom χ_{2P}^2, where P is the number of such bifrequencies in OT.

Since the area of OT is $1/48\tau^2$, P is approximately $N/48$ for $\Delta_N \approx 1/\sqrt{(N\tau)}$. CHISUM can be transformed into an approximately Gaussian statistic for detecting the signal a(t) in the window.

It is inconvenient to use a χ_n^2 statistic when n is large. A commonly used transformation of a central chi square (Kendall and Stuart, p. 371, 1958) is $(\chi_n^2/n)^{1/3}$, which is approximately Gaussian with mean $1-2/(9n)$ and variance $2/(9n)$. Thus the test statistic

$$Z = (1/9P)^{-1/2}[((CHISUM/2P)^{1/3})-1+1/9P] \qquad (3.5)$$

is approximately a <u>standardized</u> Gaussian variate (mean zero and unit variance).

The test threshold, denoted L_α, for a selected false alarm probability α is determined from $\alpha = Pr(Y > L_\alpha)$, which is tabulated in most statistics books.

This CHISUM is less than the Hinich test statistic for Gaussianity, which is the sum of the CHI(j,k) over <u>all</u> the $(j\Delta_N, k\Delta_N)$ in the principal domain grid. Estimates of the power of this test for several types of nonlinear and non-Gaussian models is given by Ashley, Hinich, and Patterson (1986). Since we show that the approximation used in the Hinich test is good for samples as small as N=256, there is no reason to doubt the validity of the large N approximation as applied to our restricted sum test for a transient signal.

If the signal is present (H_a is true) then $Y(f_k)=NA(f_k)+U(f_k)$. In this case the bispectrum of the $\{y(t)\}$ is not zero and depends on the $A(f_k)$. The detectability of the signal is analyzed in the next section.

<u>Signal Detectability</u>. The test's power function

$$\pi = Pr(Z > L_\alpha|H_a) \qquad (4.1)$$

depends upon the magnitude of $\hat{B}_y(j,k)$ for each bifrequency in the OT triangle when the signal is present. The magnitude will now be approximated under the simplifying assumption that the amplitudes $|A(f)|$ are equal to a constant level G. Thus the energy signal-to-noise ratio (SNR) is NG^2/σ^2 from (1.3).

Suppose the signal is present. Thus from Eq. (3.1) it follows that

$$NB_y^{(N)}(j,k) = N^3 A(f_j)A(f_k)A(-f_{j+k})$$

$$+ N^2[A(f_j)A(f_k)U(-f_{j+k})+A(f_j)A(-f_{j+k})U(f_k)+A(-f_{j+k})A(f_k)U(f_j)]$$

$$+ O_p(N^2) \quad , \tag{4.2}$$

where $O_p(N^2)$ denotes that the error is of order of N^2 in probability. The error is of this probabilistic order since $U(f_k)/\sqrt{N}$ is $O_p(\sqrt{1})$ from the key approximation result stated above, and thus $NU(f_j)U(f_k)$ is $O_p(N^2)$.

Note that the indices of the U terms in the three lower order cross terms in Eq. (4.2) are j,k, and $j+k$. Thus the double convolution of these cross terms are approximately unidimensional convolutions of the U terms since the phase of $A(f)$ is assumed to be slowly varying in a band of bandwidth Δ_N. The weighted average of the U terms is then approximately complex Gaussian and of order $O_p(N^{1/4})$ since the average is over $M=O(\sqrt{N})$ terms of order $O_p(\sqrt{N})$. Then it follows from Eq. (4.2) that

$$\hat{B}_y(j,k) = N^2 A(f_j)A(f_k)A(-f_{j+k}) + NG^2 O_p(N^{1/4}) + O_p(N) \quad , \tag{4.3}$$

where $G=|A(f)|$. Thus

$$N|\hat{B}_y(j,k)|^2 = N^5 G^6 + N^4 G^5 O_p(N^{1/4}) \quad . \tag{4.4}$$

Let $\rho=NG^2/\sigma^2$, the SNR for the transient model used here. Applying (4.4) to (3.4) and summing over the OT triangle, we have the following approximation for the test statistic when the signal is present:

$$CHISUM = N^2 \rho^3 (1+N^{-3/4}G^{-1}v)/24c \quad , \tag{4.5}$$

where v is a $O_p(1)$ Gaussian variate. Applying the transformation given by (3.5), the major term of the approximately Gaussian test statistic is

$$Z \approx (16/3N)^{-1/2} c^{-1/3} N^{1/3} \rho$$

$$= c^{-1/3} (3/16)^{1/2} N^{5/6} \rho \tag{4.6}$$

This approximation implies that π, the power of the test, is nearly one if $(3/16)^{1/2} N^{5/6} \rho \gg 1$ for large N. In other words, the probability of signal detection is nearly one if $(3/16)^{1/2} N^{5/6}$ is larger than $1/\rho$.

This result will now be compared with the power of "energy" detection, the standard approach for detecting a signal of unknown waveform. Consider the special case where the signal's amplitude is equal to G for all positive frequencies, $A(0)=0$, and the noise is white with known variance. The "energy" test statistic is the sample variance-to-noise variance ratio

$$S = \sum_{n=0}^{N-1} y(n\tau) - \bar{y})2/(N-1)\sigma^2 \tag{4.7}$$

where

$$\bar{y} = N^{-1} \sum_{n=0}^{N-1} (y(n\tau)).$$

Given the assumptions made about the noise in the Introduction, the distribution of the statistic S is approximately central chi square with N-1 degrees-of-freedom under H_o.

If the signal is present, then $E(S)=\rho+1$. This result implies that the power of the bispectrum test will be high when the power of the energy test will be near zero if the signal-to-noise ratio ρ is small but $N^{5/6} \rho$ is larger than $c^{-1/3}(3/16)^{1/2}$.

An Example. It is now useful to present a numerical example of the signal detectability presented above. Suppose that the transient is a one second long pulse of bandwidth 20 kHz. All the assumptions previously made about the signal and noise apply.

If the received signal is bandlimited at 20 kHz and then sampled at the Nyquist rate of 40 kHz using a sliding window of length T=1 sec, the sample size per window is $N=4\times10^4$.

Suppose that the frequency smoothing is done over a square of M^2 adjacent bifrequencies, where $M=\sqrt{N}$. The smoothing bandwidth is then 5.0 kHz, and the kernel constant is $c=1$.

Recall that the power of the test is high if $N^{5/6}(3/16)^{1/2}$ is larger than $1/\rho$. This implies that the probability of detection of the transient is high if ρ is greater than -16.4 dB.

Summary. This paper presents a method for detecting an unknown transient signal $a(t)$ of duration T in broadband noise. All that is known about the signal is that its frequency band is in the interval (f_a, f_b), and that the value of T is within some error band. Thus, matched filtering detection is not appropriate for this problem since the signal waveform is unknown. The method uses a newly discovered mathematical property of the bispectrum of a bandlimited continuous time random process. It is shown that the detectability of the signal is a function of $(Tf_o)^{2/3}\rho$ where ρ denotes the signal-to-noise ratio and $f_o=f_b-f_a$. This result implies that the bispectrum based test may be used to detect a weak signal of unknown form when standard energy detection methods will have small detection probabilities.

Acknowledgment. This research was funded by the Office of Naval Research under Contract N00014-86-K-0237.

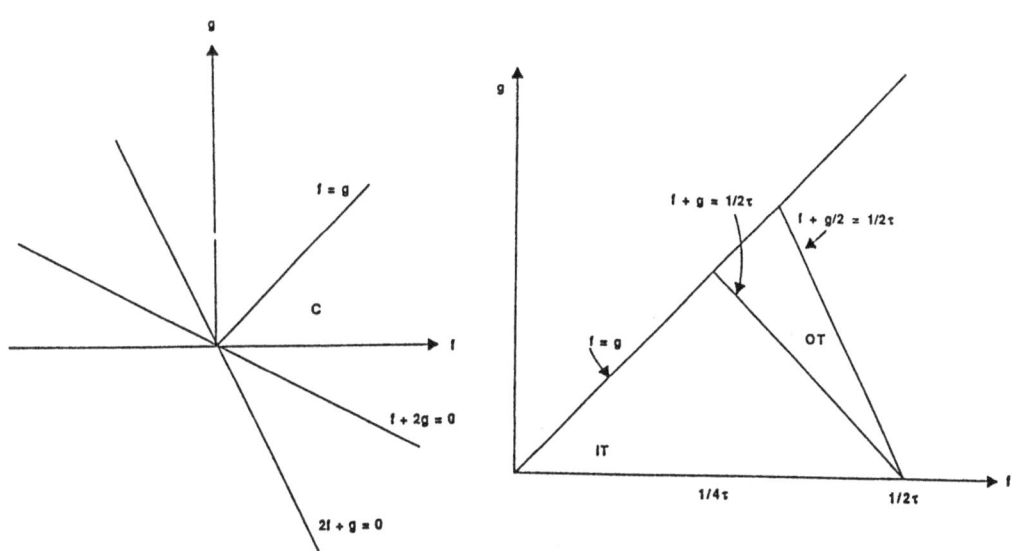

Fig. 1.
SYMMETRIES OF BISPECTRUM B(f,g)

Fig. 2.
DISCRETE-TIME PRINCIPAL DOMAIN

REFERENCES

1. Ashley, R., M. J. Hinich and D. Patterson, "A Diagnostic Test for Nonlinear Serial Dependence in Time Series Fitting Errors," J. of Time Series Analysis 7(3), 165–178, (1986).

2. Billingley, P., Probability and Measure, (John Wiley, New York, 1979) Sect. 27.

3. Brillinger, D. and M. Rosenblatt, "Asymptotic Theory of kth Order Spectra," in Spectral Analysis of Time Series, B. Harris, Ed. (John wiley, New York, 1967) pp. 153–188.

4. Brillinger, D. and M. Rosenblatt, "Computation and Interpretation of kth Order Spectra," Ibid. 189–232.

5. Brillinger, D., Time Series, Data Analysis and Theory, (Holt, Rinehart and Winston, New York, 1975) Sect. 2.7.

6. Brockett, P. L., M. J. Hinich and G. R. Wilson, "Nonlinear and Non–Gaussian Ocean Noise," J. Acoust. Soc. Am. 82(4), 1386–1394 (1987).

7. Brockett, P. L., M. J. Hinich and D. Patterson, "Bispectral Based Tests for the Detection of Gaussianity and Linearity in Times Series," To appear, J. Acoust. Soc. Am., Applications Section (1988).

8. Deutsch, R., Estimation Theory (Prentice Hall, Englewood Cliffs, 1965) Chapter 13.

9. Hinich, M. J., "Testing for Gaussianity and Linearity of a Stationary Time Series," J. Time Series Analysis 3(3), 169–176.

10. Hinich, M. J. and D. M. Patterson, "Evidence of Nonlinearity in Daily Stock Returns," J. Business and Economic Statistics 3(1), 69–77 (1985).

11. Hinich, M. J. and M. A. Wolinsky, "A Test for Aliasing Using Bispectral Analysis," J. Amer. Statistical Assoc., 83(402), Theory and Methods, 499–502, (1988).

12. Lii, K. S. and M. Rosenblatt, "Deconvolution and Estimation of Transfer Function Phase and Coefficients for NonGaussian Linear Processes," Ann. Statist. 10(4), 1195–1208 (1982).

13. Priestley, M. B., Spectral Analysis and Time Series, Vol. 2, (Academic Press, New York, 1981).

14. Kendall, M.G. and A. Stuart, The Advanced Theory of Statistics, Vol. 1, (Hafner, New York, 1958).

15. Rosenblatt, M., "Cumulants and Cumulant Spectra," in Handbook of Statistics, Vol. 3, D. Brillinger and P. Krishnaiah, eds. (North–Holland, Amsterdam, 1983), Chpt. 17.

16. Rosenblatt, M., Stationary Sequences and Random Fields, (Birkhauser, Boston, 1985).

17. Subba Rao, T. "The Bispectral Analysis of Nonlinear Stationary Time Series With Reference to Bilinear Time Series Models," in Handbook of Statistics, Vol. 3, D. Brillinger and P. Krishnaiah, eds. (North–Holland, Amsterdam, 1983) Chpt. 14.

18. Subba Rao, T. and M. Gabr, "A Test for Linearity of Stationary Time Series," J. Time Series Analysis 1, 145–158, (1980).

19. Tretter, S.A., Introduction to Discrete–Time Signal Processing, (John Wiley, New York, 1976).

AN IMPORTANT PATHOLOGY OF FINITE DIMENSIONAL APPROXIMATIONS OF STATISTICAL ESTIMATORS

by

John M. Morrison

The University of Delaware

1.Introduction. We will begin by extending a result about projection schemes found in [1]. Then we will quickly summarize some basic results on Orlicz spaces; in [2] it was shown that these provide the right mathematical envirrnment in which to deal with estimation problems using fidelity criteria more general that mean squate error. Finally we willl apply these to some important examples and allude to several others that are considered in [3].

2. An abstract property of projection schemes. Let us talke a few moments to recall some terminology. Let $\mathcal{E}$ be a Banach space and $\{P_n\}_{n\in N}$ be a sequence of projections on $\mathcal{E}$; $P_n : \mathcal{E} \to \mathcal{E}$ is a bounded linear operator on $\mathcal{E}$ and $P_n^2 = P_n$, $n \in N$. We say $\{P_n\}_{n\in N}$ that forms a *projection scheme* on if for each x in $\mathcal{E}$, $P_n(x) \to x$ and for each $n \in N$, P_n has finite rank. The following result is proved in [1] on p. 58.

Theorem 1: Let $\mathcal{E}$ be a Banach space and $\{P_n\}_{n\in N}$ be a projection scheme on $\mathcal{E}$. Then for a subset Bof $\mathcal{E}$ we have

$$\sup_{x\in B} \| P_n(x) - x \| \to 0 \text{ as } n \to \infty$$

if and only if B is a precompact subset of $\mathcal{E}$.

We will tease this result a bit to obtain a consequence that willl be central to the analysis we are presenting here. First, we will remove the requirement that the projections converge to the identity map. In exchange we will pay for this by requiring *consistency* of our projection scheme, which we define presently.

Definition: We will say that a sequence of projections is *consistent* if $P_{m\wedge n} = P_n P_m = P_m P_n$ for all m, n $\in N$, if for all x$\in$ $\mathcal{E}$ the sequence $\{P_n(x)\}_{n\in N}$ converges. An easy calculation shows that and if ker P_n contains ker P for each n $\in N$, where P denotes the pointwise limit of the P_n. Notice that we may apply the uniform boundedness principle to see that the limit of this sequence is itself a bounded projection on $\mathcal{E}$.

Theorem 2: Let be a $\{P_n\}_{n\in N}$ consistent sequence of projections on the Banach space $\mathcal{E}$ having limit P and Im P be infinite dimensional. Then for a subset B of $\mathcal{E}$

$$\sup_{x\in B} \| P_n(x) - P(x) \| \to 0 \text{ an } n \to \infty$$

if and only if there exists a compact subset C of Im P so that B lies in C + ker P.

Proof. Notice that we have the direct sum decomposition

$$\mathcal{E} = \ker P \oplus \operatorname{Im} P$$

of $\mathcal{E}$ into closed complimentary subspaces. Next observe that if we restrict to Im P, then $\{P_n\}_{n \in \mathbb{N}}$ is a consistent projection scheme. Put A = P(B); this is the projection of B onto Im P under the canonical projection in the direct sum decomposition mentioned above. Then if $x \in B$:

$$\| P_n(P(x)) - P(x) \| = \| P(P_n(x) - P(x)) \|$$

$$\leq \| P \| \, \| P_n(x) - P(x) \|$$

$$\leq \| P \| \sup_{x \in B} \| P_n(x) - P(x) \|$$

so if $P_n \to P$ uniformly on B, we have $P_n \to P$ uniformly on A. But we may now apply Theorem 1 to see that A is a compact subset of Im P and hence of $\mathcal{E}$. Take C to be the closure of A in $\mathcal{E}$. It is now clear that $P_n \to P$ uniformly on C + ker P. Finally we take note of the fact that

$$B = P(B) + (I - P)(B) \subset C + \ker P.$$

Conversely, suppose B is a subset of C + ker P, where C is a compact subset of Im P. By Theorem 1, $P_n \to P$ uniformly on C. Thus $P_n \to P$ uniformly on C + ker P.   QED

Theorem 3: Let be a $\{P_n\}_{n \in \mathbb{N}}$ consistent sequence of projections on the Banach space $\mathcal{E}$ having limit P and Im P be infinite dimensional. Then if we denote by S the unit sphere of $\mathcal{E}$ and by $\mathcal{E}_1$ the unit ball of $\mathcal{E}$ and if a subset B of $\mathcal{E}$ satisifies

$$\sup_{x \in B} \| P_n(x) - P(x) \| \to 0 \text{ as } n \to \infty$$

then $B \cap S$ and $B \cap \mathcal{E}_1$ are nowhere dense S in $\mathcal{E}_1$ and respectively.

Proof. By theorem 2, there must be a compact subset C of Im P so that C + ker P contains B. Fix a point x in the interior of the closure of C + ker P. Because of the direct sum decomposition of $\mathcal{E}$ into Im P and ker P, we have that the interior of the closure of C + ker P must contain U + V for some relative neighborhoods U of P(x) in Im P and V of x - P(x) in ker P. But this forces U to be a subset of C. Thus a point of Im P has a relative neighborhood that is locally compact; this forces Im P to be finite dimensional. $\Rightarrow \Leftarrow$. We conclude that C + ker P must be a nowhere dense subset of $\mathcal{E}$, insuring that B is nowhere dense in $\mathcal{E}$ and $\mathcal{E}_1$.

Without loss of generality we may assume that B is a closed subset of S Suppose that $B \cap S$ has a nonvoid relative interior in S. Then

$$\| P_n(\lambda x) - P(\lambda x) \| \leq \lambda \| P_n(x) - P(x) \|$$

$$\leq \| P_n(x) - P(x) \|$$

so

$$\sup_{x \in co((B \cap S) \cup \{0\})} \| P_n(x) - P(x) \| \leq \sup_{x \in B \cap S} \| P_n(x) - P(x) \| \to 0 \text{ as } n \to \infty,$$

which forces the closure of co($(B \cap S) \cup \{0\}$) to be nowhere dense in S. But $B \cap S$ has nonvoid interior in S so co($(B \cap S) \cup \{0\}$) has nonvoid interior. $\Rightarrow\Leftarrow$. 　　　　　　　　QED

Corollary 4: Let $\mathcal{E}$ be a Banach space and $\{P_n\}_{n \in N}$ be a consistent projection scheme on $\mathcal{E}$ with $P_n \to P$ and Im P be infinite dimensional. Let $\varepsilon > 0$ be fixed and S denote the unit sphere of $\mathcal{E}$. Then if we fix $x_0 \in \mathcal{E}$ and let $\{\xi_n\}_{n \in N}$ be any sequence of positive real numbers converging to zero, $\{y \in S$: for some $N \in N, n \geq N \Rightarrow \| P_n(x + \varepsilon y) - P(x + \varepsilon y) \| < \xi_n \}$ is of the first category in the complete metric space S. Also, $\{y \in \mathcal{E}$: $\| y \| \leq 1$ and for some $N \in N, n \geq N \Rightarrow \| P_n(x + \varepsilon y) - P(x + \varepsilon y) \| < \xi_n \}$ is of the first category in the closed unit ball of $\mathcal{E}$.

　　　This corollary tells us that no matter how small we choose $\varepsilon > 0$, an ε-perturbation in almost all directions in the Baire category sense from a fixed point will make as long a delay in the convergence of the projection scheme as we wish.

3. Basic facts about Orlicz spaces. For the convienence of the reader we will give a quick summary of the most important aspects of these spaces. For more careful studies of these see [2] and the references given therein. The canonical work in this area is [4].

　　　Let (Ω, Σ, P) be a probability space and suppose $\Phi: [0, \infty) \to [0, \infty)$ be a continuously differentiable strictly increasing convex function. We say that Φ satisfies the *doubling condition* if there exist positive real numbers M and C so that $x \geq M \Rightarrow \Phi(2x) \leq C\Phi(x)$. In this case, the space $L^\Phi(\Omega, \Sigma, P)$ forms a vector space; in fact the doubling condition on Φ is a necessary and sufficient condition for to $L^\Phi(\Omega, \Sigma, P)$ be a vector space. We can make $L^\Phi(\Omega, \Sigma, P)$ a Banach space if we norm it with the Luxemburg norm

$$\inf \left\{ \lambda > 0: \int_\Omega \Phi\left(\frac{|X|}{\lambda}\right) dP \leq \Phi(1) \right\}.$$

This norm is described in [2] and [4]. Moreover, if Φ satisfies the doubling condition, for any sequence of random variables $\{X_n\}_{n \in N}$ we have $X_n \to X$ if and only if

$$\int_\Omega \Phi(| X_n - X|) \, dP \to 0 \text{ as } n \to \infty.$$

If for all ε, $0 < \varepsilon < 1/4$ there exists an $R_\varepsilon > 0$ with

$$\liminf_{x \to \infty} \frac{\Psi'(x)}{\Psi'((1-\varepsilon)x)} \geq R_\varepsilon$$

then the Orlicz space L^Φ is a uniformly convex and therefore reflexive Banach space in the Luxemburg norm. Moreover, as Banach spaces, L^Φ is isomorphic to L^Ψ where Ψ is the conjugate convex function to Φ as described in [2] on p. 229.

4. Approximation of conditional expectations.

Let be a (Ω, Σ, P) probability space and $\mathcal{F}$ be a σ-subalgebra of Σ. Then recall that the conditional expectation operator is orthogonal projection of L^2 (Ω, Σ, P) onto the closed subspace $L^2(\Omega, \mathcal{F}, P)$ and that for $Y \in L^2(\Omega, \mathcal{F}, P)$ the random varialble $E(Y| \mathcal{F})$ uniquely solves

$$\min\left\{ \int_\Omega |Y - Z|^2 \, dP : Z \in L^2(\Omega, \mathcal{F}, P) \right\}.$$

If $\{\mathcal{F}_n\}_{n \in \mathbb{N}}$ is a sequence of σ-subalgebras of Σ we write $\mathcal{F}_n \Uparrow \mathcal{F}$ to denote that $\{\mathcal{F}_n\}_{n \in \mathbb{N}}$ is an increasing family of σ-subalgebras of Σ and that $\mathcal{F}$ is the smallest σ-algebra on Ω containing the $\mathcal{F}_n$. Likewise we will use $\mathcal{F}_n \Downarrow \mathcal{F}$ to denote that $\{\mathcal{F}_n\}_{n \in \mathbb{N}}$ is a decreasing family of σ-subalgebras of Σ and that $\mathcal{F}$ is the intersection of the $\mathcal{F}_n$. It is a standard result of probability theory that if $\mathcal{F}_n \Uparrow \mathcal{F}$ or $\mathcal{F}_n \Downarrow \mathcal{F}$, $1 \leq p \leq \infty$ and $Y \in L^p(\Omega, \Sigma, P)$, then $E(Y| \mathcal{F}_n) \to E(Y| \mathcal{F})$ in the L^p sense and almost surely as $n \to \infty$. We can do a similar thing for Orlicz spaces: if Φ is a convex increasing function satisfying the doubling condition, $\mathcal{F}_n$ $\Uparrow \mathcal{F}$ or $\mathcal{F}_n \Downarrow \mathcal{F}$, and $Y \in L^\Phi(\Omega, \Sigma, P)$, $E(Y| \mathcal{F}_n) \to E(Y| \mathcal{F})$ in the L^Φ sense and almost surely as $n \to \infty$. In this case we also get

$$\int_\Omega \Phi(|E(Y|\mathcal{F}_n) - E(Y|\mathcal{F})|) \, dP \to 0 \text{ as } n \to \infty.$$

Let be (Ω, Σ, P) a probability space and $\{X\}_{t \in [0,T]}$ be a random process having state space S, a separable metric space that is continuous in probability. We can think of $\{X_t\}_{t \in [0,T]}$ as being a random signal we are observing. Suppose we are trying to estimate a random variable Y by observing the process $\{X_t\}_{t \in [0,T]}$. Then if Y is a second order random variable, the mean-square best approximation of Y given $\{X_t\}_{t \in [0,T]}$ is $E(Y| X_t: t \in [0,T])$.

To deal with this sort of problem from a practical angle, we see two immediate limitations on our ability to observe $\{X_t\}_{t \in [0,T]}$. Firstly, our measuring device has limited accuracy so it quantizes the state space S in the following way that is developed in [2]. It partitions the state space Y into finitely many measurable subsets $S_1, S_2, ..., S_n$ and fixes a point $s_k \in S_k$, $1 \leq k \leq n$; if it observes the process in S_k it reports the value s_k, $1 \leq k \leq n$. Accordingly the map $Q:S \rightarrow S$ defined by $Q(s) = s_k$ if $s \in S_k$, $1 \leq k \leq n$ describes the action of this quantization. What our measuring device actually sees is $Q(X_t)$.

Definition: A sequence of maps $\{Q_n\}_{n \in N}$ defined on S is said to be a *round-off scheme* on S if each Q_n has finite range, for each the partition defined Q_n by refines that defined by Q_{n+1}, $n \in N$ and if dia $Q_n^{-1}($ $Q_n(x)) \rightarrow 0$ as $n \rightarrow \infty$.

Secondly, we are often constrained to observing the process at only finitely many times $P = \{0 = t_0 < t_1 < t_2 < ..... < t_n = T\}$. Thus the question we are inclined to ask is how well $E(Y|Q(X_t) : t \in P_m)$ approximate $E(Y| X_t: t \in [0,T])$ in the mean square sense? The following asymptotic result for this is obtained in [2].

Theorem 5: Let (Ω, Σ, P) be a probability space, $Y \in L^\Phi(\Omega, \Sigma, P)$, $T > 0$, $\{P_m\}_{m \in N}$ be a sequence of partitions on $[0, T]$ whose union is dense in $[0, T]$ and $\{Q_n\}_{n \in N}$ be a round off scheme on the separable metric space S and let $\{X_t\}_{t \in [0,T]}$ be a process defined on (Ω, Σ, P) that is continuous in probability. Then we have

$$E(Y|Q_n(X_t) : t \in P_m) \rightarrow E(Y|(X_t: t \in [0,T]) \text{ as } m, n \rightarrow \infty;$$

equivalently,

$$\lim_{m, n \rightarrow \infty} \int_\Omega \Phi(| E(Y|Q_n(X_t) : t \in P_m) - E(Y|(X_t: t \in [0,T])|) \, dP = 0,$$

provided that Φ satisfies the doubling condition.

Now notice that the σ-algebras $\sigma(Q(X_t) : t \in P_m)$ are finite so the maps $E(\cdot | Q(X_t) : t \in P_m)$ form a consistent sequence of projections scheme on $L^\Phi (\Omega, \Sigma, P)$ that satisfy the hypotheases of Corollary 4. We may now specialize Corollary 4 to obtain our main result.

Theorem 6: Let $Y \in L^\Phi(\Omega, \Sigma, P)$, $\epsilon > 0$ be fixed and let $\{\xi_n\}_{n \in N}$ be any sequence of positive real numbers converging to zero. Then only in a nowhere dense subset of $L^\Phi(\Omega, \Sigma, P)$ do we have

$$\int_\Omega \Phi(| E(Y + \epsilon Z |Q_n(X_t) : t \in P_m) - E(Y + \epsilon Z |(X_t: t \in [0,T])|) \, dP \leq \xi_n$$

for n sufficiently large.

5.Conclusion. Estimation procedures often depend on judicious applications of the projection theorem and give consistent sequences of projections that form the approximations to the quantity we wish to estimate. The discussion here shows that even a small amount of uncertainty about the quantity being estimated will almost surely in the Baire Category sense result in a very long delay in the rate of convergence of the scheme. The overarching theme here is that the norm is far too crude a measurement to deal with the delicate business of controlling rates of convergence of estimation schemes. The interested reader should see [3]; it contains several other situations where this theory will produce some eye-opening results with ease.

REFERENCES

[1]   Deimling, K. *Nonlinear Functional Analysis*, Springer-Verlag, New York, 1985.

[2]   Morrison, J. and Wise, G., "An Asymptotic Property of Nonlinear Estimators Arising as Solutions to a Certain Class of Convex Programming Problems", *Systems and Control Letters* vol. 7, No. 3, 1986, pp. 225-232.

[3]   Morrison, J., "Some Nonrobustness Problems with Kalman filtering and their Relation to a Class of Measures on Hilbert Spaces", *Proc. of the Twenty-First Annual Conference on Information Sciences and Systems*, 25-27 March 1987, Baltimore, MD. annual Allerton Conference.

[4]   Krasnoselskii, M. and Rutickii, Y., *Convex Functions and Orlicz Spaces*, Nordhoff, Gronigen, 1961.

COUNTEREXAMPLES IN DETECTION AND ESTIMATION

Gary L. Wise

Department of Electrical and Computer Engineering

and

Department of Mathematics

The University of Texas at Austin

Austin, Texas 78712

Introduction.   Detection and estimation theory is an area of engineering which sometimes seems permeated with mathematically sloppy or incorrect reasoning. In this paper I will present several counterexamples to erroneous claims which sometimes pass for fact in the general area of detection and estimation theory.

In [1] I presented some counterexamples in the general area of communication theory. To begin, we note from [1] or from [2] that conditional expectation does not minimize mean square error, even for bounded random variables. Specifically, we have the following result.

Theorem: For any positive real number B, there exists a probability space $(\Omega, \mathcal{F}, P)$, two bounded random variables X and Y defined on $(\Omega, \mathcal{F}, P)$, and a function $f:R \rightarrow R$ such that $f(X) = Y$ pointwise on Ω but $E([Y - E(Y \mid X)]^2) > B$.

I believe that this result strikes at the heart of much of what passes for rigor in detection and estimation. Theoretical subtleties do not exist merely to take up space on a page. With regard to the problem of the minimization of the mean square error, one may wish to consider [2]. While we are on the subject of conditional expectation, it might do us well to consider the following example, which is taken from [3, p. 133], where it is credited to C. Sugahara. Let U and V be independent Gaussian random variables defined on a common probability space and each having zero mean and unit variance. Let $X = U + V$ and let $Y = U - V$. Note that X and Y are independent Gaussian random variables defined on the same probability space and each has zero mean and a variance of two. Now, simply note that, off a measurable set of zero probability, we have that $E(X \mid U) = E(U + V \mid U) = E(U \mid U) + E(V \mid U) = E(U \mid U) = U = E(U \mid U) - E(V \mid U) = E(U - V \mid U) = E(Y \mid U)$. Therefore, we see that independent random variables can have conditional expectations which are nondegenerate and almost surely equal; furthermore, this phenomenon can occur when all random variables concerned are Gaussian.

Notation and Conventions.   For a topological space T, we will let $\mathcal{B}(T)$ denote the family of Borel sets on T, given by the σ-algebra generated by the open sets. We will let N denote the positive integers. We will let I_A denote the indicator function of the set A. For a measure space $(\Omega, \mathcal{F}, \mu)$, we will say that a property holds almost everywhere if there exists a set $N \in \mathcal{F}$ such that $\mu(N) = 0$ and the property of interest holds for all $\omega \in (\Omega - N)$. A subset of an underlying set is said to be cocountable if its complement is countable. We will adopt the convention that $0 \cdot \infty = 0$.

Let $(\Omega, \mathcal{F}, P)$ be a probability space. Recall that $L_0(\Omega, \mathcal{F}, P)$ denotes the set of all real valued random variables defined on $(\Omega, \mathcal{F}, P)$ modulo almost sure equivalence equipped with the topology of convergence in probability. Let $\Phi:[0, \infty) \rightarrow [0, \infty)$ be a convex, strictly increasing function such that

$\Phi(0) = 0$ and let Φ satisfy the doubling condition given by $\limsup\limits_{x \rightarrow \infty} \dfrac{\Phi(2x)}{\Phi(x)} < \infty$. Let $L_\Phi(\Omega, \mathcal{F}, P)$ be the

Orlicz space equipped with the Luxemburg norm corresponding to Φ. We recall that $L_\Phi(\Omega, \mathcal{F}, P) = \{X \in L_0(\Omega, \mathcal{F}, P): E(\Phi(|X|)) < \infty\}$, and we recall that the Luxemburg norm on $L_\Phi(\Omega, \mathcal{F}, P)$ is given by $\|X\|_{L_\Phi} = \inf \{\lambda > 0: E(\Phi(|X|/\lambda)) < \Phi(1)\}$. Note that $L_\Phi(\Omega, \mathcal{F}, P)$ is a real Banach space. As an example, if $\Phi(x) = x^p/p$ for some $p \in [1, \infty)$, then $L_\Phi(\Omega, \mathcal{F}, P) = L_p(\Omega, \mathcal{F}, P)$.

Radon-Nikodym Derivatives.   Let μ and ν be measures on the measurable space $(\Omega, \mathcal{F})$. If $\mu(S) = 0$ implies that $\nu(S) = 0$ for all $S \in \mathcal{F}$, ν is said to be absolutely continuous with respect to μ and this is denoted by $\nu \ll \mu$. A measure on $(\Omega, \mathcal{F})$ is said to be σ-finite if Ω can be covered by a countable family of measurable sets, each having finite measure. If $\nu \ll \mu$ and if μ is a σ-finite measure, then it follows from the Radon-Nikodym Theorem that there exists an $\mathcal{F}$-measurable function f taking values in $[0, \infty]$

such that $\nu(S) = \displaystyle\int_S f \, d\mu$. Furthermore, this function f is a. e. $[\mu]$ unique, and it is known as the Radon-

Nikodym derivative of ν with respect to μ. Radon-Nikodym derivatives play a key role in much of detection and estimation theory. Indeed, for several optimization criteria, a Radon-Nikodym derivative, or a function of it, is an optimal test statistic in the theory of signal detection, and conditional expectation, which often arises in estimation theory, is a Radon-Nikodym derivative.

Assume that μ and ν are measures on the measurable space $(\Omega, \mathcal{F})$ and that $\nu \ll \mu$. Then if μ is σ-finite, must ν be σ-finite? If μ is finite, must ν be σ-finite? If μ is σ-finite, can the Radon-Nikodym derivative of ν with respect to μ be chosen to be real valued? If μ is finite, can the Radon-Nikodym derivative of ν with respect to μ be chosen to be real valued?

The answer is no to each of the above questions, and I will now give a counterexample to affirmative answers. Let $\Omega = [0, 1]$ and let $\mathcal{F}$ be the σ-algebra on Ω consisting of all subsets of $[0, 1]$ which are either countable or cocountable. Let μ be Lebesgue measure restricted to $\mathcal{F}$, and let ν be the measure defined on $\mathcal{F}$ by $\nu(S) = 0$ if S is countable and $\nu(S) = \infty$ if S is cocountable. Recall that a countable union of countable sets is countable and $[0, 1]$ is not countable; thus ν is not σ-finite. For $S \in \mathcal{F}$, note that $\mu(S) = 0$ if and only if S is countable. Furthermore, if S is countable, then $\nu(S) = 0$. Thus $\nu \ll \mu$, μ is a finite measure, and ν is not σ-finite. Finally, note that the Radon-Nikodym derivative of ν with respect to μ is given by the constant ∞; that is, it straightforwardly follows that for each $S \in \mathcal{F}$, we have that

$$\nu(S) = \int_S \infty \, d\mu.$$

Once again, let μ and ν be measures on the measurable space $(\Omega, \mathcal{F})$. If ν is a finite measure that is absolutely continuous with respect to the measure μ, must a Radon-Nikodym derivative of ν with respect to μ exist? The answer to this question is no. Consider the following situation. Let $\Omega = [0, 1]$ and let $\mathcal{F} = \mathcal{B}([0, 1])$. Let ν be Lebesgue measure on $\mathcal{B}([0, 1])$, and let μ be counting measure on $\mathcal{B}([0, 1])$. Note that $\nu(\Omega) = 1$ and that μ is not σ-finite since μ assigns finite measure only to finite sets, $[0, 1]$ is uncountable, and countable unions of countable sets are countable. Further, note that $\nu \ll \mu$, since μ assigns measure zero only to the empty set. Now let f be any $\mathcal{F}$-measurable function taking values in

$[0, \infty]$, and consider the quantity $\displaystyle\int_S f \, d\mu$, where $S \in \mathcal{F}$. If f is everywhere zero, then this quantity is everywhere zero and is thus not equal to $\nu(S)$ for certain sets S. Furthermore, if f is positive at some point $x \in [0, 1]$, then by picking $S = \{x\}$, we see that the quantity represented by the integral is positive while $\nu(S) = 0$. Thus we see that in this case there does not exist a Radon-Nikodym derivative of ν with respect to μ.

Asymptotic Results. Often in detection and estimation, asymptotic results are utilized. Here I will discuss some possibly surprising consequences of some commonly used asymptotic methods.

Often one considers estimating an unknown probability density function based upon the first N terms of a sequence of mutually independent, identically distributed random variables having a common unknown probability density function. Let $k \in N$ and let f be a $\mathcal{B}(R^k)$-measurable probability density function defined on R^k. Assume that we attempt to estimate f by $\hat{f}_N(x, X_1, X_2, \ldots, X_N)$, where $x \in R^k$, $\hat{f}_N$ is a Borel measurable function of its arguments, and $X_1, X_2, \ldots, X_N$ are mutually independent random vectors with the common density function f. Let $p \in [1, \infty)$. Then for any such sequence $\{\hat{f}_N\}$ and for any sequence $\{a_n\}$ of positive numbers converging to zero, it follows from [4] that there exists an infinitely differentiable probability density function f with compact support such that

$$E(\int_R |\hat{f}_N - f|^p \, dx) > a_n \text{ infinitely often.}$$

The central limit theorem is often used in detection and estimation theory to attempt to justify a Gaussian approximation to a probability distribution of interest. The usual concept of convergence associated with the central limit theorem for mutually independent, identically distributed second order random variables is $\displaystyle\lim_{n \to \infty} \sup_{x \in R} |F_n(x) - \Phi(x)| = 0$, where F_n is the distribution function of the centered, normalized sum of mutually independent, identically distributed second order random variables and Φ is the distribution function of a zero mean, unit variance Gaussian random variable. It follows from [5] that for any sequence $\{a_n : n \in N\}$ of positive numbers converging to zero, there exists a sequence of mutually independent, identically distributed second order random variables having zero mean and unit variance and a positive integer N such that

$$\sup_{x \in R} |F_n(x) - \Phi(x)| \geq a_n \text{ for all } n > N.$$

Now, consider a sequence of random variables $\{X_n: n \in N\}$ defined on a common probability space. This sequence might converge almost surely to a real valued function defined on the same probability space which is not a random variable. Indeed, pick any probability space which is not complete; for example, consider R and $\mathcal{B}(R)$ with standard Gaussian measure. Then there exists a measurable set N of zero measure and a nonmeasurable subset $Z \subset N$. If X_n converges almost surely to a random variable X, then note that $X_n - X$ converges almost surely to I_Z, which is not a random variable. Naturally, $X_n - X$ also converges almost surely to the identically zero function.

Often in estimation theory one derives a sequence of estimators, say $\{Y_n: n \in N\}$, and one may wish to show that as $n \to \infty$, Y_n converges in an appropriate sense to a random variable Y representing possibly a best estimator. In many cases, the martingale convergence theorem arises as the technique to establish the desired convergence. For example, if X is a second order random variable and if $\{X_n: n \in N\}$ is a sequence of random variables, then $E(X \mid X_1, X_2, \ldots, X_k)$ is a martingale and the martingale convergence theorem establishes the almost sure as well as the L_2 convergence of $E(X \mid X_1, X_2, \ldots, X_k)$ as $k \to \infty$. Let $(\Omega, \mathcal{F}, P)$ be a separable probability space such that Ω cannot be written as a finite union of P-atoms, and let $\Phi:[0, \infty) \to [0, \infty)$ be a convex, strictly increasing function such that $\Phi(0) = 0$ and let Φ satisfy the doubling condition . Let S be a nonvoid open $L_\Phi(\Omega, \mathcal{F}, P)$ ball. It then follows from [6] that there exists an increasing sequence $\{\mathcal{F}_n\}$ of σ-subalgebras of $\mathcal{F}$ such that for any sequence $\{a_n: n \in N\}$ of positive numbers converging to zero, there exists $X \in S$ so that
$$\|E(X \mid \mathcal{F}_n) - \lim_{k \to \infty} E(X \mid \mathcal{F}_k)\|_{L_\Phi} > a_n \text{ infinitely often.}$$

On the other side of the proverbial coin, a martingale can converge instantaneously fast, but not to the random variable of interest. Furthermore, this can occur even when all random variables concerned are Gaussian with positive variances. In [7] an example is given where $E(X \mid X_1, X_2, \ldots, X_k) = 0$ a.s. for all $k \in N$, even when $X, X_1, X_2, \ldots$ are all Gaussian with zero mean and positive variances. Moreover, in this example, for any $k \in N$, there exists a function $f:R \to R$ such that $X = f(X_k)$ pointwise on the underlying probability space. Thus we see that the convergence guaranteed by the martingale convergence theorem can be utterly useless.

Now consider an adaptive system whose adaptation is based upon the use of a sequential estimate of an unobservable random variable in lieu of the random variable itself. Although the range of the data is often viewed as uncountably infinite, in modern systems the adaptation would likely be implemented by a digital computer, which is only capable of manipulating finite data sets. Thus, regardless of the range of the data, the presence of the digital computer would force us to be restricted to considering simple random variables. Let $\Phi:[0, \infty) \to [0, \infty)$ be a convex function with a strictly increasing continuous derivative such that $\Phi(0) = 0$, and let Φ satisfy the doubling condition. Let $(\Omega, \mathcal{F}, P)$ be a separable probability space such that Ω cannot be written as a finite union of P-atoms. It follows from [8] that there exists a sequence $\{X_n: n \in N\}$ of simple random variables defined on $(\Omega, \mathcal{F}, P)$ such that the convergence of $\{E(\cdot \mid X_1, X_2, \ldots, X_n): n \in N\}$ is nonuniform on any nonvoid open $L_\Phi(\Omega, \mathcal{F}, P)$ ball. Let $Y \in L_\Phi(\Omega, \mathcal{F}, P)$. We know that $E(Y \mid X_1, X_2, \ldots, X_n) \to E(Y \mid X_n: n \in N)$ in $L_\Phi(\Omega, \mathcal{F}, P)$ [9]. However, the work in [8] tells us that this rate of convergence is not robust in the choice of the random variable Y. More precisely, for any $\varepsilon > 0$, we know that there exists $N \in N$ such that for $n > N$,

$\|E(Y \mid X_1, X_2, \ldots, X_n) - E(Y \mid X_n: n \in N)\|_{L_\Phi} < \varepsilon$; however this N is a function of not only ε but also Y, the random variable being estimated [8].

A Singular Detection Problem. Consider a one sample detection problem modeled by the following hypothesis testing situation:

$H_0: X = N$

$H_1: X = s + N,$

where the signal s is a known constant and the noise N is a random variable whose distribution is known. Let P_0 be the probability measure on $\mathcal{B}(R)$ associated with H_0, and let P_1 be the probability measure on $\mathcal{B}(R)$ associated with H_1. We say that the detection problem is singular if there exists $B \in \mathcal{B}(R)$ such that $P_0(X \in B) = 1$ and $P_1(X \in B) = 0$. In this case of singular detection, we see that we can use the test which announces H_0 if and only if $X \in B$ and our error probability under either hypothesis is zero. It is possible for P_0 and P_1 to have precisely the same support and yet the resulting detection problem can be singular. For example, let $\{q_n: n \in N\}$ be an enumeration of the rationals, let N be a random variable such that the probability that $N = q_n$ is positive for each $n \in N$ and such that the probability that $N \in Q$ is one, and let s be any fixed irrational number. Now recall that a rational number added to an irrational number is irrational and simply note that $P_0(X \in Q) = 1$ and $P_1(X \in Q) = 0$. Thus this detection problem is singular. Further, since Q is dense in R, we see that the support of P_0 and the support of P_1 are the same and are equal to R. Thus, in this situation we have a detection problem based upon one sample where the probability measures under the two hypotheses have the same support and the detection problem is singular.

Neyman-Pearson Detection. Consider a one sample detection problem modeled by the following hypothesis testing problem:

$H_0: X = N$

$H_1: X = s + N,$

where the signal s is a known constant and the noise N is a random variable whose distribution is known.

A Neyman-Pearson detector is a detector whose false alarm probability does not exceed a set value and whose detection probability is maximized subject to the constraint on the false alarm probability. If the known signal s is positive and if the noise N has a probability density function which is unimodal and symmetric, then will a Neyman-Pearson detector announce that the signal is present when the observation is large? If the distribution of the noise were Gaussian, the answer to this question would be affirmative. However, in general, there exist situations consistent with the above question, such that the Neyman-Pearson detector announces that the signal is absent for large values of the observation.

For example, assume that s = 2 and that the density of the noise is given by $f(x) = \frac{1/\pi}{1 + x^2}$, which is unimodal and symmetric. Then $f(x - 2)/f(x) = \frac{1 + x^2}{1 + (x - 2)^2}$, which we note approaches one as $x \to \infty$, equals one at x = 1, is strictly greater than one for $x \in (1, \infty)$, and is strictly less than one for $x \in (-\infty, 1)$.

Further, note that $\int_{-\infty}^{1} f(x) \, dx = 3/4$. Thus, we see that for a Neyman-Pearson detector such that the false alarm probability is fixed to be not greater than some number $\alpha_0 < 1/4$, the threshold for the likelihood ratio will be some number greater than one, and consequently, the detector will announce the signal as absent when the observation is sufficiently large.

Fusion in Estimation. Let X be a second order random variable and let Y_1 and Y_2 be random variables, all defined on a common probability space. How might we fuse or combine $E(X \mid Y_1)$ and $E(X \mid Y_2)$ so as to approximate X in such a way as to attempt to minimize the mean square error between X and our estimate of it based on $E(X \mid Y_1)$ and $E(X \mid Y_2)$? The following example was taken from [10], where several subtleties associated with such problems can be found. Consider the probability space $(\Omega, \mathcal{F}, P)$ where $\Omega = [0, 1]$, $\mathcal{F} = \mathcal{B}([0, 1])$, and P denotes Lebesgue measure restricted to $\mathcal{B}([0, 1])$. Let A be a positive real number, $\sigma(Y_1) = \sigma([0, 1/2))$, $\sigma(Y_2) = \sigma([1/4, 3/4))$, and $X(\omega) = A$ for $\omega \in [0, 1/4) \cup [1/2, 3/4)$ and $X(\omega) = -A$ for $\omega \in [1/4, 1/2) \cup [3/4, 1]$. Then it straightforwardly follows that $E(X \mid Y_1) = E(X \mid Y_2) = 0$ a.s. and that $E(X \mid Y_1, Y_2) = X$ a.s. Recalling that $E(X \mid Y_1)$ is $\sigma(Y_1)$-measurable and that $E(X \mid Y_2)$ is $\sigma(Y_2)$-measurable, we see that $E(X \mid Y_1) = E(X \mid Y_2) = 0$ pointwise in ω; similarly, we see that $E(X \mid Y_1, Y_2) = X$ pointwise in ω. Further, any function of $E(X \mid Y_1)$ and $E(X \mid Y_2)$ is a constant function. Thus, we see that it is fruitless to attempt to approximate X based on any function of $E(X \mid Y_1)$ and $E(X \mid Y_2)$. Furthermore, note that any two of the three random variables X, Y_1, and Y_2 are independent. This might lead an unwary, overly cavalier investigator to perhaps assert that $E(X \mid Y_1, Y_2) = E(X)$ a.s., or perhaps that $E(X \mid Y_1, Y_2) = E(X \mid Y_i)$ a.s. for i = 1, 2. Each of these assertions is incorrect.

We now consider another example taken from [10]. Recall from [11] that for any integer n greater than one, there exists an underlying probability space on which n + 1 random variables X, $Y_1, \ldots, Y_n$ are defined such that the set $\{X, Y_1, \ldots, Y_n\}$ is not mutually Gaussian and not mutually independent, yet any proper subset of $\{X, Y_1, \ldots, Y_n\}$ containing at least two random variables is mutually Gaussian, mutually independent, and identically distributed with zero mean and unit variance. For any proper subset $\mathcal{D}$ of $\{Y_1, \ldots, Y_n\}$ note that $E(X \mid \mathcal{D}) = 0$ a.s. since X is independent of $\mathcal{D}$. However, for this example in [11], it can be straightforwardly shown that $E(X \mid Y_1, \ldots, Y_n) =$

$\frac{1}{2\sqrt{2}} Y_1 \cdots Y_n \exp[-((Y_1)^2 + (Y_2)^2 + \cdots + (Y_n)^2)/2]$ a.s. Thus, it would not be reasonable to attempt to estimate $E(X \mid Y_1, \cdots, Y_n)$ based upon $E(X \mid \mathcal{D})$ where $\mathcal{D}$ ranges over all proper subsets of $\{Y_1, \ldots, Y_n\}$. Note that even the ubiquitous Gaussian assumption does not come to the rescue in this case.

<u>Nonaffine Transformations of Gaussian Random Variables.</u>   Let X be a Gaussian random variable. Let λ denote Lebesgue measure on $\mathcal{B}(R)$ and let f:R→R be a Borel measurable function such that inf {λ{x ∈ R: f(x) ≠ ax + b}: a ∈ R and b ∈ R} is large. That is, the set of points where an affine function fails to coincide with f has large Lebesgue measure. If f is such a function, then is f(X) a Gaussian random variable? We know that an affine function of X is Gaussian, but is this still true if f is not even close to an affine function? The correct answer is that it can be; indeed, such a random variable f(X) can have the same distribution as X.

For example, let X be a zero mean, unit variance Gaussian random variable. For a positive number A,

let $f(x) = \begin{cases} -x \text{ if } |x| < A \\ x \text{ if } |x| \geq A \end{cases}$. Then it straightforwardly follows that f(X) is a Gaussian random variable with

zero mean and unit variance. However, by choice of A, we could make this function highly dissimilar in the above sense to any affine function.

<u>Improper Riemann Integrals.</u>   Consider a function mapping a compact interval into R. If this function is bounded, it is well known that it is Riemann integrable if and only if it is almost everywhere (with respect to Lebesgue measure) continuous (see, for instance, [12, p. 85]). Furthermore, in this case if the function is Riemann integrable, then it is Lebesgue integrable and the two integrals are equal (see, for instance, [12, pp. 83-84]). Does a similar result hold if a function maps R into R and possesses an improper Riemann integral? What if the function of interest is bounded and infinitely differentiable? The correct answer is that even this does not imply that the function is Lebesgue integrable. Consider, for instance, the function $f(x) = \frac{\sin x}{x}$ for x ≠ 0 and f(0) = 1. This function has an improper Riemann integral and is infinitely differentiable, and yet it is not Lebesgue integrable.

<u>Integrals and Expectations of Stochastic Processes.</u>   Let {X(t): t ∈ [0, 1]} be a stochastic process defined on a probability space $(\Omega, \mathcal{F}, P)$ with sample functions that are integrable over the set [0, 1]. Assume that E(|X(t)|) < ∞ for each t ∈ [0, 1]. How might $\int_0^1 E(X(t))\, dt$ compare to $E(\int_0^1 X(t)\, dt)$? Consider the following example, which is adapted from [13, p. 124]. Let $\Omega = [0, 1]$, $\mathcal{F} = \mathcal{B}([0, 1])$, and P equal Lebesgue measure restricted to $\mathcal{B}([0, 1])$. Now let A be a positive real number and define

$$X(t, \omega) = \begin{cases} A\omega^{-2} \text{ if } 0 < t < \omega < 1 \\ -At^{-2} \text{ if } 0 < \omega < t < 1 \\ 0 \text{ otherwise for } t \in [0, 1] \text{ and } \omega \in [0, 1] \end{cases}$$

Note that $E(|X(t)|) = \dfrac{2A}{t} - A$ for $t \neq 0$ and $t \neq 1$; and $E(|X(0)|) = 0 = E(|X(1)|)$. Then we have that

$$E(X(t)) = -A \text{ and } \int_0^1 X(t)\, dt = A. \text{ Therefore, we get that } \int_0^1 E(X(t))\, dt = -A \text{ and } E(\int_0^1 X(t)\, dt\,) = A.$$

Now, pick A to be large and look at what happens. In the general context of this situation, one may wish to carefully consider Fubini's theorem.

Filtrations of Sigma Algebras.   Let $\{X(t): t \in [0, \infty)\}$ be a stochastic process defined on the probability space $(\Omega, \mathcal{F}, P)$. By a P-null set we will mean a subset of a measurable set which has probability zero. The canonical filtration of the stochastic process $\{X(t): t \in [0, \infty)\}$ is given by the family of σ-subalgebras $\mathcal{F}_t = \sigma(X(s): s \in [0, t]) \vee (\text{P-null sets})$ for $t \in [0, \infty)$. Define $\mathcal{F}_{0-} = \mathcal{F}_0$; and for

$t \in (0, \infty)$, define $\mathcal{F}_{t-} = \bigvee_{s<t} \mathcal{F}_s$. For $t \in [0, \infty)$, define $\mathcal{F}_{t+} = \bigcap_{s>t} \mathcal{F}_s$. We say that the filtration $\{\mathcal{F}_t\}$ is left continuous at t if $\mathcal{F}_t = \mathcal{F}_{t-}$, it is right continuous at t if $\mathcal{F}_t = \mathcal{F}_{t+}$, and it is continuous at t if $\mathcal{F}_{t-} = \mathcal{F}_{t+}$. We say that the filtration $\{\mathcal{F}_t\}$ is left continuous, right continuous, or continuous if it is left continuous, right continuous, or continuous, respectively, at t for all $t \in [0, \infty)$.

In dealing with martingales and with stochastic differential equations, we often encounter hypotheses stipulating right continuity of filtrations of σ-algebras. Indeed, this is a blanket assumption made by many in the French and Soviet schools of stochastic process theory.

It might be tempting and pleasing to the intuition to believe that the regularity of the sample paths of a stochastic process and the continuity of its canonical filtration are closely related. For example, separable Brownian motion has continuous sample paths and with the aid of the Blumenthal 0-1 Law, it can be shown that it has a continuous canonical filtration (see, for example, [14, pp. 11-17]). On the other side of the proverbial coin, martingales with respect to right continuous filtrations have versions that are almost surely cadlag (see, for example, [15, p. 33]). The following counterexamples are taken from [16].

Let $\Omega = [0, 1]$, $\mathcal{F} = \mathcal{B}([0, 1])$, and P equal Lebesgue measure on $\mathcal{B}([0, 1])$. Pick a function $f \in C^\infty([0, \infty))$ such that $f(x) = 0$ for $x \leq 1/2$ and so that f is strictly increasing on $[1/2, \infty)$. Note that such functions abound (see, for example, [17, p. 39]). Now for $t \in [0, \infty)$, define $X(t, \omega) = \omega f(t)$. Then we see that this stochastic process has infinitely differentiable sample paths, and yet it has a discontinuous

canonical filtration since $\sigma(X(s): s \in [0, t]) = \begin{cases} \{\emptyset, \Omega\} & \text{for } t \leq 1/2 \\ \mathcal{B}([0, 1]) & \text{for } t > 1/2 \end{cases}$.

Now let $f:[0, \infty) \to R$. Consider a one point probability space and define the stochastic process $X(t, \omega) = f(t)$ for $\omega \in \Omega$ and for $t \in [0, \infty)$. Since this stochastic process takes values in $\overline{R}$, which is compact, the separability theorem (see, for example, [18, pp. 166-168]) implies that $\{X(t): t \in [0, \infty)\}$ has a separable version $\{Y(t): t \in [0, \infty)\}$, with $Y(t):\Omega \to \overline{R}$. But $Y(t) = X(t) = f(t)$ for $t \in [0, \infty)$, and thus we see that $\{X(t): t \in [0, \infty)\}$ was separable. Now note that for all $t \in [0, \infty)$,

$\sigma(X(s): s \in [0, t]) \vee (P\text{-null sets})$ is P-trivial; that is, for $A \in \sigma(X(s): s \in [0, t]) \vee (P\text{-null sets})$, $P(A)$ equals zero or one. Thus we see that there exists a separable stochastic process with a continuous canonical filtration and arbitrary sample paths.

It is well known that if $f: R \rightarrow R$ is monotone, then its only discontinuities are jump discontinuities and they occur at only countably many points. Notice that the canonical filtration of a stochastic process is monotone. Does the canonical filtration of a stochastic process have discontinuities at only countably many points? In [16] an example of a stochastic process $\{X(t): t \in [0, \infty)\}$ taking values in $[0, 1]$ is given such that its canonical filtration $\{\mathcal{F}_t\}$ is discontinuous at every positive t. In the context of the continuity of filtrations of σ-subalgebras, one may wish to consider [16].

Sufficiency. Let $(\Omega, \mathcal{F})$ be a measurable space, and let $\mathcal{P}$ be a family of probability measures on $(\Omega, \mathcal{F})$. Thus, for each $P \in \mathcal{P}$, $(\Omega, \mathcal{F}, P)$ is a probability space. In this case $(\Omega, \mathcal{F}, \mathcal{P})$ is sometimes called a probability structure. If $\mathcal{G}$ is a σ-subalgebra of $\mathcal{F}$, we say that $\mathcal{G}$ is sufficient if for each $\mathcal{F}$-measurable bounded real valued function f defined on Ω, there exists a $\mathcal{G}$-measurable bounded real

valued function g defined on Ω such that $\int_B f \, dP = \int_B g \, dP$ for each $B \in \mathcal{G}$ and for all $P \in \mathcal{P}$; that is, g is

a.s.[P] equal to the conditional expectation of f conditioned on $\mathcal{G}$ when P is the relevant probability measure, and g is not dependent on P, but the set of P-measure zero might depend on P. Intuitively, there may be sets in $\mathcal{F}$ which are not in $\mathcal{G}$, but these are irrelevant to drawing inferences about the unknown probability measure $P \in \mathcal{P}$. A real valued sufficient statistic is a random variable X defined on $(\Omega, \mathcal{F})$ such that $\sigma(X) = X^{-1}(\mathcal{B}(R))$ is sufficient.

If $(\Omega, \mathcal{F}, \mathcal{P})$ is a probability structure, if $\mathcal{G}_1$ and $\mathcal{G}_2$ are σ-subalgebras of $\mathcal{F}$ such that $\mathcal{G}_1 \subset \mathcal{G}_2$, and if $\mathcal{G}_1$ is sufficient, then is $\mathcal{G}_2$ sufficient? One may be tempted to suspect that an affirmative answer is correct, since any set in $\mathcal{G}_1$ is contained in $\mathcal{G}_2$ and $\mathcal{G}_1$ is sufficient. However, the correct answer is negative; see [19] for a counterexample.

If $(\Omega, \mathcal{F}, \mathcal{P})$ is a probability structure, if X_1 and X_2 are random variables such that X_1 is a sufficient statistic and if there exists a function $F: R \rightarrow R$ such that $X_1 = F(X_2)$, then is X_2 a sufficient statistic? Once again, the answer is negative; see [19] for a counterexample.

Consider the probability structure $(\Omega, \mathcal{F}, \mathcal{P})$. The family of probability measures $\mathcal{P}$ is said to be dominated if there exists a σ-finite measure μ on $(\Omega, \mathcal{F})$ such that $P \ll \mu$ for each $P \in \mathcal{P}$. In the counterexamples referenced above, the family of probability measures was not dominated. In [20] it is shown that if the family of probability measures $\mathcal{P}$ is dominated, then any σ-subalgebra of $\mathcal{F}$ which includes a sufficient σ-subalgebra of $\mathcal{F}$ is sufficient.

Conditional Probability. Let $(\Omega, \mathcal{F}, P)$ be a probability space and let $\mathcal{A}$ be a σ-subalgebra of $\mathcal{F}$. For $F \in \mathcal{F}$, consider the conditional probability $P(F \mid \mathcal{A})$. For F fixed, this is an $\mathcal{A}$-measurable random variable. For all ω off a measurable set of zero probability, is $P(\cdot \mid \mathcal{A})(\omega)$ a probability measure on $\mathcal{F}$?

Consider the following example, which is adapted from [21, p. 464, problem 33.13]. Let $\Omega = [0, 1]$, and let H be a subset of [0, 1] such that the inner Lebesgue measure of H is zero and the outer Lebesgue measure of H is 1. Let $\mathcal{F} = \{(H \cap B_1) \cup (H^c \cap B_2): B_1 \text{ and } B_2 \text{ belong to } \mathcal{B}([0, 1])\}$. It follows that $\mathcal{F}$ is a σ-algebra on Ω. Further, it follows that $P: \mathcal{F} \rightarrow [0, 1]$ via $P((H \cap B_1) \cup (H^c \cap B_2)) = [\lambda(B_1) + \lambda(B_1)]/2$, where λ denotes Lebesgue measure on $\mathcal{B}([0, 1])$, is a probability measure (see, for example, [7]). Note that $P(H) = 1/2$ and that for $B \in \mathcal{B}([0, 1])$, $P(B) = \lambda(B)$ and $P(H \cap B) = \lambda(B)/2$. Thus, H and $\mathcal{B}([0, 1])$ are statistically independent. Now, let N be an event of zero probability and assume that for $\omega \notin N$, $P(\cdot \mid \mathcal{B}([0, 1]))(\omega)$ is a probability measure on $\mathcal{F}$. Let the set $\{A_n \subset [0, 1]: n \in N\}$ be such that for any positive integers i and j there exists a positive integer m such that $A_i \cap A_j = A_m$, and such that the σ-algebra on [0, 1] generated by $\{A_n \subset [0, 1]: n \in N\}$ is $\mathcal{B}([0, 1])$. Note that for $n \in N$, $P(A_n \mid \mathcal{B}([0, 1])) = I_{A_n}$ a.s. since $A_n \in \mathcal{B}([0, 1])$. For $n \in N$, let $C_n = \{\omega \in \Omega: P(A_n \mid \mathcal{B}([0, 1]))(\omega)$

$= I_{A_n}(\omega)\}$. Note that, for all $n \in N$, $C_n \in \mathcal{F}$ and $P(C_n) = 1$. Consequently, $C = \bigcap_{n \in N} C_n \cap N^c$ is

such that $P(C) = 1$. Note that the function which, for a fixed $\omega \in C$, maps $B \in \mathcal{B}([0, 1])$ to $I_B(\omega)$ is a probability measure on $\mathcal{B}([0, 1])$ which agrees with $P(B \mid \mathcal{B}([0, 1]))(\omega)$ when $B \in \{A_n \subset [0, 1]: n \in N\}$. It is then a consequence of the Dynkin System Theorem [22, p. 169] that for $\omega \in C$, $P(B \mid \mathcal{B}([0, 1]))(\omega)$ is uniquely determined for $B \in \mathcal{B}([0, 1])$ to be $I_B(\omega)$. Thus $\omega \in C$ implies that $P(B \mid \mathcal{B}([0, 1]))(\omega) = I_B(\omega)$ for all $B \in \mathcal{B}([0, 1])$. Thus, for $\omega \in C$, $P(\{\omega\} \mid \mathcal{B}([0, 1]))(\omega) = 1$. Now, recalling our assumption that for $\omega \notin N$, $P(\cdot \mid \mathcal{B}([0, 1]))(\omega)$ is a probability measure on $\mathcal{F}$, we see that $\omega \in H \cap C$ implies that $P(H \mid \mathcal{B}([0, 1]))(\omega) \geq P(\{\omega\} \mid \mathcal{B}([0, 1]))(\omega) = 1$ and that $\omega \in H^c \cap C$ implies that $P(H \mid \mathcal{B}([0, 1]))(\omega) \leq P(\Omega - \{\omega\} \mid \mathcal{B}([0, 1]))(\omega) = 0$. Thus $P(C) = 1$ and $\omega \in C$ implies that $P(H \mid \mathcal{B}([0, 1]))(\omega) = I_H(\omega)$. But, H and $\mathcal{B}([0, 1])$ are statistically independent. Thus, $P(H \mid \mathcal{B}([0, 1])) = 1/2$ a.s., yielding a contradiction. Therefore, $P(\cdot \mid \mathcal{B}([0, 1]))$ is not almost surely a probability measure on $\mathcal{F}$.

We will now consider one further aspect of conditional probability. This example is taken from [21, pp. 458-459]. Let the probability space be given by [0, 1], $\mathcal{B}([0, 1])$, and Lebesgue measure restricted to $\mathcal{B}([0, 1])$. Consider the σ-subalgebra $\mathcal{G}$ given by the family of countable and cocountable subsets of [0, 1]. Let $B \in \mathcal{B}([0, 1])$ and consider the conditional probability $P(B \mid \mathcal{G})$. Now, since $\mathcal{G}$ contains all singleton sets, to know which elements of $\mathcal{G}$ contain ω is to know ω itself. Thus, an overly cavalier approach to conditional probability might lead one to suspect that $P(B \mid \mathcal{G})(\omega) = I_B(\omega)$. However, this is incorrect; indeed, if B is neither countable nor cocountable, then $I_B(\cdot)$ is not $\mathcal{G}$-measurable. Note that every set in $\mathcal{G}$ has probability either zero or one. Thus, we see that $P(B \mid \mathcal{G}) = P(B)$ a.s.

Conclusion. Some counterexamples to claims which seem to commonly arise in the literature of detection and estimation theory have been presented.

Acknowledgement. This research was supported by the Air Force Office of Scientific Research under Grant AFOSR-86-0026.

Also, I would like to thank Prof. Donald L. Burkholder of the University of Illinois at Urbana-Champaign for pointing out to me difficulties that may arise in the context of sufficient σ-subalgebras.

References

[1] G. L. Wise, "Some common misconceptions in communication theory," *Proceedings of the Twenty-Sixth Midwest Symposium on Circuits and Systems,* Puebla, Puebla, Mexico, August 15-16, 1983, pp. 33-37.

[2] G. L. Wise, "A note on a common misconception in estimation," *Systems and Control Letters,* Vol. 5, pp. 355-356, April 1985.

[3] J. P. Romano and A. F. Siegel, *Counterexamples in Probability and Statistics,* Wadsworth & Brooks/Cole Advanced Books and Software, Monterey, California, 1986.

[4] L. Devroye, "On arbitrarily slow rates of global convergence in density estimation," *Z. Wahrscheinlichkeitstheorie verw. Gebiete,* Vol 62, pp. 475-483, 1983.

[5] V. K. Matskyavichyus, "A lower bound for the convergence rate in the central limit theorem," *Theory of Probability and Its Applications,* Vol 23, pp. 596-601, December 1986 (Soviet version published in December 1985).

[6] J. M. Morrison and G. L. Wise, "The martingale convergence theorem, with tears," *Systems and Control Letters,* Vol. 9, pp. 275-279, September 1987.

[7] A. E. Wessel and G. L. Wise, "On estimation of random variables via the martingale convergence theorem," *Systems and Control Letters,* Vol. 11, pp. 61-64, July 1988.

[8] J. M. Morrison and G. L. Wise, "Some comments on the convergence rate for adaptive schemes," to appear in *Proceedings of the Thirty-First Midwest Symposium on Circuits and Systems,* St. Louis, Missouri, August 10-12, 1988.

[9] J. M. Morrison and G. L. Wise, "An asymptotic property of nonlinear estimators arising as solutions to a certain class of convex programming problems," *Systems and Control Letters,* Vol. 7, pp. 225-232, June 1986.

[10] E. B. Hall, A. E. Wessel, and G. L. Wise, "On fusion in estimation theory," to appear in *Proceedings of the Twenty-Sixth Annual Allerton Conference on Communication, Control, and Computing,* Monticello, Illinois, September 28-30, 1988.

[11] D. A. Pierce and R. L. Dykstra, "Independence and the normal distribution," *The American Statistician,* vol. 23, no. 4, 1969.

[12] R. L. Wheeden and A. Zygmund, *Measure and Integral An Introduction to Real Analysis*. Marcel Dekker, New York, 1977.

[13] B. R. Gelbaum and J. M. H. Olmsted, *Counterexamples in Analysis*. Holden-Day, San Francisco, 1964.

[14] R. Durrett, *Brownian Motion and Martingales in Analysis*. Wadsworth, Belmont, CA, 1984.

[15] N. Ikeda and S. Watanabe, *Stochastic Differential Equations and Diffusion Processes*. North Holland, Amsterdam, 1981.

[16] J. M. Morrison and G. L. Wise, "Continuity of filtrations of sigma algebras," *Statistics and Probability Letters,* Vol. 6, pp. 55-60, September 1987.

[17] W. M. Boothby, *An Introduction to Differentiable Manifolds and Riemannian Geometry*. Academic Press, New York, 1977.

[18] R. B. Ash and M. F. Gardner, *Topics in Stochastic Processes*. Academic Press, New York, 1975.

[19] D. L. Burkholder, "Sufficiency in the undominated case," *Annals of Mathematical Statistics,* Vol. 32, pp. 1191-1200, December 1961.

[20] R. R. Bahadur, "Sufficiency and statistical decision functions," *Annals of Mathematical Statistics,* Vol. 25, pp. 423-462, September 1954.

[21] P. Billingsley, *Probability and Measure,* second edition. John Wiley and Sons, New York, 1986.

[22] R. B. Ash, *Real Analysis and Probability*. Academic Press, New York, 1972.

On the Convergence of the Projection Method for an Autoregressive Process and a Matched DPCM Code

Morteza Naraghi-Pour   David L. Neuhoff
Louisiana State University   University of Michigan

1. **Introduction:** The analysis of differential pulse code modulation (DPCM) involves the study of the steady state behavior of the code, the evaluation of the distortion incurred between the input process and its reproduction, and the optimization of the code parameters so as to minimize its distortion. The key step in such study is to find the stationary probability distribution of a random process in terms of which the code distortion can be evaluated. Due to the nonlinear nature of the quantizer and the presence of the feedback loop analytical solutions to this problem have only been found in special cases. However, a well known approximation technique has been frequently applied to obtain an appropriate probability distribution. The validity of this technique however, has not been previously justified. In this paper we establish a framework in which the approximation technique can be rigorously justified.

Let $\{X_n\}$ be a stationary Gaussian first-order autoregressive source, i.e., $X_{n+1} = \rho X_n + W_{n+1}$, where $\{W_n\}$ is an IID Gaussian innovation process with mean zero and variance one, and where ρ is a correlation coefficient, $|\rho| < 1$. Each X_n is Gaussian with mean zero and variance $\sigma_X^2 = 1/(1 - \rho^2)$ and $E[X_n X_{n+1}] = \rho \sigma_X^2$. Consider a DPCM code with quantizer $Q(\cdot)$ and first-order predictor with prediction coefficient a. Such a code produces Y_n as the reproduction for X_n according to the rule

$$Y_n = aY_{n-1} + Q(X_n - aY_{n-1}) \quad , \tag{1}$$

where Y_{n-1} is the state of the code, aY_{n-1} is the prediction of X_n, and where the initial state Y_0 may be a constant or a random variable with some known distribution.

When the prediction coefficient a equals ρ, the code is said to be *matched* to the source, as opposed to *mismatched* for the case $a \neq \rho$. When a equals one, the code is said to have a perfect integrator as opposed to a leaky integrator for $a < 1$. The prediction error process $\{Z_n\}$ is given by

$$Z_n = X_n - aY_{n-1} = \rho X_{n-1} + W_n - aY_{n-1} \quad , \tag{2}$$

and the reproduction error process $\{E_n\}$ is given by

$$E_n = X_n - Y_n = Z_n - Q(Z_n) \quad . \tag{3}$$

The primary goal in the study of these codes is to compute the average distortion incurred in reproducing the sequence $\{Y_n\}$ instead of $\{X_n\}$. With respect to the squared-error distortion measure, this distortion is

$$D \stackrel{\Delta}{=} \limsup_{N \to \infty} \frac{1}{N} \sum_{n=0}^{N-1} E[X_n - Y_n]^2 \quad . \tag{4}$$

Using (3) the average distortion can be expressed in terms of the prediction error process $\{Z_n\}$. In the case of matched predictor ($\rho = a$), it follows from (2) and (3) that

$$Z_n = \rho [Z_{n-1} - Q(Z_{n-1})] + W_n \quad ,$$

from which it is easy to see that the prediction error process $\{Z_n\}$ is first-order Markov. It follows from the results of Gersho [8] that $\{Z_n\}$ is asymptotically stationary in the sense that there is a unique stationary distribution v^* such that the distribution of Z_n converges weakly to v^* irrespective of the initial distribution of the code state Y_0. The average distortion can then be written in terms of the stationary distribution v^* as $D = E_{v^*}[Z - Q(Z)]^2$, where Z is a random variable with distribution v^*. It is straightforward to show that v^* has a probability density function. Let $q(z)$ denote this density. Then $q(z)$ satisfies the following Chapman-Kolmogorov equation

$$q(z) = \int p_W(z - \rho \, \varepsilon(x)) \, q(x) \, dx \quad , \tag{5}$$

where $p_W(w)$ is the probability density function of W_n and $\varepsilon(x) = x - Q(x)$. In order to compute D one needs to solve this integral equation for the unknown probability density function $q(z)$. Unfortunately, however, exact solution of such equations is a formidable if not impossible, task. To our knowledge analytical solutions have only been found in special cases [6,9]. Due to the inherent difficulties in obtaining analytical solutions for such equations, one generally resorts to efficient numerical techniques.

Numerical solution of integral equations has been the subject of extensive research in the mathematical community [11]. Davisson [4] was the first to use such techniques to solve equation (5). He was followed by Slepian [16], Arnstein [2], Janardhanan [10] and Farvardin [5] who applied the same techniques to DPCM codes for Gauss Markov input processes. More recently, a two dimensional version of this technique has been applied to evaluate the performance of the mismatched DPCM for an autoregressive input process [13].

In what follows we describe a general formulation of this approximation technique which is known as the projection or Galerkin method [11].

2. Projection Method: Let H be a given separable Hilbert space with inner product $\langle \, , \, \rangle_H$. Let $K : H \to H$ be a bounded linear operator and suppose we would like to solve the eigenvalue/eigenfunction (e.v./e.f.) problem given by the equation

$$\lambda y = K y \quad . \tag{6}$$

The projection method for the solution of this equation is as follows: Let $\{M_n\}_{n=0}^{\infty}$ be an increasing sequence of subspaces of H such that $\bigcup_{n=0}^{\infty} M_n$ is dense in H. Let $\{P_n\}_{n=0}^{\infty}$ be the sequence of (orthogonal) projections such that $P_n : H \to M_n$. Replace equation (6) by the following approximate equation

$$P_n K P_n y = \lambda_n P_n y \quad .$$

Equivalently let $y_n = P_n y$, where $y_n \in M_n$, then the approximate equation becomes

$$P_n K y_n = \lambda_n y_n \quad , \quad y_n \in M_n \quad . \tag{7}$$

We now search for a solution (λ_n, y_n) to this last equation, where $y_n \in M_n$ is considered an approximation to y. We have replaced λ in (6) by λ_n in (7) to emphasize the fact that λ may no longer be an e.v. of the operator $P_n K : M_n \to M_n$ in equation (7).

Suppose we have a complete orthonormal sequence $\{\phi_i\}_{i=0}^{\infty}$ in H (since H is separable, one always exists). Then a candidate often used for the subspace M_n is the span of the set $\{\phi_i\}_{i=0}^{n-1}$. Then P_n becomes the orthogonal projection of H onto M_n, i.e., $P_n y \stackrel{\Delta}{=} \sum_{i=0}^{n-1} \langle y, \phi_i \rangle_H \phi_i$. In this case the solution to equation (7) is

$$y_n = \sum_{i=0}^{n-1} y^{(i)} \phi_i \qquad\qquad (8)$$

where $y = (y^{(0)}, y^{(1)}, ..., y^{(n-1)})^T$ is the solution to the matrix equation

$$\lambda_n y = A y \ , \qquad\qquad (9)$$

and where A is an $n \times n$ matrix with

$$A_{ij} = \langle K \phi_j , \phi_i \rangle_{H} \ , \qquad i,j = 0, 1, 2,...,n-1 \ .$$

Note that $y^{(i)} = \langle y_n, \phi_i \rangle_{H}$.

When one applies the projection method to equation (6), the following questions should be dealt with:

(i)   Suppose that (λ , y) is an e.v. / e.f. pair for the operator K. Does the operator $P_n K$ in equation (7) have an e.v. / e.f. pair for each value of n?

(ii)   If the answer to (i) is in the affirmative, is this e.v. / e.f. pair unique for each fixed n?

(iii)   More importantly, does any such solution approximate (λ , y) in an appropriate sense?

To our knowledge to date answers to all of these questions can be given only if the operator K is compact [11]. For this case the following theorem summarizes the results of interest to us [11;12].

Theorem 1

Let H and $\{P_n\}_{n=0}^{\infty}$ be as before. Let $K : H \longrightarrow H$ be a compact linear operator, and consider the e.v. / e.f. problem of equation (6). Then

(i)   There exists n_0 such that for all $n \geq n_0$, the operator $P_n K$ in equation (7) has an e.v. λ_n.

(ii)   Let $\{\lambda_n\}$ be a sequence of such eigenvalues (λ_n is an e.v. of $P_n K$). Then any limit point of $\{\lambda_n\}$ is an e.v. of the operator K.

(iii)   For any e.v. λ_0 of K, there exists a sequence $\{\lambda_n\}$, where λ_n is an e.v. of $P_n K$, such that $\lambda_n \longrightarrow \lambda_0$.

(iv)   If $\{(\lambda_n , y_n)\}$ is a sequence of e.v. / e.f.'s for the sequence of operators $\{P_n K\}$ in (7) with the e.v.'s λ_n converging to λ_0, and with the e.f.'s y_n having norm one, then

    (a)   The sequence $\{y_n\}$ contains a convergent subsequence.

    (b)   The limit of any convergent subsequence $\{y_{n_k}\}$ is an e.f. of the operator K, corresponding to the e.v. λ_0.

(v)   If $\{(\lambda_n , y_n)\}$ is a sequence of e.v. / e.f.'s for the sequence of operators $P_n K$ in (7), with the e.v.'s λ_n converging to λ_0, then for all n sufficiently large

$$\dim N(\lambda_n - P_n K) \leq \dim N(\lambda_0 - K) \ ,$$

where $N(\lambda - K)$ is the null space of the operator $\lambda I - K$.

3.   **The projection method applied to (5):** Equation (5) can be written in operator notation as $Tq = q$, where T is the integral operator given by

$$(Tf)(z) = \int p_W(z - \rho \, \varepsilon(x)) \, f(x) \, dx \quad . \tag{10}$$

In this light, we see that the sought after stationary density q (the solution to (5)) is the solution to an e.v./e.f. problem in the form of (6) with eigenvalue one and eigenfunction q. Accordingly, one may use the projection method and to do so one needs to choose the Hilbert space H and the subspaces M_n. Ideally, H would be chosen so that T was compact and, consequently, so that Theorem 1 could be applied to show convergence of the approximate solutions to the true.

Janardhanan [10] has applied the projection method to equation (5) with the space L_2 (the space of square integrable functions with respect to the Lebesgue measure) as the Hilbert space H. Gerr [7] has also used L_2 to study the convergence of the approximation method in the case of adaptive DPCM. However, as discussed later, the operator T is not a compact operator on L_2. Therefore, we now describe two other choices for the space H, one of which was utilized by Davisson, Arnstein and Farvardin [4,2,5], and will, accordingly, be adopted for our developments here. This choice makes the operator T compact.

Let $\Psi(x)$ be a probability density function and denote by S, the support of Ψ, i.e.,

$$S \overset{\Delta}{=} \{ \, x : \Psi(x) > 0 \, \} \quad .$$

Denote by $H(\Psi^{-1})$ the space of Lebesgue measurable functions $f : S \longrightarrow R$, that are square integrable with respect to the weight function Ψ^{-1}. That is

$$H(\Psi^{-1}) = \{ \, f : \int_S |f(x)|^2 \, \Psi^{-1}(x) \, dx < \infty \, \} \quad . \tag{11}$$

Similarly let $H(\Psi)$ be the space of measurable functions $f : S \longrightarrow R$, that are square integrable with respect to the weight function $\Psi(x)$, i.e.,

$$H(\Psi) = \{ \, f : \int |f(x)|^2 \, \Psi(x) \, dx < \infty \, \} \quad . \tag{12}$$

It is clear that $f \in H(\Psi)$ if and only if $f\Psi \in H(\Psi^{-1})$. Let us define the following inner products on the two spaces $H(\Psi)$ and $H(\Psi^{-1})$: For f and $g \in H(\Psi)$ let

$$\langle \, f \, , g \, \rangle_\Psi = \int f(x) \, g(x) \, \Psi(x) \, dx \quad , \tag{13}$$

and for f and $g \in H(\Psi)$ let

$$\langle \, f \, , g \, \rangle_{\Psi^{-1}} = \int f(x) \, g(x) \, \Psi^{-1}(x) \, dx \quad . \tag{14}$$

Equipped with these inner products $H(\Psi)$ and $H(\Psi^{-1})$ become separable Hilbert spaces. The norms corresponding to the inner products in (13) and (14) are given by $\| f \|^2_\Psi = \int |f(x)|^2 \, \Psi(x) \, dx$, and $\| f \|^2_{\Psi^{-1}} = \int |f(x)|^2 \, \Psi^{-1}(x) \, dx$.

It is well known that corresponding to $\Psi(x)$ there is a unique complete orthonormal sequence of polynomials $\{\Gamma_i(x)\}_{i=0}^\infty$ in $H(\Psi)$, such that $\Gamma_i(x)$ has degree i [3]. More explicitly,

$$\int \Gamma_i(x) \, \Gamma_j(x) \, \Psi(x) \, dx = \delta_{ij} \quad ,$$

where $\delta_{ij} = 0$ for $i \neq j$ and $\delta_{ii} = 1$. It can also be easily shown that $\{\phi_i(x)\}_{i=0}^\infty$, where $\phi_i(x) = \Psi(x)\Gamma_i(x)$, is a complete orthonormal sequence for $H(\Psi^{-1})$.

Henceforth, $H(\Psi^{-1})$ will be utilized for the study of the projection method. The issues that influence this choice are speed of convergence, compactness of the operator T and the generality of the space so as it contains the unknown probability density function $q(z)$. These are discussed in detail in [14]. This space with a particular choice of the density function $\Psi(x)$ was also selected by Davisson, Amstein and Farvardin [10,1,5]. We will discuss the choice of $\Psi(x)$ in the next section.

To apply the projection method, we now let M_n be the span of the set $\{\phi_i(x)\}_{i=0}^{n-1}$ where $\phi_i(x) = \Psi(x)\Gamma_i(x)$. The approximate solution to $q(z)$ is

$$q_n(z) = \sum_{i=0}^{n-1} q^{(i)}\Psi(z)\Gamma_i(z) \quad , \tag{15}$$

where $q = (q^{(1)}, q^{(2)}, \ldots, q^{(n-1)})^T$ is the solution to the matrix equation $\lambda_n q = \Lambda q$, where A is an $n \times n$ matrix with

$$A_{ij} = \langle T\phi_j, \phi_i \rangle_{\Psi^{-1}} = \iint p_W(z - \rho\,\epsilon(x))\,\Psi(x)\,\Gamma_i(z)\,\Gamma_j(x)\,dz\,dx \quad . \tag{16}$$

For this method to work, we need to show that q belongs to $H(\Psi^{-1})$. In the next section we consider this question and establish sufficient conditions on the probability density function $\Psi(\cdot)$ which guarantee that q belongs to $H(\Psi^{-1})$. We will also prove the convergence of the projection method for this problem.

4. Approximation Technique for a Matched Autoregressive Process: In the following we assume that the input process $\{X_n\}$ is autoregressive with an IID zero mean, unit variance Gaussian innovation process $\{W_n\}$. We also assume that the code is matched to the source in the sense that $\rho = a$. If the code operates with small distortion, the error incurred in the encoding process is small, and therefore the prediction error $\{Z_n\}$ is close in distribution to the innovation process $\{W_n\}$. Therefore $q(\cdot)$ must be approximately Gaussian. This assertion, which is based on intuition, is supported by the results of O'Neal [15], Stroh [17] and Janardhanan [12]. Therefore it seems appropriate to choose a Gaussian probability density function for $\Psi(\cdot)$. Accordingly, let

$$\Psi(z) = g_\sigma(z) \triangleq \frac{1}{\sqrt{2\pi}\,\sigma}\, e^{-\frac{z^2}{2\sigma^2}} \quad .$$

The complete orthonormal sequence of polynomials corresponding to $g_\sigma(z)$ is the sequence of Hermite polynomials denoted by $\{\Phi_i\}_{i=0}^{\infty}$. $H(g_\sigma^{-1})$ is the space of measurable functions $f : \mathbb{R} \longrightarrow \mathbb{R}$ such that

$$\sqrt{2\pi}\,\sigma \int |f(z)|^2\, e^{\frac{z^2}{2\sigma^2}}\, dz < \infty \quad . \tag{17}$$

We would like to point out that in order to compute an approximation to the stationary probability density function q, Davisson [10], Amstein [1] and Farvardin [5] have applied the projection method to equation (10) in the space $H(g_\sigma^{-1})$ with $\sigma = 1$.

In an attempt to justify that the stationary probability density function $q(\cdot)$ belongs to $H(g_\sigma^{-1})$, Amstein [1, Appendix D] showed that the moments of q are bounded. Unfortunately however, for q to belong to $H(g_\sigma^{-1})$, bounded moments are necessary but not sufficient.

The following theorem whose proof is given in [14] shows that for an appropriate choice of σ, the stationary probability density function $q(\cdot)$ belongs to the space $H(g_\sigma^{-1})$.

Theorem 2

If the quantizer $Q(\cdot)$ has a finite number of output levels, $Q(z) \geq 0$ for $z \geq 0$, $Q(z) < 0$ for $z < 0$, and $\sigma > (2 - 2\rho^2)^{-\frac{1}{2}}$, then $q(\cdot) \in H(g_\sigma^{-1})$.

Remark

This theorem shows that if the quantizer satisfies some mild conditions and if $\sigma > (2 - 2\rho^2)^{-\frac{1}{2}}$, then $q \in H(g_\sigma^{-1})$. Since $q(z)$ is bounded, it can be seen from equation (17) that the fact that $q(z)$ belongs to $H(g_\sigma^{-1})$ does not depend on the nature of $q(z)$ on any bounded set, but rather on the tails of $q(z)$ as $|z| \to \infty$. Let $p_X(\cdot)$ denote the probability density function of the input process $\{X_n\}$. Then

$$p_X(x) = \frac{1}{\sqrt{2\pi}\,\sigma_X}\, e^{-\frac{x^2}{2\sigma_X^2}} \quad ,$$

where $\sigma_X = (1 - \rho^2)^{-\frac{1}{2}}$. Now from the fact that the reproduction process is bounded and since $Z_k = X_k - aY_{k-1}$, it follows that the tails of $q(\cdot)$ are the same as those of $p_X(\cdot)$. Furthermore, it follows from (17) that $p_X \in H(g_\sigma^{-1})$, if and only if $\sigma > \dfrac{\sigma_X}{2}$, or equivalently if and only if $\sigma > (2 - 2\rho^2)^{-\frac{1}{2}}$. Therefore for values of $\sigma \leq (2 - 2\rho^2)^{-\frac{1}{2}}$, the sufficient condition (17) fails for the probability density function $q(z)$, and thus the lower bound on σ given in Theorem 2 cannot be weakened.

We now turn our attention to the operator T and study its properties on the space $H(g_\sigma^{-1})$. We know there exists a unique probability density function $q \in H(g_\sigma^{-1})$ such that $Tq = q$. Therefore T has an e.f. $q \in H(g_\sigma^{-1})$, corresponding to e.v. $\lambda = 1$. However, the reader may note that we have not yet established the fact that T is an operator from $H(g_\sigma^{-1})$ into $H(g_\sigma^{-1})$. Although we know that if $f(\cdot)$ is a probability density function, then $(Tf)(\cdot)$ given by (10) is also a probability density function, this does not imply that $Tf \in H(g_\sigma^{-1})$. In other words we need to prove that the range of the operator T is a subset of $H(g_\sigma^{-1})$. Also note that the operator T conceivably might have eigenvalues other than $\lambda = 1$, and corresponding to $\lambda = 1$, T conceivably might have eigenfunctions that are not a scalar multiple of q. Theorems 3 and 4 whose proofs are given in [14] establish the fact that the range of T is a subset of $H(g_\sigma^{-1})$, and that T is compact. Theorem 5 whose proof is also given in [14] shows that $\lambda = 1$ is the only eigenvalue of T, while q is the only eigenfunction of T to within a scalar multiple.

Theorem 3

If $\sigma > (2 - 2\rho^2)^{-\frac{1}{2}}$, then the integral operator T in equation $Tq = q$ is a bounded linear operator from $H(g_\sigma^{-1})$ into $H(g_\sigma^{-1})$, and

$$\| T \| \leq \left[\frac{1}{2\pi} \iint \exp - [z - \rho \varepsilon(x)]^2 \, \exp \frac{z^2 - x^2}{2\sigma^2} \, dz \, dx \right]^{\frac{1}{2}} < \infty \quad , \tag{18}$$

where $\| T \|$ denotes the norm of T.

Theorem 4

If $\sigma > (2 - 2\rho^2)^{-\frac{1}{2}}$, then the operator T is compact.

Theorem 5

(i)　If $q(\cdot)$ is the unique probability density function that satisfies the Chapman-Kolmogorov equation (5), i.e., $q(\cdot)$ is the stationary probability density function, then

$$N(1-T) = \{ h \in H(g_\sigma^{-1}) : h = rq \text{ where } r \in \mathbb{R} \} ,$$

i.e., $\dim N(1-T) = 1$.

(ii)　The only e.v. of the operator T is $\lambda = 1$.

Remark

　　We have selected the space $H(g_\sigma^{-1})$ as the Hilbert space in which the eigenvalue problem is studied. However, from the integral operator point of view, the Hilbert space L_2 might seem more natural. As noted by Gerr [7] in the study of a similar problem, the main difficulty with such a choice is that the operator T is not compact with respect to the corresponding norm of L_2. This is caused by the fact that the integral operator is similar to a convolution integral where the range of integration is the entire real line. These types of operators considered in the space L_2 are not generally compact. For example it can be shown that in the space L_2, the sequence of functions given by

$$f_n(t) = \begin{cases} 1 & n < t \le n+1 \\ 0 & \text{otherwise} \end{cases}$$

converges weakly to zero, whereas the sequence Tf_n does not converge strongly to zero, and therefore T is not compact in L_2.

　　Gerr and Cambanis [7] select L_2 for the study of the projection method, and since the integral operators they have studied are not compact in L_2, they have only obtained partial results, in that they have not shown that the truncated equations have any solutions, or whether solutions, if they exist, converge in L_2. However, they have shown that if the truncated equations have solutions and if the sequence of solutions has a limit point, then that limit point is the solution of the original integral equation.

　　We now demonstrate the convergence of the projection method when applied to the Chapman-Kolmogorov equation. We would like to solve the equation $Tq = q$, where T is the integral operator in (10) and q is the unknown stationary probability density function of the error process $\{Z_n\}$. We view equation $Tq = q$ as an e.v. / e.f. problem in the space $H(g_\sigma^{-1})$, where σ is chosen so that $\sigma > (2-2\rho^2)^{-\frac{1}{2}}$. We have shown before that any solution of equation (10) is a scalar multiple of q. Therefore our approach is to first find the unique solution f such that $\|f\| = 1$, and then normalize f by the proper factor, namely $r = \int f$ to obtain q. The following procedure will find a sequence of functions $\{f_n\}$ for which f is the only limit point.

　　Let M_n be the span of the set $\{g_\sigma \Phi_i\}_{i=0}^{n-1}$, where $\{\Phi_i\}_{i=0}^{\infty}$ is the sequence of Hermite polynomials corresponding to g_σ. Also let $\{P_n\}_{n=0}^{\infty}$ be the sequence of projections, such that $P_n : H(g_\sigma^{-1}) \longrightarrow M_n$. Solve the sequence of equations

$$P_n T f_n = \lambda_n f_n , \tag{19}$$

for $f_n \in M_n$ and $\|f_n\| = 1$ (by reduction to a matrix equation). Choose a limit point of the sequence $\{f_n\}$. This will be the desired solution.

The following theorem justifies this approach.

Theorem 6

(i) There exists n_0 such that for each $n \geq n_0$, equation (19) has a solution (λ_n, f_n).

(ii) For any such sequence of solutions $\{(\lambda_n, f_n)\}$,

$\quad$ (a) $\quad \lambda_n \longrightarrow \lambda = 1$ as $n \to \infty$.

$\quad$ (b) $\quad$ For a given λ_n, f_n is unique.

$\quad$ (c) $\quad f$ is the only limit point of the sequence $\{f_n\}$.

Proof: See [14].

We have shown that if the parameter σ in the Hermite polynomial expansion of $q(z)$ is large enough $(\sigma > (2 - 2\rho^2)^{-\frac{1}{2}})$, then the projection method yields a sequence of solutions that approximate $q(z)$ arbitrarily closely. In applying the projection method Arnstein [2] and Davisson [4] have chosen $\sigma = 1$, which implies convergence for all $\rho < .7$. Arnstein was able to obtain convergence for all values of $\rho < .9$. We believe this discrepancy can be explained as follows. Our results give a sufficient condition for the convergence of the projection method. While $\sigma = 1$ implies convergence for all $\rho < .7$, it does not suggest that convergence may not be achieved for values of $\rho > .7$. Moreover, the convergence in our study is in the sense of the norm in (17), whereas Arnstein has obtained convergence of distortion. For a symmetric probability density function such as $q(z)$ convergence in our norm implies convergence in distortion. Therefore, our result, being stronger, applies to a smaller set of ρ values.

In [14] it is shown that not only is f the only strong limit point of $\{f_n\}$ (as proved in Theorem 6), but it is also the only weak limit point of $\{f_n\}$.

References

[1] Arnstein, D.S., "Predictive Source Coding of Correlated Gaussian Processes," Ph.D. dissertation, University of California, Los Angeles, 1970.

[2] Arnstein, D.S., "Quantization Error in Predictive Coders," *IEEE Trans. on Commun.*, Vol.COM-23, pp.423-429, April 1975.

[3] Beckman, P., *Orthogonal Polynomials for Engineers and Physicists*, Boulder, CO: The Golem Press, 1973.

[4] Davisson, L.D., "Information Rates for Data Compression." *IEEE WESCON*, 1968, Session 8, Paper 1.

[5] Farvardin, N., "Rate-Distortion Performance of Optimum Entropy-Constrained Quantizers," Ph.D. dissertation, Rensselaer Polytechnic Institute, 1983.

[6] Fine, T.L., "The Response of a Particular Nonlinear System with Feedback to Each of Two Random Processes," *IEEE Trans. on Inform. Theory*, Vol.IT-14, pp.255-264, March 1968.

[7] Gerr, N.L., "Exact Analysis of a Delayed Delta Modulator and an Adaptive Differential Pulse-code Modulator," Ph.D. dissertation, University of North Carolina, 1982.

[8] Gersho, A., "Stochastic Stability of Delta Modulation," *BSTJ*, Vol.51, No.4, pp.821-841, April 1972.

[9] Hayashi, A., "An Exact Analysis of Some Digital Transmission Systems with Gaussian Inputs," Ph.D. dissertation, University of Hawaii, 1976.

[10] Janardhanan, E., "Differential PCM Systems," *IEEE Trans. on Communications*, Vol.COM-27, pp.82-93.

[11] Krasnosel'skii, M.A., *Approximate Solution of Operator Equations*, Wolters-Noordhoff, 1972.

[12] Lo, W.S., "Spectral Approximation Theorems for Bounded Linear Operators," *Bull. Austral. Math. Soc.*, Vol.8, pp.279-287, 1973.

[13] Naraghi-Pour, M. and D.L. Neuhoff, "Mismatched DPCM Encoding of Autoregressive Processes," submitted for publication to IEEE Trans. on Information Theory.

[14] Naraghi-Pour, M., "Predictive Quantization of Autoregressive and Composite Random Processes," Ph.D. dissertation, The University of Michigan, 1987.

[15] O'Neal, J.B., Jr., "Delta Modulation Quantizing Noise — Analytic and Simulation Results for Gaussian on Television Input Signals," *IEEE Trans. on Communication Technol. (corresp.)*, Vol.COM-19, pp.568-569, Aug. 1971.

[16] Slepian, D., "On Delta Modulation," *Bell System Technical Journal*, Vol.51(10), pp.2101-2137, Dec. 1972.

[17] Stroh, R.W., "Optimum and Adaptive Differential Pulse Code Modulation," Ph.D. dissertation, Polytechnic Institute of Brooklyn, NY, 1970.

Morteza Naraghi-Pour is with the Department of Electrical and Computer Engineering, Louisiana State University, Baton Rouge, LA 70803.

David L. Neuhoff is with the Department of Electrical Engineering and Computer Science, The University of Michigan, Ann Arbor, MI 48109.

Tree-Structured Vector Quantization for Progressive Transmission Image Coding

Robert M. Gray   Eve A. Riskin

1 Introduction

Tree-structured vector quantization (TSVQ) provides an approach to image coding that is capable of high fidelity at low bit rates and modest complexity. (See, e.g., [1] for a survey of vector quantization in general and tree-structured vector quantizers in particular.) When the tree-structured codes are designed using the splitting method [2, 3], the resulting codes have a successive approximation character that is well suited to progressive transmission image coding, codes in which an image is progressively improved as additional bits arrive [4].

Many data sources, including speech and images, can be coded with higher fidelity at the expense of added complexity by using variable rate codes, codes which assign more bits to active regions of input signal and fewer bits to relatively quiescent portions. The performance gains available can outweigh the added complexity incurred by the use of variable rate systems and their requirement for buffering and treatment of overflow and underflow. Traditionally, most such variable rate schemes have been designed by concatenating a fixed rate data compression system (such as vector quantization or a transform code) with a variable rate noiseless code such as a Huffman, arithmetic, or Lempel-Ziv code. An alternative is to use a pruned tree-structured vector quantizer (PTSVQ) designed by "pruning" a large TSVQ so as to optimally trade off average rate (or average number of bits) with average distortion. An algorithm for optimally pruning a TSVQ was developed in [5] by generalizing an algorithm of Breiman, Friedman, Olshen, and Stone [6] for optimally pruning binary decision trees.

The goal of this presentation is to summarize the basic code design technique and to demonstrate its performance using medical images, specifically magnetic resonance brain scans. Medical images implicitly require higher fidelity because of the importance of accurate diagnosis (and the potential legal issues). In some applications the full accuracy of the original digitized image may be required. As these images may need to be communicated over a slow communication link (phone lines) or stored in a limited capacity disk reservoir, progressive transmission provides a means of providing good fidelity fast with perfect fidelity available as needed.

2 TSVQ Design

This section presents an informal review of the TSVQ design algorithm. The algorithm assumes the following quantities are given:

1. A training sequence $T = \{X_n; n = 1, 2, ..., L\}$ of vectors, e.g., 2×2 pixel blocks from a collection of sample images.

2. A distortion measure $d(X_n, \hat{X}_n)$ between input X_n and corresponding reproduction $\hat{X}_n$ such as mean square error.

3. A threshold ϵ.

4. A design rate R (assumed to be a factor of 2 for simplicity).

TSVQ Design Algorithm

Initialization Form the centroid of the entire training sequence: $C_0 = centroid(T)$. (If a mean square error is used, this is simply the Euclidean centroid of the entire training sequence.) This is the rate 0 codebook. Set the rate variable $r = 0$.

Splitting Each of the 2^r terminal leaves in the tree-structured codebook has associated with it a codeword or a "label." The codeword is split into a new two-word codebook by including the given codeword and adding a new one formed by slightly perturbing the original codeword. Two new branches are then extended from the formerly terminal leaf to form two new terminal leaves and the leaves are labeled by the new two codewords. The extended tree now has 2^{r+1} terminal leaves, each corresponding to a codeword.

Lloyd Algorithm 1. Encode the training sequence using the current TSVQ; that is, trace through the tree from the root to the terminal leaf, at each point selecting the node corresponding to the codeword which is the closest in a minimum distortion sense to the input vector. This encoding partitions the input into 2^{r+1} subsequences.

2. Replace the codeword for each terminal node by the centroid of the subsequence mapping into that node. Compute the total distortion $\sum_{n=1}^{L} d(X_n, \hat{X}_n)$ for the resulting encoding. If the ratio of the decrease in distortion to the current distortion is less than the threshold ϵ, then go to the rate increase step. Otherwise go to the Lloyd algorithm Step 1.

Rate Increase If $r = R$, then quit with the final TSVQ. Otherwise go to the Splitting step.

The principal advantage of a TSVQ is that a codebook of 2^R vectors can be searched to find a good codeword by making only $2R$ distortion computations and R comparisons instead of 2^R distortion computations and one R-ary comparison as in full search VQ. The disadvantage is that the search may not find the best codeword, that is, the minimum distortion codeword among all the terminal leaf labels. Furthermore, the terminal leaf codewords designed in this fashion will usually not be as good an overall codebook for the training sequence as one designed with a full search Lloyd algorithm and encoded with a full search encoder. Experience has shown, however, that in many applications TSVQ performs nearly as well as the full search VQ, and the reduction of search complexity from exponential to linear more than compensates for this slight loss in performance. An additional disadvantage of TSVQ is the approximate doubling of required memory. As with most VQ strategies, however, it is assumed that extra memory will be tolerable if the computational complexity can be significantly reduced.

Another approach to TSVQ design is proposed by Makhoul et al. [7]. They propose a different splitting algorithm wherein at the first level of the tree, only one of the two codewords is split: compute the distortion contribution of each codeword and then split only the worst codeword, that is, the one contributing the most to the average distortion. The tree is designed in this fashion, only splitting the single worst codeword in the entire tree, until the leaves hit a predetermined final depth. At this point they are split no longer.

There exist several approaches to designing tree searches matched to a given codebook, possibly designed using a full search Lloyd algorithm. These approaches, however, have not yet proved successful for large codebooks [8].

A powerful advantage of the splitting design technique is that the code is specifically designed for the given tree structure and each new codebook is optimized for the tree labels up to that level. In other words, a TSVQ is a form of successive approximation VQ: each additional bit provides improved performance, at least on average. One application of the successive approximation character is to a progressive transmission image coding system. The basic idea is this: suppose an 8 bit TSVQ is used to encode 2×2 pixel blocks. The overall bit rate is then $8/4 = 2$ bits per pixel (bpp). If each subblock of the image is first encoded only one level deep in the tree and this single bit for each subblock is sent, the decoder can construct a 1/4 bpp reproduction of the entire image. Next if each subblock is encoded one layer deeper and the bits are sent, the intermediate reproduction will be at 1/2 bpp. With each successive group of bits received, the image quality will improve. Thus if the image is being communicated over a slow link, the image can be gradually constructed and improved rather than waiting for the entire bit stream to produce the reproduction. Obvious bit savings are realized in telebrowsing: if the wrong image is sent, the transmission is aborted and the next image will be sent. The appropriateness of TSVQ for progressive transmission image coding was first observed by Tzou [4].

3   Pruned TSVQ

Ordinary TSVQ is a fixed rate VQ in that every input vector is coded with an equal number of bits. A variable rate system has the advantage of being able to use more bits for active or important regions of an image and fewer for low activity or unimportant (e.g., background) regions. Until recently most variable rate schemes were constructed by following a fixed rate compression scheme with a variable rate noiseless code. Such combinations, however, lose the successive approximation characteristic of the fixed rate TSVQ. An alternative approach is to use a PTSVQ. Given a fixed rate TSVQ, a variable rate code can be formed by simply removing or "pruning" terminal leaves. The encoder then traces the best (minimum distortion) path through the tree until a terminal node is reached, which can now be at any depth in the tree. By appropriate choice of binary symbols (e.g., a prefix code), the decoder can still uniquely decode the path map without additional information. The design goal now becomes finding the subtree that is optimum in the sense of minimizing the average distortion for a given average bit rate. In principle this can be found by exhaustive search of all possible subtrees, but this is not in general practicable. An alternative and computationally feasible approach is developed in [5] where it is shown that an extension of an algorithm for pruning binary decision trees due to Breiman, Friedman, Olshen, and Stone [6] can be used to optimally prune TSVQ's. This algorithm, which we shall call the generalized BFOS algorithm, can be summarized as follows.

Let T be a complete tree-structured codebook or, simply, a binary tree with labels on all its leaves. The labels of the terminal leaves constitute the final codebook. Let S be a subtree of that tree, and let us write for this $S \preceq T$. The generalized BFOS algorithm minimizes the tree functional, $J(S) = \delta(S) + \lambda \ell(S)$; here, $\delta(S)$ is the average distortion and $\ell(S)$ is the expected codeword length that would be measured from using S as a TSVQ on the given training sequence. The parameter λ can be interpreted as a Lagrange multiplier that trades off distortion for rate, and it is equal to the magnitude of the ratio of the increase in distortion to the decrease in rate due to pruning off a branch of the tree. The algorithm starts with a large tree and prunes off the branches in order of increasing λ (and decreasing rate) to find the lower convex hull of the operational distortion-rate function

$$\hat{D}_T(R) = \min_{S \preceq T} \{\delta(S) | \ell(S) \le R\}.$$

(Recall that on a lower convex hull of a distortion-rate curve, the slope always decreases in magnitude as the rate increases.) It is proved in [5] that the optimal subtrees of the original tree are nested.

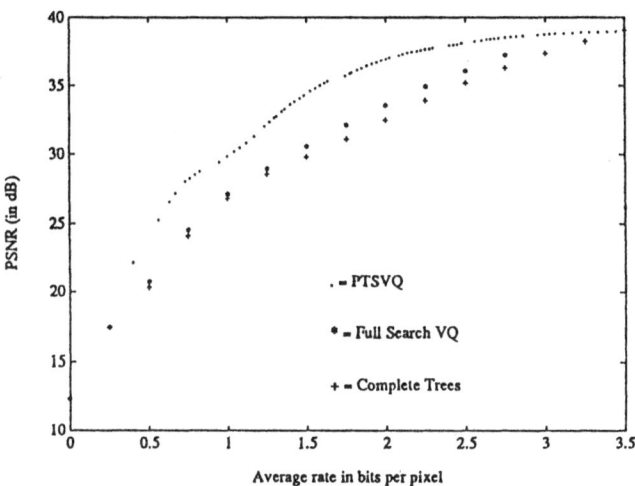

Figure 1: PSNR vs. average rate for medical image sequence.

Hence it is possible to start with the full tree and prune back to the root. This will produce a list of nested subtrees which trace out the vertices of the lower boundary of the convex hull.

An alternative approach proposed by R. Lindsay of Unisys Corp. is to use the Makhoul splitting algorithm and then simply remove terminal leaves in the reverse order in which they are added. This design algorithm is less complex, but it does not have the property of the BFOS algorithm of yielding the provably optimal subtree.

4 PTSVQ Image Coding

PTSVQ experiments have been performed using locally generated data from a General Electric Signa machine. The training sequence consists of 20 magnetic resonance brain scans and the test sequence is made up of 5 magnetic resonance brain scans not in the training sequence. The distortion measure is mean square error.

The vector dimension used is $k = 4$ corresponding to subblocks of 2×2 pixels. The height of the final tree is 14, yielding a maximum rate of 3.5 bpp. The PSNR vs. average rate of the complete TSVQ and PTSVQ has been measured and plotted in Fig.1 along with the corresponding performance of full search VQ for comparison.

Fig. 2 is the original image, a sagittal slice through the brain of a healthy subject, displayed here at 8 bpp. In Fig. 3, this image is quantized with full search VQ at 1.75 bpp. It is quantized at 1.75 bpp with PTSVQ in Fig. 4. The overall quality and the edge reproduction of the image are improved over full search VQ by using PTSVQ.

To use the PTSVQ as a progressive transmission coder, at least three strategies are natural. In the first, one can bit slice flat across the tree, that is, send at a fixed rate until the regions with the lowest instantaneous rate have been completely transmitted. After this, bits are sent only where the instantaneous bit rate is higher until all the bits of the image have been sent. A better way would be to "trace out" the tree at the beginning. In other words, at the beginning, transmit bits

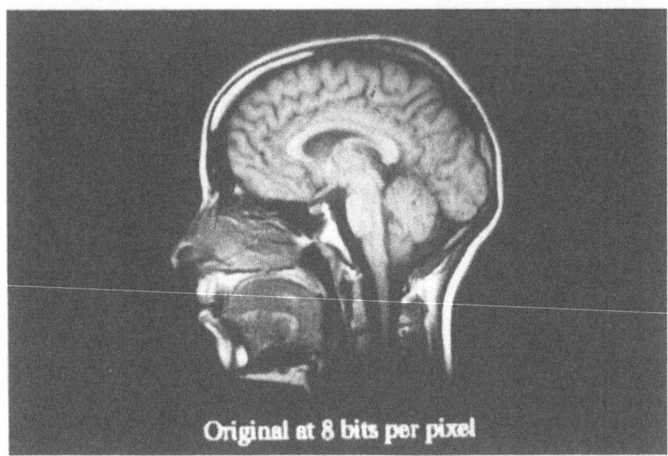

Figure 2: Sagittal image at 8 bpp.

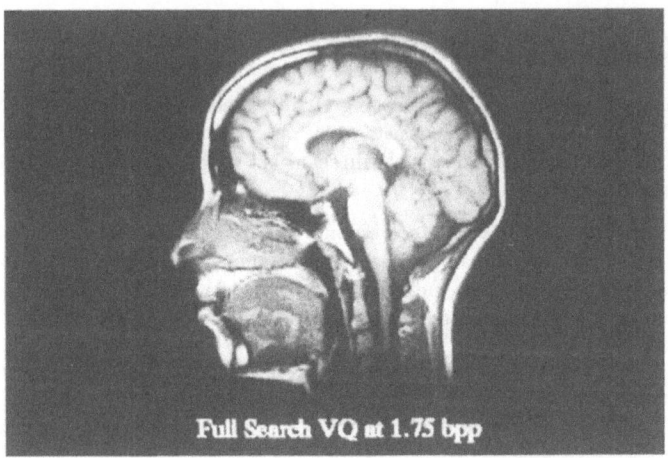

Figure 3: Sagittal image compressed to 1.75 bpp using full search VQ.

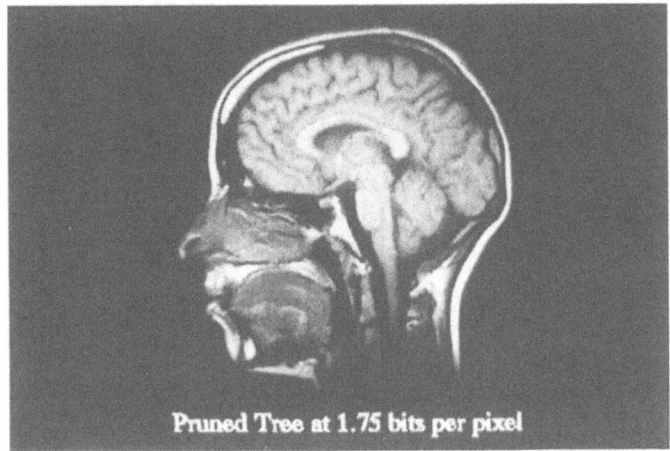

Figure 4: Sagittal image compressed to 1.75 bpp using PTSVQ.

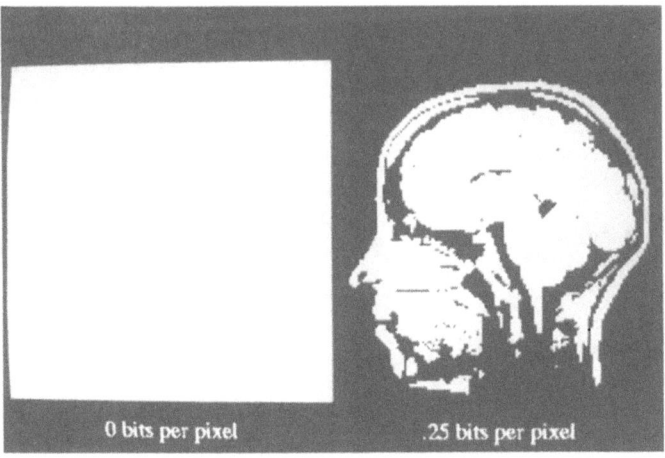

Figure 5: Sagittal image at 0.0 (left) and 0.25 (right) bpp using PTSVQ.

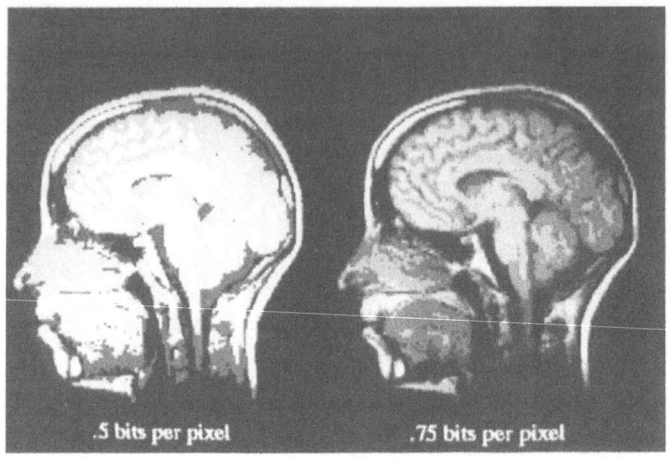

Figure 6: Sagittal image at 0.5 (left) and 0.75 (right) bpp using PTSVQ.

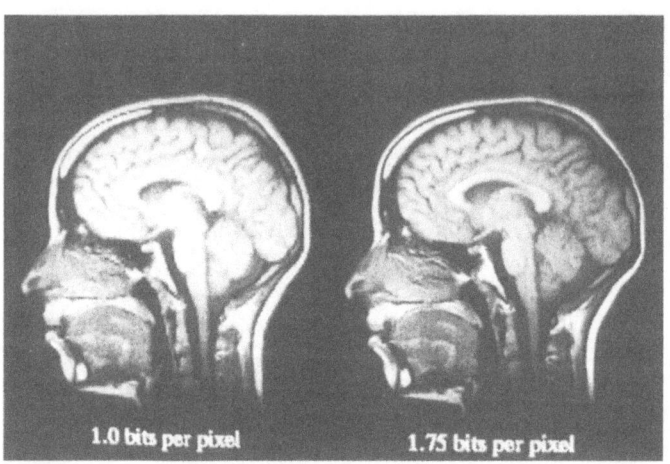

Figure 7: Sagittal image at 1.0 (left) and 1.75 (right) bpp using PTSVQ.

mostly where the instantaneous bit rate is highest and then towards the end, send bits at an equal rate for the entire image. This has the advantage of improved intermediate approximations of the image at the receiver.

The best way to use PTSVQ for progressive transmission is that proposed by Philip A. Chou to the authors. Use the sequence of subtrees generated by the pruning algorithm. As noted before, the sequence of subtrees is nested and each one gives the optimal performance for its given average rate. Therefore, one could select a subset of these trees. Start with the smallest subtree as the encoder. To obtain the next group of bits, begin where the first subtree left off (its terminal nodes) and continue to code the input vectors through to the terminal nodes of the next subtree, and so on. Each intermediate reproduction will be the best possible for the number of bits needed to reproduce it because of the optimal distortion-rate tradeoff used to prune the trees.

For the images shown in this paper, the first method is used because it is simplest to implement.

In Figures 5, 6, and 7, a sequence of progressive reproductions of the image is displayed. The rates shown are 0.0, 0.25, 0.5, 0.75, 1.0, and 1.75 bpp. Clearly, the image quality improves as the bit rate increases, but notice that the image is recognizable even at an early stage.

5 Conclusion

We have reviewed the design and properties of tree-structured vector quantization (TSVQ) and its amenability to progressive transmission image coding systems. By optimally pruning a TSVQ to trade off average distortion for average rate, one obtains a variable rate (or variable length) tree-structured code. This encoder can send more (fewer) bits for more (less) active or important subblocks. The variable rate code can be used in a progressive transmission scheme by sending the first bit for each subblock codeword, then the second (if there is one), and so on. If lossy transmission is required, the VQ code can be followed by a noiseless coding of the residual error to eventually provide a perfect reproduction.

6 Acknowledgement

The authors are with the Information Systems Laboratory, Durand Building, Stanford University, Stanford, CA 94305. They wish to thank Philip A. Chou and Tom Lookabaugh for their help with the applications of the BFOS algorithm and Peter Webb and David Parish for providing the medical images. This work was supported by ESL, Inc., Rockwell International Semiconductor Products Division, and by a Doctoral Fellowship from the American Electronics Association.

References

[1] R. M. Gray, "Vector quantization," *IEEE ASSP Magazine*, vol. 1, pp. 4–29, Apr. 1984.

[2] A. Buzo, A. H. Gray Jr., R. M. Gray, and J. D. Markel, "Speech coding based upon vector quantization," *IEEE Transactions on Information Theory*, vol. 28, pp. 562–574, Oct. 1980.

[3] R. M. Gray, "Tree-searched block codes," in *Proceedings of the 1980 Allerton Conference*, (Allerton, IL), Oct. 1980.

[4] K. Tzou, "Progressive image transmission: a review and comparison of techniques," *Optical Engineering*, vol. 26, pp. 581–589, July 1987.

[5] P. A. Chou, T. Lookabaugh, and R. M. Gray, "Optimal pruning with applications to tree structured source coding and modeling," *IEEE Transactions on Information Theory*, 1988. To appear.

[6] L. Breiman, J. H. Friedman, R. A. Olshen, and C. J. Stone, *Classification and Regression Trees. The Wadsworth Statistics/Probability Series*, Belmont, California: Wadsworth, 1984.

[7] J. Makhoul, S. Roucos, and H. Gish, "Vector quantization in speech coding," *Proceedings of the IEEE*, vol. 73, pp. 1551–1588, 1985.

[8] D. Cheng and A. Gersho, "A fast codebook search algorithm for nearest neighbor pattern matching," in *Proceedings of ICASSP*, pp. 265–268, IEEE Acoustics Speech and Signal Processing Society, 1986.

PATH MAP SYMBOL RELEASE ALGORITHMS AND
THE EXPONENTIAL METRIC TREE

W. W. Chang and Jerry D. Gibson
Texas A&M University

Introduction. Tree encoding is known to be capable of performing arbitrarily close to the rate distortion bound for any memoryless source and single–letter fidelity criterion. It employs a multipath search which pursues some or all of the paths in the code tree and chooses among them the best path at some search depth L. Designs of such a source encoder usually involve choosing a suitable code generator, an efficient tree search algorithm, and an appropriate distortion measure. In addition, it is also important to specify a rule to release the path map symbols to the channel.

This paper addresses the problem of path map symbol release algorithm performance for the exponential metric tree. We present simulation results of the average single–letter distortion performance of various symbol release algorithms with exhaustive search and compare these numerical results with those given by theoretical analyses.

The organization of this paper is as follows. Section II gives a general description of tree coding and defines some terminology. Section III presents the theoretical analyses and simulation results of various path map symbol release algorithms, including incremental encoding, block encoding, and two variable symbol release rules. Section IV concludes that the single symbol release algorithm performs the best.

Conventional Tree Coders. Tree coders employ encoding delay to provide a multipath search capability, and all of their possible quantized output sequences can be envisioned as a tree structure which consists of branches and nodes. There is a fixed number of N branches emanating from each node, each of which will terminate in another node in one sampling interval. In a code tree, each branch is labelled with β branch letters chosen from a reconstruction alphabet. We define the encoding rate in bits per sample as $(1/\beta) \log_2 N$. A path is a sequence of connected branches as the tree encoding progresses in time and can be uniquely located by the path map symbols that specify which of the N emanating branches is to be followed to the node at the next level. Instead of transmitting branch letters, the tree coders transmit these path map symbols. To gain insight into path map symbol release rules, the metric tree is introduced to be used in analyzing tree coder performance. It differs from the code tree only in that its branch is labelled with the metric increment which quantifies the distortion associated with the reconstruction value. The path metric is defined as the cumulative metric increment through the path.

A tree coder has four elements: a code generator, a tree search algorithm, a distortion measure, and a path map symbol release algorithm, as shown in Fig. 1. Given an L source letter sequence $\mathbf{s} = s_1, s_2, \ldots, s_L$, the code generator will map all possible path maps to the branch letter sequence $\hat{\mathbf{s}} = \hat{s}_1, \hat{s}_2, \ldots, \hat{s}_L$. The code generators can be classified to have deterministically populated or stochastically populated tree codes. The tree codes associated with conventional source encoders such as pulse code modulation (PCM), differential PCM (DPCM), and adaptive DPCM (ADPCM), belong to the class of deterministic tree codes. They are simple to implement, but they also provide the least gains over the conventional source encoder. The stochastic tree coders use appropriately chosen random variables to populate the reconstruction samples and asymptotically in L provide better performance.

The distortion measure calculates the path metric $e(\mathbf{s}, \hat{\mathbf{s}})$ that quantifies how well the reconstruction for a given path map approximates the source letters. The most popular single–letter fidelity criterion is the mean squared error. The tree search algorithm finds the path with the least path metric by sequentially feeding different path maps to the code generator and using the distortion measure to evaluate the reconstructed $\hat{\mathbf{s}}$ until the best match is found. Some popular tree search algorithms are the Viterbi algorithm (exhaustive search), the (M, L) algorithm, the stack algorithm, and the 2–cycle algorithm. The most basic of all is the Viterbi algorithm, which sets the upper bound on the performance for all other search algorithms. It searches all possible branches of the code tree to depth L, and hence the best $\hat{\mathbf{s}}$ is always found.

The path map symbol release algorithm specifies the rule to release the path map symbols to the channel. Most of the previous research has involved incremental encoding (or block encoding) in that no matter how deep the tree is searched, a single path map symbol (or the whole block) is released at a time. Most previous investigations are concerned with the selection of a suitable tree code generator and an efficient tree search algorithm for some chosen fixed symbol release rule. We concentrate on the path map symbol release algorithm performance here.

Path Map Symbol Release Algorithms. The theoretical analysis of path map symbol release algorithm performance is complicated by the fact that even an elementary memoryless data source and a simple code generator have a complex metric tree structure. To permit a theoretical analysis, we apply various path map symbol release algorithms to a binary metric tree in which the per–level metric increase μ is distributed as the exponential density. As shown in Appendix A, this exponential metric tree corresponds to the actual situation of applying binary, rate–1/2 tree encoding to a memoryless Gaussian source. In this paper, we study various symbol release algorithms on a binary L–depth metric tree in which the increase in path metric per level is distributed as the exponential density e^{-x}, $x \geq 0$, also studied by Bodie [2]. We define λ_k as the least path metric attainable with an exhaustive search from level k to level L. Some density functions of λ_k are derived in Appendix B.

<u>Fixed symbol release algorithm.</u> First, we study the fixed symbol release algorithm in which no matter how deep the tree is searched, a fixed number of path map symbols are released after each search. Both incremental and block tree encoding are special cases of this algorithm, which release 1 and L symbols, respectively. Some theoretical analyses of the fixed symbol release

algorithm performance are presented in Appendix C. The data shown in Table C.2 are for the L–depth exponential metric tree using the exhaustive tree search. The parameter D_j is the average per–letter distortion attainable with j number of path map symbols released. The leftmost column ($j = 1$) and rightmost column ($j = L$) correspond to incremental and block encoding, respectively. Notice that none of the multiple symbol release algorithms is able to outperform the single symbol release algorithm. Simulation results for blocklength from 1 to 12 are shown in Table 1. Compared with Table C.2, the theoretical and experimental values of average distortion agree very closely.

Actually, the performance is degraded with the increase of the fixed symbol release number. The reason for this is that some path switching, which occurs whenever the coder finds a better path than the one it is pursuing, is desirable to have better performance. The fixed symbol release algorithm does not recognize this condition and misses the opportunity to switch to a better path [3]. Also, the block tree coding resets the decoder to the zero state before each search and hence has the problem of locally large distortion near the edges of the block.

Variable Symbol Release Algorithm.   As noted by Gray [1], a variable symbol release algorithm should be employed to stay on a good path long enough to achieve the promised long–term fidelity. The logic behind this approach is that the first step of the good path, which has the least path metric, may be a poor one with large sample distortion. Two variable symbol release rules are investigated. One rule releases the path map symbols on the best path until the running average distortion is less than or equal to the long–term average distortion, while the other releases until the level which has the least running average distortion. Goris [3] suggested that we should have the maximum number of symbols released constrained to $J = \lfloor L/2 \rfloor$, the largest integer that less than $L/2$, and we adopt this rule here. The following procedures are used.

1) Exhaustively search for the best L–depth path with the minimum path metric, calculate the long–term average distortion E_{mmse} and the running average distortion at jth level E_j, $j = 1, 2, \ldots, J$,

$$E_{mmse} = \frac{1}{L} \sum_{i=1}^{L} (s_i - \hat{s}_i)^2$$

and

$$E_j = \frac{1}{j} \sum_{i=1}^{j} (s_i - \hat{s}_i)^2.$$

2) Select one variable symbol release rule,

> Rule 1: Release the path map symbols until the ith level whose running average distortion E_i is less than or equal to E_{mmse} [3].
>
> Rule 2: Search for the minimum running average distortion E_i at level i. If E_i is less than E_{mmse}, then release i path map symbols. If E_i is greater than E_{mmse}, then release

only one symbol.

3) If the number of path map symbols released in Step 2 reaches the constraint J, then stop sending.

We have analyzed both of the variable symbol release rules for blocklength 4 and 5 in Appendix D. As shown in Table D.1, both of the variable symbol release algorithms are unable to outperform the single symbol release algorithm, and Rule 2 has better performance than Rule 1. This implies that the first step on the best path has a good sample distortion. Table 2 shows the simulation results for blocklengths from 4 to 12. Compared with Table D.1, the theoretical prediction and simulation results are very close.

Conclusion.   We have shown that the single symbol release algorithm performs the best with exhaustive search on the exponential metric tree. Fixed multiple symbol release algorithm performance is degraded with the increase of released symbols by the fact that it may miss desirable path switching and has the problem of locally large distortion near the edges of the block. Variable symbol release algorithms fail to outperform the single symbol release algorithm, but their performance difference is small. Hence, one possible use of the variable symbol release algorithm is in the reduction of computational load when selecting an exhaustive search.

APPENDIX A
ONE CODE TREE WITH EXPONENTIAL METRIC TREE

To simplify the theoretical analyses, we have assumed a binary metric tree in which the metric increase μ per level is distributed as the exponential density. There exists a practical situation where tree encoding an actual data source will generate the exponential metric tree. This occurs when we apply tree encoding to a memoryless Gaussian source, whose source letters are distributed independently with normal density $N(0, \frac{1}{2})$, and use a half–rate code tree structure with two branches emanating from each node, each branch associated with two branch letters, which are assumed to be chosen from the same $N(0, \frac{1}{2})$ as the source letters. We denote the two reconstruction letters on one branch as $\hat{S}_1$ and $\hat{S}_2$, corresponding to the source letters S_1 and S_2, and denote X_i as the difference between S_i and $\hat{S}_i$. We also define the metric increase per level $\mu = X_1^2 + X_2^2$.

Since both source letters and branch letters have the identical independent density function $N(0, \frac{1}{2})$, their difference X_i has the density $N(0, 1)$. It is known that if $X_1, \ldots, X_n$ is a random sample from a normal distribution with mean ξ and variance σ^2, then $U = \sum_{i=1}^{n}(X_i - \xi)^2/\sigma^2$ has a chi–square distribution with n degrees of freedom [4]. Hence the metric increase per level μ will be distributed as χ_2^2, chi–square density with 2 degrees of freedom. This is also the exponential density with parameter $\lambda = \frac{1}{2}$

$$f_\mu(\mu) = \frac{1}{2}e^{-\frac{1}{2}\mu} \qquad \mu \geq 0.$$

APPENDIX B
EXHAUSTIVE SEARCH ON THE EXPONENTIAL METRIC TREE

We assume a binary L-depth metric tree in which the increase in path metric per level (μ, the metric increase) is distributed as the exponential density. We define λ_k as the least path metric attainable with an exhaustive search from level k to level L. Extending the search one level further back involves choosing for each node the branch which contributes to the smaller path metric from level $k-1$ to level L. These path metrics are the sum of two independent components λ_k and μ, and their density functions equal the convolution of their respective density functions,

$$f_{\lambda_k'}(x) = f_{\lambda_k}(x) * f_\mu(x).$$

Then the λ_{k-1}, the least path metric from level $k-1$ to level L, has the density of the minimum of two random variables which are distributed as $f_{\lambda_k'}(x)$,

$$f_{\lambda_{k-1}}(x) = 2f_{\lambda_k'}(x)[1 - F_{\lambda_k'}(x)].$$

Since we always have $\lambda_L = 0$, $f_{\lambda_L} = \delta(x)$ at the last level of the tree. Starting from the last level, we may iterate the above two equations to obtain the density function of λ_k at any level.

Thus,

$$f_{\lambda_L}(x) = \delta(x);$$

$$f_{\lambda_{L-1}}(x) = 2e^{-2x};$$

$$f_{\lambda_{L-2}}(x) = 8e^{-2x} - 12e^{-3x} + 4e^{-4x};$$

$$f_{\lambda_{L-3}}(x) = 22.2e^{-2x} - 80e^{-3x} + 117.3e^{-4x} - 91.1e^{-5x} + 40e^{-6x}$$
$$- 9.3e^{-7x} + 0.9e^{-8x};$$

$$f_{\lambda_{L-4}}(x) = 52.6e^{-2x} - 341.8e^{-3x} + 1040.7e^{-4x} - 1982.8e^{-5x} + 2650.6e^{-6x}$$
$$- 2629.5e^{-7x} + 1992e^{-8x} - 1167.7e^{-9x} + 531.1e^{-10x} - 186.1e^{-11x}$$
$$+ 49.4e^{-12x} - 9.6e^{-13x} + 1.3e^{-14x} - 0.1e^{-15x};$$

$$f_{\lambda_{L-5}}(x) = 113.5e^{-2x} - 1188.1e^{-3x} + 6197e^{-4x} - 21507.5e^{-5x} + 55789.2e^{-6x}$$
$$- 114968.3e^{-7x} + 195238.6e^{-8x} - 279807.2e^{-9x} + 344013e^{-10x}$$
$$- 367097e^{-11x} + 342883.2e^{-12x} - 282050.3e^{-13x} + 205210.4e^{-14x}$$
$$- 132438e^{-15x} + 75941.9e^{-16x} - 38712.1e^{-17x} + 17534.9e^{-18x}$$
$$- 7047.2e^{-19x} + 2506.6e^{-20x} - 786.2e^{-21x} + 216.4e^{-22x}$$
$$- 51.9e^{-23x} + 10.8e^{-24x} - 1.9e^{-25x} + 0.3e^{-26x}.$$

APPENDIX C

FIXED SYMBOL RELEASE ALGORITHM PERFORMANCE

We denote $\bar{\lambda}_k$ as the expected value of λ_k, the least path metric attainable with an exhaustive search from level k to level L. From the density f_{λ_k} given in Appendix B, we can calculate both

λ_k and D_j. See Tables C.1 and C.2.

$\bar{\lambda}_k$: expected value of the path metric from level k to level L.

D_j: the least per–letter average distortion attainable with the exhaustive search and j fixed symbol release algorithm.

$$\bar{\lambda}_k = E(\lambda_k) = \int_{-\infty}^{+\infty} x f_{\lambda_k}(x)\, dx;$$
$$D_j = (\,\bar{\lambda}_0 - \bar{\lambda}_j\,)\,/\,j.$$

APPENDIX D

VARIABLE SYMBOL RELEASE ALGORITHM PERFORMANCE

A variable symbol release algorithm is employed to stay on a good path long enough to achieve the promised long–term fidelity. Two variable symbol release rules are investigated here.

Rule 1: Release the path map symbols until the running average distortion is less than or equal to the long–term average distortion, or reaches the release constraint $\lfloor L/2 \rfloor$.

Rule 2: Release the path map symbols until the level which has the smallest running average distortion, or the release constraint $\lfloor L/2 \rfloor$ is reached. Only release one symbol when the smallest running average distortion is greater than the long–term average distortion.

A. Depth–4 Exponential Metric Tree

1). Performance of Rule 1:

We denote R_i as the event that i path map symbols are decided to be released and set $u_1 = 3\lambda_0, u_2 = 4\lambda_1$, and $y = u_1 - u_2$.

$$P(R_1) = P[\,(\lambda_0 - \lambda_1) \le \frac{\lambda_0}{4}\,]$$
$$= P[\,3\lambda_0 - 4\lambda_1 \le 0\,] = P[\,(u_1 - u_2) \le 0\,]$$
$$= \int_{-\infty}^{0} \int_{-3y}^{+\infty} f_{u_1,u_2}(u_1, u_1 - y)\, du_1\, dy = 0.535347,$$

where

$$f_{u_1,u_2} = f_{u_1|u_2} \cdot f_{u_2}$$
$$= e^{\frac{-2u_1}{3}}[\,3.7 - 13.3e^{\frac{-u_2}{4}} + 19.6e^{\frac{-u_2}{2}} - 15.2e^{\frac{-3u_2}{4}} + 6.7e^{-u_2}$$
$$- 1.6e^{\frac{-5u_2}{4}} + 0.15e^{\frac{-3u_2}{2}}\,],$$
$$u_1 \ge 0, \quad u_2 \ge 0, \quad u_1 \ge \frac{3}{4}u_2.$$

The least average path metric attainable with Rule 1 is

$$D = D_1 \cdot P(R_1) + D_2 \cdot P(R_2)$$
$$= 0.3462 \times 0.535347 + 0.3598 \times 0.464653 = 0.3525.$$

2). Performance of Rule 2:

$$P(R_2) = P[\; \frac{\lambda_0 - \lambda_2}{2} \leq \lambda_0 - \lambda_1 \quad \text{and} \quad \frac{\lambda_0 - \lambda_2}{2} \leq \frac{\lambda_0}{4}\;]$$

$$= P[\; (\lambda_0 - 2\lambda_1 + \lambda_2) \geq 0 \quad \text{and} \quad -\lambda_0 + 2\lambda_2 \geq 0\;]$$

$$= \int\int\int f_{\lambda_0,\lambda_1,\lambda_2}(x,y,z)\, dz\, dy\, dx$$

where

$$f_{\lambda_0,\lambda_1,\lambda_2} = f_{\lambda_0|\lambda_1,\lambda_2} \cdot f_{\lambda_1|\lambda_2} \cdot f_{\lambda_2}$$

$$= [2e^{-2(\lambda_0 - \lambda_1)}] \cdot [2e^{-2(\lambda_1 - \lambda_2)}] \cdot [8e^{-2\lambda_2} - 12e^{-3\lambda_2} + 4e^{-4\lambda_2}]$$

$$= [32e^{-2\lambda_0} - 48e^{-2\lambda_0 - \lambda_2} + 16e^{-2\lambda_0 - 2\lambda_2}], \qquad \lambda_0 \geq \lambda_1 \geq \lambda_2 \geq 0.$$

The upper and lower limits in the integral must be chosen to satisfy the following three constraints :

i). $\quad x \geq y \geq z \geq 0$;

ii). $\quad -x + 2z \geq 0 \qquad \text{or} \quad z \geq \frac{x}{2}$;

iii). $\quad x - 2y + z \geq 0 \qquad \text{or} \quad z \geq (2y - x)$.

There exist two possible conditions:

$$Event\; A : z \geq (2y - x) \geq \frac{x}{2} \qquad \text{and} \quad x \geq y \geq z \geq 0;$$

$$Event\; B : z \geq \frac{x}{2} \geq (2y - x) \qquad \text{and} \quad x \geq y \geq z \geq 0.$$

Hence

$$P(event\; A) = P[\; (2\lambda_1 - \lambda_0) \geq \frac{\lambda_0}{2}\;]$$

$$= P[\; (3\lambda_0 - 4\lambda_1) \leq 0\;] = 0.535347,$$

$$P(event\; B) = 1 - P(event\; A) = 0.464653,$$

$$P(R_2) = P(\; R_2|event\; A\;) \cdot P(event\; A) + P(\; R_2|event\; B\;) \cdot P(event\; B)$$

$$= [\; \int_0^{+\infty} \int_{\frac{2x}{4}}^{x} \int_{2y-x}^{y} f_{\lambda_0,\lambda_1,\lambda_2}(x,y,z)\, dz\, dy\, dx\;] \cdot P[\; (2y - x) \geq \frac{x}{2}\;]$$

$$+ [\; \int_0^{+\infty} \int_{\frac{x}{2}}^{\frac{3x}{4}} \int_{\frac{x}{2}}^{y} f_{\lambda_0,\lambda_1,\lambda_2}(x,y,z)\, dz\, dy\, dx\;] \cdot P[\; \frac{x}{2} \geq (2y - x)\;]$$

$$= 0.1283 \times 0.535347 + 0.1075 \times 0.464653 = 0.118635.$$

The least average path metric attainable with Rule 2 is

$$D = D_1 \cdot P(R_1) + D_2 \cdot P(R_2)$$

$$= 0.3462 \times 0.881365 + 0.3598 \times 0.118635 = 0.3478.$$

B. Depth–5 Exponential Metric Tree

1). Performance of Rule 1:

We set $u_1 = 4\lambda_0$, $u_2 = 5\lambda_1$, and $y = u_1 - u_2$, so

$$P(R_1) = P[\,(\lambda_0 - \lambda_1) \leq \frac{\lambda_0}{5}\,]$$
$$= P[\,4\lambda_0 - 5\lambda_1 \leq 0\,] = P[\,(u_1 - u_2) \leq 0\,]$$
$$= \int_{-\infty}^{0} \int_{-4y}^{+\infty} f_{u_1,u_2}(u_1, u_1 - y)\,du_1\,dy = 0.46916,$$

where

$$f_{u_1,u_2} = f_{u_1|u_2} \cdot f_{u_2}$$
$$= e^{\frac{-u_1}{2}} \cdot [5.3 - 34.2e^{\frac{-u_2}{5}} + 104e^{\frac{-2u_2}{5}} - 198.3e^{\frac{-3u_2}{5}}$$
$$+ 265e^{\frac{-4u_2}{5}} - 263e^{-u_2} + 199.2e^{\frac{-6u_2}{5}} - 116.8e^{\frac{-7u_2}{5}}$$
$$+ 53e^{\frac{-8u_2}{5}} - 18.5e^{\frac{-9u_2}{5}} + 5e^{-2u_2} - 1.0e^{\frac{-11u_2}{5}} + 0.13e^{\frac{-12u_2}{5}}$$
$$- 0.01e^{\frac{-13u_2}{5}}], \qquad u_1 \geq 0, \quad u_2 \geq 0, \quad u_1 \geq \frac{4}{5}u_2.$$

The least average path metric attainable with Rule 1 is

$$D = D_1 \cdot P(R_1) + D_2 \cdot P(R_2)$$
$$= 0.3275 \times 0.46916 + 0.3369 \times 0.53084 = 0.3325.$$

2). Performance of Rule 2:

$$P(R_2) = P[\,\frac{\lambda_0 - \lambda_2}{2} \leq \lambda_0 - \lambda_1 \quad \text{and} \quad \frac{\lambda_0 - \lambda_2}{2} \leq \frac{\lambda_0}{5}\,]$$
$$= P[\,(\lambda_0 - 2\lambda_1 + \lambda 2) \geq 0 \quad \text{and} \quad (-3\lambda_0 + 5\lambda_2 \geq 0)\,]$$
$$= \int \int \int f_{\lambda_0,\lambda_1,\lambda_2}(x, y, z)\,dz\,dy\,dx,$$

where

$$f_{\lambda_0,\lambda_1,\lambda_2} = f_{\lambda_0|\lambda_1,\lambda_2} \cdot f_{\lambda_1|\lambda_2} \cdot f_{\lambda_2}$$
$$= [2e^{-2(\lambda_0 - \lambda_1)}] \cdot [2e^{-2(\lambda_1 - \lambda_2)}] \cdot [22.2e^{-2\lambda_2} - 80e^{-3\lambda_2} + 117.3e^{-4\lambda_2}$$
$$- 91.1e^{-5\lambda_2} + 40e^{-6\lambda_2} - 9.3e^{-7\lambda_2} + 0.9e^{-8\lambda_2}]$$
$$= e^{-2\lambda_0} \cdot [88.8 - 320e^{-\lambda_2} + 469.3e^{-2\lambda_2} - 364.4e^{-3\lambda_2} + 160e^{-4\lambda_2}$$
$$- 37.2e^{-5\lambda_2} + 3.6e^{-6\lambda_2}].$$

The upper and lower limits in the integral must be chosen to satisfy the following three constraints:

i). $x \geq y \geq z \geq 0$;

ii). $-3x + 5z \geq 0$ or $z \geq \dfrac{3x}{5}$;

iii). $x - 2y + z \geq 0$ or $z \geq (2y - x)$.

There exist two possible conditions:

$$Event\ A : z \geq (2y - x) \geq \frac{3x}{5} \quad \text{and} \quad x \geq y \geq z \geq 0;$$

$$Event\ B : z \geq \frac{3x}{5} \geq (2y - x) \quad \text{and} \quad x \geq y \geq z \geq 0.$$

Hence

$$P(R_2) = P(R_2|event\ A) \cdot P(event\ A) + P(R_2|event\ B) \cdot P(event\ B)$$

$$= [\int_0^{+\infty} \int_{\frac{4x}{5}}^{x} \int_{2y-x}^{y} f_{\lambda_0,\lambda_1,\lambda_2}(x,y,z)\, dz\, dy\, dx\ ] \cdot P[\ (2y - x) \geq \frac{3}{5}x\ ]$$

$$+ [\int_0^{+\infty} \int_{\frac{3x}{5}}^{\frac{4x}{5}} \int_{\frac{3x}{5}}^{y} f_{\lambda_0,\lambda_1,\lambda_2}(x,y,z)\, dz\, dy\, dx] \cdot P[\ (2y - x) \leq \frac{3}{5}x\ ]$$

$$= 0.1313 \times 0.46916 + 0.1023 \times 0.53084 = 0.1159.$$

The least average path metric attainable with Rule 2 is

$$D = D_1 \cdot P(R_1) + D_2 \cdot P(R_2)$$

$$= 0.3275 \times 0.8841 + 0.3369 \times 0.1159 = 0.3285.$$

REFERENCES

[1] R. M. Gray, "Time–invariant trellis encoding of ergodic discrete–time sources with a fidelity criterion," IEEE Trans. Inform. Theory, vol. IT–23, pp. 71–83, Jan. 1977.

[2] J. B. Bodie, "Multi-path tree encoding for analog data sources," Commu. Res. Lab., Mc-Master Univ., Hamilton, O.N., Canada, CRL Internal Rept. Series CRL–20, June 1974.

[3] A. C. Goris and J. D. Gibson, "Incremental Tree Coding of Speech," IEEE Trans. Inform. Theory, vol. IT–27, pp. 511–516, July 1981.

[4] A. M. Mood and F. A. Graybill, Introduction to the Theory of Statistics, Second edition, McGraw–Hill, New York, 1963.

TABLE 1

AVERAGE PER-LETTER DISTORTION FOR
FIXED SYMBOL RELEASE ALGORITHM (Simulation Results)

L	1	2	3	4	5	6	7	8	9	10	11	12
D_1	0.5014	0.4229	0.3799	0.3544	0.3312	0.3232	0.3100	0.2987	0.3030	0.2876	0.2913	0.2857
D_2		0.4659	0.3950	0.3653	0.3416	0.3267	0.3183	0.2964	0.2998	0.2951	0.2915	0.2857
D_3			0.4325	0.3855	0.3557	0.3363	0.3206	0.3091	0.3049	0.2987	0.2940	0.2868
D_4				0.4147	0.3727	0.3454	0.3237	0.3141	0.3070	0.2969	0.2954	0.2908
D_5					0.4049	0.3571	0.3365	0.3223	0.3099	0.3048	0.3014	0.2901
D_6						0.3838	0.3478	0.3298	0.3172	0.3137	0.3009	0.2945
D_7							0.3638	0.3446	0.3227	0.3119	0.3069	0.2952
D_8								0.3598	0.3330	0.3188	0.3117	0.3027
D_9									0.3567	0.3282	0.3173	0.3029
D_{10}										0.3470	0.3213	0.3109
D_{11}											0.3372	0.3158
D_{12}												0.3342

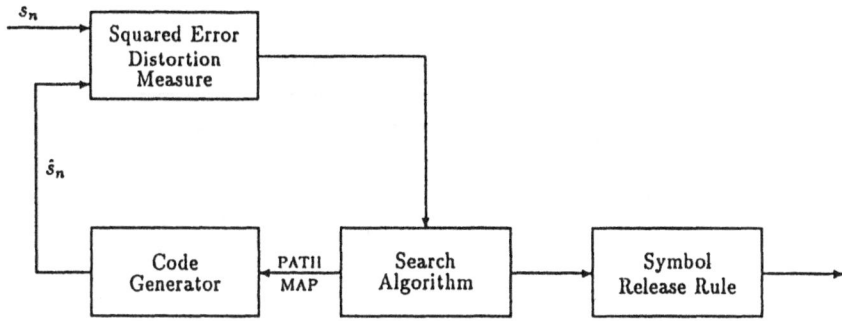

Fig. 1. Tree Coder

TABLE 2

AVERAGE PER–LETTER DISTORTION FOR
VARIABLE SYMBOL RELEASE ALGORITHM
(Simulation Results)

L	Incremental	Block	Rule 1	Rule 2
4	0.3544	0.4147	0.3628	0.3575
5	0.3312	0.4049	0.3396	0.3353
6	0.3232	0.3838	0.3282	0.3302
7	0.3100	0.3638	0.3141	0.3110
8	0.2987	0.3598	0.3063	0.3033
9	0.3030	0.3567	0.3039	0.3031
10	0.2876	0.3470	0.2967	0.2964
11	0.2913	0.3372	0.2947	0.2925
12	0.2857	0.3342	0.2907	0.2870

TABLE C.1

EXPECTED VALUE OF PATH METRIC

L	λ_0	λ_1	λ_2	λ_3	λ_4	λ_5
1	0.5000	0.0000				
2	0.9167	0.5000	0.0000			
3	1.2901	0.9167	0.5000	0.0000		
4	1.6363	1.2901	0.9167	0.5000	0.0000	
5	1.9638	1.6363	1.2901	0.9167	0.5000	0.0000

TABLE C.2

AVERAGE PER–LETTER DISTORTION
(Theoretical Results)

L	D_1	D_2	D_3	D_4	D_5
1	0.5000				
2	0.4167	0.4584			
3	0.3734	0.3951	0.4300		
4	0.3462	0.3598	0.3788	0.4091	
5	0.3275	0.3369	0.3490	0.3660	0.3928

TABLE D.1

AVERAGE PER–LETTER DISTORTION
(Theoretical Results)

L	Incremental	Block	Rule 1	Rule 2
4	0.3462	0.4090	0.3525	0.3478
5	0.3275	0.3928	0.3325	0.3285

STRUCTURE SIMPLIFICATION OF 2-D DIGITAL FILTERS: AN OVERVIEW

W.-S. Lu   E. B. Lee

Dept. of Electrical & Computer Eng.   Dept. of Electrical Engineering
University of Victoria   University of Minnesota
Victoria, B.C., CANADA V8W 2Y2   Minneapolis, MN 55455

I. Introduction

Simplifying the structure of a digital filter is of considerable importance since it could lead to a less expensive hardware implementation as well as higher processing speed. The simplification may be carried out at two different stages of a synthesis procedure, namely during the design and during the realization. When the order of a digital filter obtained by a certain design technique is too high, a re-design may be carried out with less restrictive constraints (design specifications) to get a lower-order, yet satisfactory, filter. An obvious alternative instead of redesigning is to find a lower-order approximation of the original transfer function. This alternative is especially suitable if one wants to design an IIR filter satisfying certain specifications in amplitude response while maintaining linear-phase characteristics over a large frequency range [1]. Once the design is accomplished, further simplication is possible at the realization level. Typically a digital filter is realized using a parallel/cascade structure. Such a structure has been proven successful in reducing the filter's round-off noise power and suitable for parallel processing as well as for VLSI chip implementation. Through a singular-value analysis, it is shown that a parallel/cascade realization for a given digital filter can be constructed such that the significance of each parallel channel in terms of its contribution to the desired frequency response can be appreciated by comparing "its" singular value with others. Consequently, the structure may be simplified by deleting those least significant channels.

While both 1-D and 2-D digital filters can be treated as discrete dynamic systems, 2-D discrete systems are of infinite-dimension and therefore technically more difficult to deal with. It is only recently that results on order reduction and section reduction have begun to appear ([2]-[10] and references therein).

Here we intend to provide the reader with an up-to-date survey on various issues regarding structure simplification of 2-D discrete systems at modelling as well as realization levels. Particular attention will be given to stability of the simplified filters, approximation error bounds, and computational issues. In the next section, order reduction techniques such as balanced approximation and the use of Davis-Kahan-Wienberger theorem for FIR and IIR 2-D filters are addressed. Attractive features of the balanced approximation include computational efficiency, good phase characterization and stability of the reduced filter for the FIR case as well as the quadrantally symmetric IIR case. The DKW theorem is found useful in finding a lower-order FIR filter that approximates the original FIR filter. In section III, the task of reducing realization channels for a given 2-D filter is considered. For the FIR case, the classic Eckart-Young-Mirsky theorem [1] is used to show that the SVD reduction is the best in L_2-sense and that the same method also minimizes the L_∞-norm of an upper bound of the reduction error. For the IIR case, it is pointed out that an L_2 suboptimal reduction may be obtained through an error analysis of a relevant infinite-dimensional matrix in conjunction with certain 1-D model reduction techniques.

II. Order Reduction

A balanced realization has been found useful in reducing the order of a 1-D dynamic system model [12]. In the 2-D case, the concept of balancing a system model may be considered either globally [13] or locally [7]. Attractive features of locally balanced 2-D systems models include computational efficiency and its application to analysis and synthesis of 2-D digital filters.

Like the 1-D case, balancing a 2-D system model is carried out in a 2-D state-space setting e.g. using Roesser's local state-space model [14] where a system is represented as by the two-dimensional difference equation (discrete-time)

$$\begin{bmatrix} x^h(i+1,j) \\ x^v(i,j+1) \end{bmatrix} = \begin{bmatrix} A_1 & A_2 \\ A_3 & A_4 \end{bmatrix} \begin{bmatrix} x^h(i,j) \\ x^v(i,j) \end{bmatrix} + \begin{bmatrix} b_1 \\ b_2 \end{bmatrix} u(i,j) \overset{\Delta}{=} Ax + bu \quad (1a)$$

$$y(i,j) = [\; C_1 \quad C_2 \;] \begin{bmatrix} x^h(i,j) \\ x^v(i,j) \end{bmatrix} + du(i,j) \overset{\Delta}{=} cx + du . \quad (1b)$$

Here $x^h(i,j) \epsilon R^{n_1}$ and $x^v(i,j) \epsilon R^{n_2}$ form the local state for the system at (i,j). The 2-D z transform of above state equations yields the transfer function

$$Q(z_1,z_2) = c(I(z_1,z_2)-A)^{-1} b + d \overset{\Delta}{=} \frac{n(z_1,z_2)}{p(z_1,z_2)} \quad (2)$$

where $I(z_1,z_2) = z_1 I_{n_1} \oplus z_2 I_{n_2}$, $p(z_1,z_2) = det[I(z_1,z_2)-A]$. Throughout the paper we assume that

$$p(z_1,z_2) \neq 0 \quad \text{for} \quad (z_1,z_2) \epsilon \{(z_1,z_2): \; |z_1| > 1, \; |z_2| > 1\} \quad (3)$$

to guarantee the BIBO stability of the system modeled as equation (1).

Let

$$f(z_1,z_2) = [I(z_1,z_2) - A]^{-1} b,$$

and

$$g(z_1,z_2) = c[I(z_1,z_2) - A]^{-1} ,$$

then the generalized reachability and observability gramians are defined as

$$K = \frac{1}{(2\pi j)^2} \oint_{|z_1|=1} \oint_{|z_2|=1} f(z_1,z_2) f^*(z_1,z_2) \frac{dz_1}{z_1} \frac{dz_2}{z_2} \quad (4a)$$

and

$$W = \frac{1}{(2\pi j)^2} \oint_{|z_1|=1} \oint_{|z_2|=1} g^*(z_1,z_2)g(z_1,z_2) \frac{dz_1}{z_1} \frac{dz_2}{z_2} \tag{4b}$$

respectively. Further denote the n_1-dimensional upper left blocks and the n_2-dimensional lower left blocks of K and W_2 by K_1,K_2 and W_1,W_2, respectively. It is known that the eigenvalues of K_1W_1 and K_2W_2 are invariant udner 2-D similarity transformations, and $K_i(W_i)$ for i = 1,2 are positive definite if system model (1) is locally reachable (observable) [7].

System model (1) is said to be (locally) balanced if

$$K_{11} = W_{11} = \text{diag}(\sigma_{11},\sigma_{12},\ldots,\sigma_{1n_1})$$

and

$$K_{22} = W_{22} = \text{diag}(\sigma_{21},\sigma_{22},\ldots,\sigma_{2n_2})$$

with $\quad \sigma_{11} > \cdots > \sigma_{1n_1} > 0 \quad , \quad \sigma_{21} > \cdots > \sigma_{2n_2} > 0 \quad .$

If system model (1) is not locally balanced, then a 2-D balancing transformation $T = T_1 \oplus T_2$ can be found (assuming (1) is locally reachable and observable) such that $(T^{-1}AT, T^{-1}b, CT, d)$ is locally balanced. A balancing transformation T can be computed by using K_{ii} and W_{ii} (i=1,2) through any reliable algorithm for 1-D balancing transformation [15]. Once a balanced realization, say $(\hat{A},\hat{b},\hat{c},d)$, is found, a reduced state-space model (A_r,b_r,c_r) of order (r_1,r_2) can be obtained by a subpartitioning of $(\hat{A},\hat{b},\hat{c})$ as

$$\hat{A} = \left[\begin{array}{cc|cc} A_{1r} & * & A_{2r} & * \\ \hline * & * & * & * \\ \hline A_{3r} & * & A_{4r} & * \\ \hline * & * & * & * \end{array}\right]\begin{array}{l}\}r_1 \\ \\ \}r_2 \\ \\ \end{array} \quad , \quad \hat{b} = \left[\begin{array}{c} b_{1r} \\ \hline * \\ \hline b_{2r} \\ \hline * \end{array}\right]\begin{array}{l}\}r_1 \\ \\ \}r_2 \\ \end{array} \quad \text{and} \quad \hat{c} = [\begin{array}{cc|cc} c_{1r} & * & c_{2r} & * \end{array}] \tag{5}$$

and then taking as the approximation the triple

$$A_r = \left[\begin{array}{c|c} A_{1r} & A_{2r} \\ \hline A_{3r} & A_{4r} \end{array}\right] \quad , \quad b_r = \left[\begin{array}{c} b_{1r} \\ \hline b_{2r} \end{array}\right] \quad , \quad \text{and} \quad c_r = [\begin{array}{c|c} c_{1r} & c_{2r} \end{array}]. \tag{6}$$

In what follows particular attention will first be given to the application of this reduction approach to FIR 2-D filters. A discussion on the stability issue for the IIR case will then follow to introduce the concept of 2-D bounded-real-balancing.

2.1 The FIR Case

Let $Q(z_1,z_2)$ be the transfer function of a FIR filter of order (n_1,n_2) i.e.

$$Q(z_1,z_2) = \sum_{j=0}^{n_2} \sum_{i=0}^{n_1} f_{ij}z_1^{-i}z_2^{-j} . \tag{7}$$

In Roesser's local state-space, (7) can be modeled by equation (1) with

$$A = \begin{bmatrix} 0 & 1 & 0 & \ldots & . & 0 & \vdots & f_{1n_2} & \ldots & \ldots & f_{11} \\ . & . & . & & . & . & \vdots & & & & \\ 0 & 0 & 0 & . & . & 1 & \vdots & . & . & . & . \\ 0 & 0 & 0 & . & . & 0 & \vdots & f_{n_1 n_2} & \ldots & \ldots & f_{n_1 1} \\ \hdashline & & & & & & \vdots & 0 & 1 & 0 & \ldots & . & 0 \\ & & 0 & & & & \vdots & . & . & . & . \\ & & & & & & \vdots & 0 & 0 & 0 & . & . & 1 \\ & & & & & & \vdots & 0 & 0 & 0 & . & . & 0 \end{bmatrix} , \quad b = \begin{bmatrix} f_{10} \\ . \\ f_{n_1 0} \\ \hdashline 0 \\ \vdots \\ 0 \\ 1 \end{bmatrix} , \tag{8}$$

$$c = [1 \ 0 \ \ldots \ 0 \ \vdots \ f_{0n_2} \ \ldots \ f_{01}], \quad \text{and} \quad d = f_{00} .$$

Note that in (8) $A_3 = 0$ meaning that the filter is separable. In addition, both A_1 and A_4 are nilpotent since $A_1^{n_1} = 0$ and $A_4^{n_2} = 0$, which considerably simplifies the procedure of solving the two Lyapunov equations described below.

To compute K_{ii} and W_{ii} $(i=1,2)$, we write $Q(z_1,z_2)$ as

$$Q(z_1,z_2) = \sum_{i=0}^{n_1} (\sum_{j=0}^{n_2} f_{ij} z_2^{-j}) z_1^{-i} \overset{\Delta}{=} \sum_{i=0}^{n_1} b_{1i}(z_2) z_1^{-i}$$

$$= \sum_{j=0}^{n_2} (\sum_{i=0}^{n_1} f_{ij} z_1^{-i}) z_2^{-j} \overset{\Delta}{=} \sum_{j=0}^{n_2} c_{2j}(z_1) z_2^{-j} \tag{9}$$

and define

$$A_1(z_2) = A_1 + A_2(z_2 I - A_4)^{-1} A_3$$
$$b_1(z_2) = b_1 + A_2(z_2 I - A_4)^{-1} b_2$$
$$c_1(z_2) = c_1 + c_2(z_2 I - A_4)^{-1} A_3$$

$$\tag{10}$$

$$A_2(z_1) = A_4 + A_3(z_1 I - A_1)^{-1} A_2$$
$$b_2(z_1) = b_2 + A_3(z_1 I - A_1)^{-1} b_1$$
$$c_2(z_1) = c_2 + c_1(z_1 I - A_1)^{-1} A_2 .$$

It is known [7] that K_{11} can be found through evaluation of the integral

$$K_{11} = \frac{1}{2\pi j} \oint_{|z_2|=1} K_1(z_2) \frac{dz_2}{z_2} \tag{11}$$

where $K_1(z_2)$ is the positive-definite Hermitian solution of the Lyapunov equation

$$A_1(z_2) K_1(z_2) A_1^*(z_2) - K_1(z_2) = -b_1(z_2) b_1^*(z_2) . \tag{12}$$

For the FIR case, $A_1(z_2) = A_1$ and

$$b_1(z_2) = \begin{bmatrix} b_{11}(z_2) \\ \\ b_{1n_1}(z_2) \end{bmatrix}$$

where $b_{1i}(z_2)$ are defined by (9). Hence (12) becomes

$$A_1 K_1(z_2) A_1^T - K_1(z_2) = -b_1(z_2) b_1^*(z_2) \quad .$$

Since A_1 is nilpotent, the solution $K_1(z_2)$ is given by

$$K_1(z_2) = \sum_{i=0}^{n_1-1} A_1^i b_1(z_2) b_1^*(z_2) (A_1^T)^i \quad .$$

Therefore

$$K_{11} = \sum_{i=0}^{n_1-1} A_1^i P_1 (A_1^T)^i \tag{13a}$$

with

$$P_1 = \frac{1}{2\pi j} \oint_{|z_2|=1} b_1(z_2) b_1^*(z_2) \frac{dz_2}{z_2} \quad .$$

A bit of computation shows that

$$P_1 = F_b F_b^T \tag{13b}$$

where F_b is defined as

$$F_b = \begin{bmatrix} f_{10} & f_{11} & \cdots & f_{1n_2} \\ f_{20} & f_{21} & \cdots & f_{2n_2} \\ \cdot & \cdot & \cdot & \cdot & \cdot \\ f_{n_10} & f_{n_11} & \cdots & f_{n_1n_2} \end{bmatrix} \quad . \tag{13c}$$

Furthermore, since both $A_2(z_1) = A_4$ and $b_2(z_1) = b_2$ are independent of z_1 and of the filter coefficients, K_{22} is the positive-definite solution of the Lyapunov equation [7]:

$$A_4 K_{22} A_4^T - K_{22} = -b_2 b_2^T$$

i.e.

$$K_{22} = \sum_{j=0}^{n_2} A_4^j b_2 b_2^T (A_4^T)^j = I_{n_2} \quad . \tag{13d}$$

Similarly it can be shown that

$$W_{11} = I_{n_1} \tag{14a}$$

and

$$W_{22} = \sum_{j=0}^{n_2} (A_4^T)^j P_2 A_4^j \tag{14b}$$

with

$$P_2 = F_c^T F_c \tag{14c}$$

and

$$F_c = \begin{bmatrix} f_{01} & f_{02} & \cdots & f_{0n_2} \\ f_{11} & f_{12} & \cdots & f_{1n_2} \\ \cdot & \cdot & \cdot & \cdot \\ f_{n_11} & f_{n_12} & \cdots & f_{n_1n_2} \end{bmatrix} . \tag{14d}$$

Note also that matrices F_b and F_c are formed using a part of the filter parameters $\{f_{ij}, 0 < i < n, 0 < j < n_2\}$ and can be obtained by properly truncating the whole coefficient matrix $F = (f_{ij}), 0 < j < n_2$ as shown in diagram below:

$$F = \begin{bmatrix} f_{00} & f_{01} & f_{02} & \cdots & f_{0n_2} \\ f_{10} & f_{11} & f_{12} & \cdots & f_{1n_2} \\ f_{20} & f_{21} & f_{22} & \cdots & f_{2n_2} \\ \cdot & \cdot & \cdot & \cdot & \cdot \\ f_{n_10} & f_{n_11} & f_{n_12} & \cdots & f_{n_1n_2} \end{bmatrix} . \tag{15}$$

Besides the computation efficiency as demonstrated through (13) and (14) there are several additional features when reducing the order of a 2-D FIR filter by the balanced approximation approach. First, the reduced system model is usually of IIR type but always has separable denominator. This is simply because $A_3 = 0$ which implies $A_{3r} = 0$. Second, the reduced IIR filter is always stable [10] and locally balanced [16].

2.2 The Quadrantally Symmetrical IIR Case

An 2-D filter defined by transfer (rational) function $Q(z_1,z_2)$ is said to be quadrantally symmetric if

$$|Q(e^{j\omega_1},e^{j\omega_2})| = |Q(e^{-j\omega_1},e^{j\omega_2})| = |Q(e^{j\omega_1},e^{-j\omega_2})| = |Q(e^{-j\omega_1},e^{-j\omega_2})| . \tag{16}$$

Practically useful 2-D linear filters such as circularly symmetric and fan filters satisfy (16) and are therefore quandrantally symmetric. It has been known that the transfer function of a quadrantally symmetric filter has separatable denominator polynomial [17], and such a $Q(z_1,z_2)$ of order (n_1,n_2) can be realized in Roesser's state-space by (1) where $A_1 \epsilon R^{n_1 \times n_1}$, $A_4 \epsilon R^{n_2 \times n_2}$, and $A_3 = 0$ [14].

With $A_2(z_1) = A_4$ and $b_2(z_1) = b_2$, K_{22} can be found by solving the 1-D discrete Lyapunov equation

$$A_4 K_{22} A_4^T - K_{22} = -b_2 b_2^T . \tag{17}$$

With $A_1(z_2) = A_1$ and $b_1(z_2) = b_1 + A_2(z_2 I - A_4)^{-1} b_2$, K_{11} can be found by solving the Lyapunov equation

$$A_1 K_{11} A_1^T - K_{11} = -\frac{1}{2\pi j} \oint_{|z_2|=1} b_1(z_2) b_1^*(z_2) \frac{dz_2}{z_2}$$

where

$$\frac{1}{2\pi j}\oint_{|z_2|=1} b_1(z_2)b_1^*(z_2)\frac{dz_2}{z_2} = b_1b_1^T + A_2(\frac{1}{2\pi j}\oint_{|z_2|=1}(z_2 I - A_4)^{-1}b_2 b_2^T(\bar{z}_2 I - A_4)^{-T}\frac{dz_2}{z_2}) A_2^T$$

$$= b_1b_1^T + A_2 K_{22}A_2^T \ .$$

Namely K_{11} satisfies the 1-D discrete Lyapunov equation

$$A_1 K_{11}A_1^T - K_{11} = -(b_1b_1^T + A_2 K_{22}A_2^T) \ . \tag{18}$$

Similarly, W_{11} and W_{22} can be computed through the following 1-D type Lyapunov equations:

$$A_1^T W_{11}A_1 - W_{11} = -c_1^T c_1 \quad \text{and} \tag{19}$$

$$A_4^T W_{22}A_4 - W_{22} = -(c_2^T c_2 + A_2^T W_{11}A_2) \ . \tag{20}$$

Note that the Lyapunov equations (17)-(20) were derived by Kawamata and Higuchi [18] through a different approach in a study of roundoff noise properties of 2-D separable digital filters. Like the FIR case, it can be shown that the reduced system is separable, locally balanced, and stable provided that $Q(z_1,z_2)$ is stable.

2.3 The FIR Case Again: The DKW Theorem

Consider the task of approximating an IIR 2-D filter again; this time we would like to carry it out by finding an appropriate lower-order FIR filter rather than an IIR filter.

Let $Q(z_1,z_2)$ defined by (7) be the transfer function of the original FIR filter of order (n_1,n_2). We seek a transfer function $Q_r(z_1,z_2)$ of the form

$$Q_r(z_1,z_2) = \sum_{j=0}^{n_{2r}} \sum_{i=0}^{n_{1r}} r_{ij}z_1^{-i}z_2^{-j} \tag{21}$$

such that the approximation error $e \triangleq |Q(z_1,z_2) - Q_r(z_1,z_2)|$ is minimized in a certain sense. Note that $Q-Q_r$ can be written as a 'quadratic' form

$$Q(z_1,z_2) - Q_r(z_1,z_2) = Z_1^T E Z_2 \tag{22}$$

where

$$Z_1^T = [\ 1 \quad z_1^{-1} \quad \ldots \quad z_1^{-n_1}\ ], \quad Z_2^T = [\ 1 \quad z_2^{-1} \quad \ldots \quad z_2^{-n_2}\ ], \tag{23}$$

and

$$E = \begin{bmatrix} (f_{ij}) & (f_{ij}) \\ \hline (f_{ij}) & (f_{ij}-r_{ij}) \end{bmatrix} \equiv \begin{bmatrix} F_{11}-R & F_{12} \\ \hline F_{21} & F_{22} \end{bmatrix}$$

with $F_{11} \in R^{(n_{r_1}+1)\times(n_{r_2}+1)}$, $F_{12} \in R^{(n_{r_1}+1)\times(n_2-n_{r_2})}$, $F_{21} \in R^{(n_1-n_{r_1})\times(n_{r_2}+1)}$, and $F_{22} \in R^{(n_1-n_{r_1})\times(n_2-n_{r_2})}$.

Denoting the L_2-norm of the approximation error by $\|e\|_2$, a bit of computation leads to

$$\|e\|_2^2 = \frac{1}{4\pi^2} \int_0^{2\pi} \int_0^{2\pi} |Q(e^{j\theta_1}, e^{j\theta_2}) - Q_r(e^{j\theta_1}, e^{j\theta_2})|^2 \, d\theta_1 d\theta_2$$

$$= tr(EE^T) = \|E\|_F^2 = \|F_{11}-R\|_F^2 + \|F_{12}\|_F^2 + \|F_{21}\|^2 + \|F_{22}\|^2$$

where $\|\cdot\|_F$ represents the Froebenius norm. Therefore, the minimizing R in the L_2-norm is given by

$$R = F_{11} \tag{24}$$

with

$$F_{11} = \begin{bmatrix} f_{00} & \cdot & \cdot & \cdot & f_{0n_{r_2}} \\ f_{10} & \cdot & \cdot & \cdot & f_{1n_{r_2}} \\ \cdot & \cdot & \cdot & \cdot & \cdot \\ f_{n_{r_1}0} & \cdot & \cdot & \cdot & f_{n_{r_1}n_{r_2}} \end{bmatrix} .$$

Another widely used error norm is the L_∞-norm $\|e\|_\infty$ defined by

$$\|e\|_\infty \overset{\Delta}{=} \max_{|z_1|=|z_2|=1} |Q(z_1,z_2) - Q_r(z_1,z_2)| . \tag{25}$$

Finding a coefficient matrix R of size $(n_{r_1}+1)\times(n_{r_2}+1)$ which minimizes (25) is rather challenging and computationally involved. Here we seek a sub-optimal solution which minimizes an upper bound of the approximation-error as follows.

By (22),

$$|Q(z_1,z_2) - Q_r(z_1,z_2)| < \|E\|_2 |z_1|^2 |z_2|^2 .$$

Notice that on the unit bicircle $T^2 = \{(z_1,z_2): |z_1| = |z_2|=1\}$,

$$\|Z_1\|_2 = \sqrt{1 + n_1} , \quad \|Z_2\|_2 = \sqrt{1 + n_2}$$

hence

$$\|e\|_\infty < \sqrt{(1+n_1)(1+n_2)} \|E\|_Z .$$

It therefore follows that a suboptimal solution to the task of minimizing $\|e\|_\infty$ may be achieved if a matrix R can be found that minimizes $\|E\|_2$. The optimization

$$\min_R \|E\|_2 = \min_R \left\| \begin{bmatrix} F_{11}-R & F_{12} \\ F_{21} & F_{22} \end{bmatrix} \right\|_2 \tag{26}$$

has a complete solution as stated in the following theorem.

Theorem 1 ([19][20])

(1) Let

$$\gamma_0 = \min_R |E|_2$$

then

$$\gamma_0 = \max\left(\left\|\begin{bmatrix} F_{12} \\ F_{22} \end{bmatrix}\right\|_2, \ |[F_{21} \ \ F_{22}]|_2\right), \qquad (27)$$

(2) Suppose $\gamma > \gamma_0$. All solutions R such that

$$\left\|\begin{bmatrix} F_{11}-R & F_{12} \\ F_{21} & F_{22} \end{bmatrix}\right\|_2 < \gamma$$

are given by

$$R = F_{11} + GF_{22}^T H - \gamma(I - GG^T)^{1/2}S(I - H^TH)^{1/2} \qquad (28a)$$

where S is an arbitrary contraction ($|S|_2 < 1$) and G and H solve the linear equations

$$G(\gamma^2 I - F_{22}^T F_{22})^{1/2} = F_{12} \qquad (28b)$$

and

$$resp(\gamma^2 I - F_{22}F_{22}^T)^{1/2}H = F_{21} \ . \qquad (28c)$$

The first part of above theorem is due to Parrott [19] while its second part is a contribution of Davis, Kahan and Weinberger [20]; which shall be referred to as the DKW theorem. Choosing S = 0, we obtain the simplest solution R which achieves the minimum in (26) as

$$R - F_{11} + G_0 F_{22}^T H_0 \overset{\Delta}{=} F_{11} + C_0 \qquad (29a)$$

where

$$C_0 = F_{12}(\gamma_0^2 I - F_{22}^T F_{22})^{-1/2} F_{22}^T(\gamma_0^2 I - F_{22}F_{22}^T)^{-1/2} F_{21} \qquad (29b)$$

with γ_0 determined by (27). On comparing (29) with (24), term C_0 reflects the difference between the L_2-norm and the L_∞-norm approximations.

III. Channel Reduction

Channel reduction of a given 2-D digital filter is a structure-simplification procedure that is carried out at the realization stage where a parallel-cascade realization scheme is used. The significance of considering this task is primarily due to the fact that any 2-D FIR filters as well as quadrantally symmetric 2-D IIR filters can be expressed as

$$Q(z_1,z_2) = \sum_{i=1}^{k} f_i(z_1)g_i(z_2) \qquad (30)$$

where f_i and g_i are causal 1-D transfer functions and consequently it can be realized using a k-channel scheme where each channel consists of two cascaded 1-D filters. A hardware implementation of filter (30) can be accomplished by adopting k parallely connected processors, each of which implements a subfilter $f_i(z_1)g_i(z_2)$. A channel reduction then simply means a savings of k-r processors. If filter (30) is implemented on a digital computer in terms of a set of programs, then reducing channel number k implies a savings of time required to execute the programs involved. In what follows the task of reducing the number of parallel channels is considered in detail for both FIR and IIR filters.

3.1 The FIR Case

Consider the FIR filter described by (7) and rewrite its expression as

$$Q(z_1,z_2) = Z_1^T \; F \; Z_2 \tag{31}$$

where Z_1 and Z_2 are defined by (23) and F is defined by (15). The singular value decomposition (SVD) of F decomposes F as

$$F = U^T \Sigma V \tag{32}$$

where $U = (u_{ij})$ and $V = (v_{ij})$ are orthogonal matrices of dimension n_1+1 and n_2+1 respectively, and (assuming $n_1 > n_2$)

$$\Sigma = \begin{bmatrix} \sigma_1 & & 0 & \vdots \\ & \ddots & & 0 \\ 0 & & \sigma_{n_1+1} & \vdots \end{bmatrix}, \quad \sigma_1 > \cdots > \sigma_k > \sigma_{kH} = \cdots = \sigma_{n_1+1} = 0,$$

with $k = \text{rank}(\Sigma)$. Note that for many practically useful FIR filters such as those with linear phase response [21] rank k is often much less than $\min(n_1+1,n_2+1)$ (in linear phase case, for example, $k < 1 + \min(n_1,n_2)/2$). Taking this into account, $Q(z_1,z_2)$ can be written as

$$Q(z_1,z_2) = \sum_{i=1}^{k} f_i(z_1)g_i(z_2) \tag{33}$$

where $f_i(z_1)$ and $g_i(z_2)$ are 1-D FIR transfer functions defined by

$$f_i(z_1) = \sum_{j=1}^{n_1+1} \sigma_i^{1/2} u_{ij} z_1^{-(j-1)}$$

$$g_i(z_2) = \sum_{j=1}^{n_2+1} \sigma_i^{1/2} v_{ij} z_2^{-(j-1)} \; .$$

If $\sigma_r \gg \sigma_{r+1}$, then the r-channel FIR filter $Q_r(z_1,z_2)$ defined by

$$Q_r(z_1,z_2) = \sum_{i=1}^{r} f_i(z_1)g_i(z_2) \tag{34}$$

approximates $Q(z_1,z_2)$ with the error

$$\varepsilon(z_1,z_2) = |Q(z_1,z_2) - Q_r(z_1,z_2)| = |\sum_{i=r+1}^{k} f_i(z_1)g_i(z_2)| = Z_1^T E r Z_2 \tag{35}$$

where

$$E_r = U^T \begin{bmatrix} 0 & | & 0 & | & 0 \\ \hline & & \sigma_{r+1} & & \\ 0 & | & & \ddots & \sigma_R & | \\ \hline 0 & | & 0 & | & 0 \end{bmatrix} \qquad V = U^T \Sigma_{r+1} V \qquad . \tag{36}$$

Similar to what was done in Section 2.3, the L_2-norm of $\varepsilon(z_1,z_2)$ is found as

$$|\varepsilon|_2 = tr(E_r E_r^T)^{1/2} = (\sum_{i=r+1}^{k} \sigma_i^2)^{1/2} = |\Sigma_{r+1}|_F \qquad . \tag{37}$$

As far as its L_∞-norm is concerned, note that by (36)

$$|E_r|_2 = |\Sigma_{r+1}|_2 = \sigma_{r+1} , \tag{38}$$

that leads (35) to

$$|\varepsilon|_\infty < \overline{\sqrt{(1+n_1)(1+n_2)}} \ \sigma_{r+1} \qquad . \tag{39}$$

Consequently if $|\Sigma_{r+1}|_F$ or $n\sigma_{r+1}$, where $n = 1+\max(n_1,n_2)$, is reasonably small, then the r-channel filter $Q_r(z_1,z_2)$ in (34) represents a good approximation of $Q(z_1,z_2)$ in L_2 or in L_∞ sense.

One may ask whether their exists a r-channel approximation that is better than (34) in the sense of L_2-norm or L_∞-norm. The Eckart-Young-Mirsky theorem quoted below answers this question.

Theorem 2 (Eckart-Young-Mirsky [11]) Let $|\cdot|$ be a unitarily invariant matrix norm (i.e. a matrix norm satisfying $|U^T X V| = |X|$ for all unitary matrices U and V) and let F be given by (32), then

$$\inf_{rank(F) < r} |F - \hat{F}| = |F - F_r| \tag{40}$$

where $F_r = U^T \Sigma_r V$ with

$$\Sigma_r = \Sigma - \Sigma_{r+1} = \begin{bmatrix} \sigma_1 & & & | & 0 \\ & \ddots & & | & \\ & & \sigma_r & | & \\ \hline 0 & & & | & 0 \end{bmatrix} \qquad .$$

Note that any r-channel FIR filter $Q_r(z_1,z_2)$ can be expressed as $Z_1^T \hat{F} Z_2$ for some $\hat{F}$ with $rank(\hat{F}) < r$. Since the Froebenius norm is unitarily invariant, it follows from (40) that

$$|Q(z_1,z_2) - Q_r(z_1,z_2)|_2 = |F - \hat{F}|_F > |F - F_r|_F = |\Sigma_{r+1}|_F \qquad .$$

In other words $Q_r(z_1,z_2)$ given by (34) is the best r-channel approximation of $Q(z_1,z_2)$ in L_2 sense.

Further notice that the 2-norm is also unitarily invariant and

$$\| Q(z_1,z_2) - \hat{Q}_r(z_1,z_2) \|_\infty < \sqrt{(1+n_1)(1+n_2)} \; \| F - \hat{F} \|_2 \quad .$$

Therefore, if we consider $\sqrt{(1+n_1)(1+n_2)} \; \| F - \hat{F} \|_2$ as a reasonable measure for the L_∞ approximation error, then (4)) simply means that $Q_r(z_1,z_2)$ given by (34) is also the best r-channel approximation of $Q(z_1,z_2)$ in L_∞ sense.

3.2 The IIR Case

Only quadrantally symmetric IIR filters will be considered here since the denominators of their transfer functions are separable. Let

$$Q(z_1,z_2) = \frac{n(z_1,z_2)}{d_1(z_1)d_2(z_2)} \tag{41}$$

and write its numerator polynomial as

$$n(z_1,z_2) = Z_1^T N Z_2 \quad .$$

Let $\text{rank}(N) = k$ and the SVD of N be given by

$$N = U^T \Sigma V \quad .$$

Denote $U = [\, u_1 \; \cdots \; u_{n_1+1} \,]$, $V = [\, v_1 \; \cdots \; v_{n_2+1} \,]$,

$$f_i(z_1) = \frac{\sigma_i^{1/2} u_i^T Z_1}{d_1(z_1)} \quad \text{and} \quad g_i(z_2) = \frac{\sigma_i^{1/2} v_i^T Z_2}{d_2(z_2)} \;, \quad 1 < i < k \quad . \tag{42}$$

$Q(z_1,z_2)$ can then be expressed by (30). Both IIR 1-D transfer functions f_i and g_i may be described in terms of their Markov parameters i.e.

$$f_i(z_1) = \sum_{j=0}^{\infty} p_{ij} z_1^{-j} \;, \quad g_i(z_2) = \sum_{j=0}^{\infty} g_{ij} z_2^{-j} \quad .$$

Now define

$$Z_1 = [\, 1 \; z_1^{-1} \; \cdots \;]^T \;, \quad Z_2 = [\, 1 \; z_2^{-1} \; \cdots \;]^T$$

$$p_i = [\, p_{io} \; p_{i1} \; \cdots \;]^T \;, \quad \text{and} \quad q_i = [\, q_{io} \; q_{i1} \; \cdots \;]^T .$$

$Q(z_1,z_2)$ can then formally be written as

$$Q(z_1,z_2) = Z_1^T \phi Z_2 \tag{43}$$

where ϕ is an infinite-dimensional matrix defined by

$$\phi = \sum_{i=1}^{k} p_i q_i^T \quad .$$

Since all 1-D filters described by $f_i(z_1)$ and $g_i(z_2)$ are stable, p_i, $q_i \in \ell_1$ (absolutely summable). Partitioning ϕ as

$$\Phi = \begin{bmatrix} \Phi_{11} & \Phi_{12} \\ \Phi_{21} & \Phi_{22} \end{bmatrix}$$

where $\Phi_{11} \epsilon R^{m \times m}$, $\Phi_{12} \epsilon R^{m \times \infty}$, $\Phi_{21} \epsilon R^{\infty \times m}$ and $\Phi_{22} \epsilon R^{\infty \times \infty}$, then with m large enough Φ_{12}, Φ_{21} and Φ_{22} are all "small" in a certain sense. Moreover, by defining

$$Q_m(z_1, z_2) = \mathcal{Z}_1^T \tilde{\Phi}_m \mathcal{Z}_2 \qquad (44)$$

with

$$\tilde{\Phi}_m = \begin{bmatrix} \Phi_{11} & 0 \\ 0 & 0 \end{bmatrix}$$

and

$$e_m(z_1, z_2) = Q(z_1, z_2) - Q_m(z_1, z_2) \qquad (45)$$

it is easy to see that

$$|e_m(z_1, z_2)|_2 = (|\Phi_{12}|_F^2 + |\Phi_{21}|_F^2 + |\Phi_{22}|_F^2)^{1/2} .$$

Since $p_i, q_i \epsilon \ell_1$ imply $p_i, q_i \epsilon \ell_2$, it follows that for any prescribed $\delta > 0$, there exists an integer m such that $(|\Phi_{12}|_F^2 + |\Phi_{21}|_F^2 + |\Phi_{22}|_F^2) < \delta^2$, i.e.

$$|e_m(z_1, z_2)|_2 < \delta . \qquad (46)$$

$Q_m(z_1, z_2)$ defined by (44) is actually a FIR 2-D filter of order (m,m):

$$Q_m(z_1, z_2) = Z_{1m}^T \Phi_{11} Z_{2m} \qquad (47)$$

where

$$Z_{1m} = [1 \quad z_1^{-1} \quad \cdots \quad z_1^{-m}]^T , \qquad Z_{2m} = [1 \quad z_2^{-1} \quad \cdots \quad z_2^{-m}]^T .$$

Note that rank(Φ_{11}) < k (with a large m, it is most likely to have rank(Φ_{11}) = k), so one still needs k channels to realize (47) where each one contains two high-order 1-D FIR filters. There are at this point two ways to reduce the number of channels required. One may simply apply the method of Section 3.1 to yield a r-channel FIR realization but the order of each subfilter could be as high as (m,m). The second approach is to keep the r-channel realization configuration but the subfilter in each channel is an IIR filter of the form $f_i(z_1)g_i(z_2)$ with an order much lower than (m,m). To do this, the SVD of Φ_{11}, i.e. $\Phi_{11} = U_1^T \Sigma_1 V_1$ with $U_1 = [u_1 \cdots u_m]$, $V_1 = [v_1 \cdots v_m]$, and $\hat{\Sigma}_1 = \text{diag}\{\sigma_1, \ldots, \sigma_m\}$, is first performed to obtain

$$Q_m(z_1, z_2) = \sum_{i=1}^{m} \hat{u}_i(z_1)\hat{v}_i(z_2)$$

where $\hat{u}_i(z_1) = \sigma_i^{1/2} u_i^T Z_{1m}$, $\hat{v}_i(z_2) = \sigma_i^{1/2} v_i^T Z_{2m}$ are 1-D FIR filters or order m. Now let

$$Q_r(z_1,z_2) = \sum_{i=1}^{r} \hat{u}_i(z_1)\hat{v}_i(z_2) \ ,$$

it is easy to see that

$$|Q_m(z_1,z_2) - Q_r(z_1,z_2)|_2 = (\sum_{i=r+1}^{m} \sigma_i^2)^{1/2} \ . \tag{48}$$

Further notice that for $\hat{u}_i(z_1)$ and $\hat{v}_i(z_2)$ it is possible to find lower-order IIR approximations $f_i(z_1)$ and $\hat{g}_i(z_2)$ respectively through their balanced realization [1]. Therefore a lower-order, r-channel realization of an IIR 2-D filter can be characterized by

$$\hat{Q}_r(z_1,z_2) = \sum_{i=1}^{r} f_i(z_1) \ \hat{g}_i(z_2) \tag{49}$$

with the reduction error

$$|e(z_1,z_2)|_2 = |Q_m(z_1,z_2) - \hat{Q}_r(z_1,z_2)|_2$$
$$< |Q(z_1,z_2) - Q_m(z_1,z_2)|_2 + |Q_m(z_1,z_2) - Q_r(z_1,z_2)|_2$$
$$+ |Q_r(z_1,z_2) - \hat{Q}_r(z_1,z_2)|_2 < \delta + (\sum_{i=r+1}^{m} \sigma_i^2)^{1/2}$$

$$+ |Q_r(z_1,z_2) - \hat{Q}_r(z_1,z_2)|_2 \tag{50}$$

where the error $|Q_r - \hat{Q}_r|_2$ can be estimated in terms of 1-D balanced approximation error bounds provided by [22].

Acknowledgements

This work is supported in part by the NSERC of Canada, in part by a Research Grant from the University of Victoria, and in part by the National Science Foundation under Grants No. DMS 8607687 and DMS 8722402.

References

[1] H. Kimura and Y. Honoki, "Balanced approximation of digital FIR filter with linear phase characteristic," Proc. ISCAS 1985, pp. 283-286.

[2] P. N. Paraskevopoulos, "Padé type order reduction of two-dimensional systems," IEEE Trans. Circuits Syst., vol. CAS-27, pp. 413-416, May 1980.

[3] N. K. Bose and S. Basu, "Two-dimensional matrix Padé approximations: Existence, Non-uniqueness, and recursive computation," IEEE Trans. Automat. Contr., vol. AC-25, pp. 509-514, June 1980.

[4] B. Lashgari, L. M. Silverman, and J. F. Abramatic, "Approximation of 2-D separable in denominator filters," IEEE Trans. Circuits Syst., vol. CAS-30, pp. 107-121, Feb. 1983.

[5] E. B. Lee, W.-S. Lu, and N. E. Wu, "Approximation of 2-D digital filters," Proc. 19th Conf. on Information Science and Systems, pp. 468-472, March 1985.

[6] W.-S. Lu, E. B. Lee, and Q.-T. Zhang, "Model reduction for 2-D systems," Proc. 1986 IEEE ISCAS, pp. 79-82, May 1986.

[7] W.-S. Lu, E. B. Lee, and Q.-T. Zhang, "Balanced approximation of two-dimensional and delay-differential systems," Proc. 25th CDC, Athens, pp. 917-922, Dec. 1986.

[8] E. I. Jury and K. Premaratne, "Model reduction of two-dimensional discrete systems," IEEE Trans. Circuits Syst., vol. CAS-33, pp. 558-562, May 1986.

[9] A. Kumar, F. W. Fairman, and J. R. Svensson, "Separately balanced realization and model reduction of 2D separable denominator transfer functions from input output data," IEEE Trans. Circuits Syst., vol. CAS-34, pp. 233-239, March 1987.

[10] K. Premaratne and E. I. Jury, "Model reduction of two-dimensional discrete systems via balanced realization," Proc. 21st Asilomar Conf. on Signals, Systems and Computers, Nov. 1987.

[11] G. H. Golub, A. Hoffman, and G. W. Stewart, "A generalization of the Eckart-Young-Mirsky matrix approximation theorem," Linear Algebra and Its Applications, vol. 88/89, pp. 317-327, 1987.

[12] B. C. Moore, "Principal component analysis in linear systems: Controllability, observability, and model reduction," IEEE Trans. Automat. Contr., vol. AC-26, pp. 17-32, Feb. 1981.

[13] K. Glover, R. F. Curtain, and J. R. Partington, "Realization and approximation of linear infinite dimensional systems with error bounds," Univ. of Cambridge, Dept. Engineering, Report CUED/F-CAMS/TR258, 1986.

[14] S. Y. Kung, B. C. Lévy, M. Morf, and T. Kailath, "New results in 2-D system theory, Part II: 2-D state-space models - Realization and notions of controllability, observability, and minimality," Proc. IEEE vol. 65, pp. 945-961, 1977.

[15] A. Laub, M. T. Heath, C. C. Paige, and R. C. Ward, "Computation of system balancing transformations and other applications of simultaneous diagonalization algorithm," Trans. Automat. Contr. vol. AC-32, pp. 115-122, Feb. 1987.

[16] W.-S. Lu, "Design of two-dimensional digital filters via balancing," to appear in Proc. 1988 Canadian Conf. on Electrical and Computer Engineering.

[17] P. K. Rajan and M. N. S. Swamy, "Quadrantal symmetry associated with two-dimensional digital transfer functions," IEEE Trans. Circuits Syst., vol. CAS-25, pp. 340-343, June 1983.

[18] M. Kawamata and T. Higuchi, "Synthesis of 2-D separable denominator digital filters with minimum roundoff noise and no overflow oscillations," IEEE Trans. Circuits Syst., vol. CAS-33, pp. 365-372, April 1986.

[19] S. Parrott, "On a quotient norm and the Sz-Nagy Foias lifting theorem," J. of Functional Analysis, vol. 30, pp. 311-328, 1978.

[20] C. Davis, W. M. Kahan, and H. F. Weinberger, "Norm-preserving dilations and their applications to optimal error bounds," SIAM J. Numer. Anal., vol. 19, pp. 445-469, 1982.

[21] W.-S. Lu, H.-P. Wang, and A. Antoniou, "Design of two-dimensional FIT digital filters by using the singular value decomposition," submitted for publication, 1988.

[22] K. Glover, "All optimal Hankel-norm approximations of linear multivariable systems and their L^{∞}-error bounds," Int. J. Control, vol. 39, pp. 1115-1193, 1984.

FACTORIZATION METHOD FOR INHERENTLY STABLE 2-D
RECURSIVE DIGITAL FILTERS

J.J. Soltis   T.J. Kent   M.A. Sid-Ahmed
University of Windsor   Bell Northern Research   University of Windsor

Introduction: Two-Dimensional (2-D) digital filters are required for signals which are inherently two-dimensional such as medical X-rays, seismic records, etc. Digital filters are of two types; non-recursive or finite-impulse response (FIR) and recursive or infinite-impulse response (IIR). IIR filters have stability problems which are non-existent for FIR filters. However, IIR filters can generally affect lower-order realizations than their non-recursive counterparts [1], [2]. Their major drawback is their computational intensive design techniques.

In this paper the design of 2-D IIR filters is considered.

The design of 2-D IIR filters to approximate general magnitude and phase responses is a complex problem, requiring in general non-linear minimization techniques involving stability constraints. Various approaches have been adopted in the literature, and these can be broadly categorized in three main categories [3]:

1.   Transformation methods [4], [5]; which produce relatively high-order filter polynomials.

2.   Iterative optimization techniques (eg. [1], [6-10]). These lack guarantee of convergence of the optimization algorithm to a global optimum with a minimum computational effort. They are usually computationally intensive.

3.   Linear programming approaches [3], [11]. These allow for an effective resolution of some of the inherent problems in the first two approaches. Convergence of the algorithm is verified to an optimum solution, based on the error criterion selected, with an overall less computational cost than the two previous methods. A review of the various design approaches for IIR filters through linear programming is provided in [10], along with an improved design method.

A direct design technique requiring neither the use of non-linear or linear programming approaches is developed in this paper to provide for inherently stable filters with relatively low-order. The methods presented in this study employ the development of multi-dimensional, z-domain mathematical arguments for use in the Chebyshev functions to yield filter transfer functions.

Other Chebyshev rational approximation schemes have been presented in the literature. A method requiring linear-programming is presented in [10] for approximating given magnitude response. Although, the method does not consider phase response, it can be used effectively to design 2-D IIR filters with prescribed magnitude specifications. The methods presented in the following sections provide:

1. Analytical solutions to the design problem of quarter-plane low-pass 2-D IIR filters.
2. Inherently stable low-order filters.
3. Provide approximate linear phase in most of the pass-band, a requirement for most digital image processing applications [12].

Chebyshev Low-Pass 2-D Digital Filters: A generalized form for the transfer function of a 2-D recursive filter with no numerator singularities can be stated as follows,

$$H(z_1,z_2) = \frac{1}{B(z_1,z_2)}$$ (1)

where,

$B(z_1,z_2)$ = finite order polynomial in z_1,z_2

Taking the square of the magnitude response of the above function and expressing it in terms of the Chebyshev function, we get,

$$\left| H(e^{j\omega_1 T}, e^{j\omega_2 T}) \right|^2 = \frac{1}{1 + (\epsilon \, C_n(Arg))^2}$$ (2)

where,

$$\epsilon = (10^{(0.1 \, Amax)} - 1)^{0.5}$$ (3)

$$Amax = \text{Passband Ripple [dB]}$$

$$C_n(Arg) = \cos(n \, \cos^{-1}(Arg))$$ (4)
$$n = \text{filter order}$$

$$Arg = \text{function in } \cos(\omega_1 T), \, \cos(\omega_2 T)$$

The development of a 2-D digital Chebyshev argument, Arg, follows. By applying the appropriate restrictions on the Chebyshev function argument the following can be stated,

$$Arg = f(\cos(\omega_1 T), \, \cos(\omega_2 T))$$

such that
$$|Arg| = < 1.0 \text{ in the filter's passband}$$

$$|Arg| > 1.0 \text{ in the filter's stopband}$$

Consider the relationship in the two digital spectral variables,

$$f(\omega_1 T, \, \omega_2 T) = \frac{(\omega_1 T)^2 + (\omega_2 T)^2}{(\gamma \, \pi)^2}$$ (5)

This function is a circularly symmetric function centered at,

$$\omega_1 T = \omega_2 T = 0.0$$

where,

$|f(w_1, w_2T)| < = 1.0$ for $(w_1T)^2 + (w_2T)^2 < = (\gamma \pi)^2$

representing the passband region and,

$|f(w_1T, w_2T)| > 1.0$ for $(w_1T)^2 + (w_2T)^2 > (\gamma \pi)^2$

representing the stopband region, where,

γ = normalized passband size parameter

$0.0 < \gamma = < 1.0$

The Taylor Series for the cosine function is given by:

$$\cos(x) = 1 - \frac{x^2}{2!} + \frac{x^4}{4!} - \frac{x^6}{6!} + \ldots \tag{6}$$

or approximately stated as,

$$\cos(x) = 1 - \frac{x^2}{2!}$$

or, $x^2 = 2 - 2\cos(x)$ \hfill (7)

By using the result of (7) in (5), the following can be derived,

$$f(\cos(w_1T), \cos(w_2T)) = \frac{2 - \cos(w_1T) - \cos(w_2T)}{1 - \cos(\gamma \pi)} \tag{8}$$

or equivalently,

$$f(\cos(w_1T), \cos(w_2T)) = \frac{2 - \cos(w_1T) - \cos(w_2T)}{2\sin^2\left(\dfrac{\gamma \pi}{2}\right)} \tag{9}$$

It should be clear at this point why $(\gamma\pi)^2$ was selected for the denominator of eqn.(5) instead of $\gamma\pi$. The approximation carried-out in the denominator could help in off-setting the error due to the approximation carried-out in the numerator. This function represents a suitable Chebyshev argument in the variables $\cos(w_1T)$ and $\cos(w_2T)$.

Since,

$$\cos(w_iT) = \frac{z_i + z_i^{-1}}{2} \tag{10}$$

then by substituting (10) into (9) an appropriate function in the z-domain variables is attained.

$$\text{Arg} = \frac{4 - z_1 - z_1^{-1} - z_2 - z_2^{-1}}{4\sin^2\left(\dfrac{\gamma \pi}{2}\right)} \tag{11}$$

It is now necessary to find the denominator roots of the transfer function given by eqn.(2). Setting that expression equal to zero yields the following,

$$1 + (\epsilon \, Cn(Arg))^2 = 0 \qquad (12)$$

letting

$$\cos^{-1} (Arg) = w = u + jv \qquad (13)$$

produces the following;

$$1 + (\epsilon \cos (n(u + jv)))^2 = 0 \qquad (14)$$

or,

$$\cos\left[n(u + jv)\right] = \pm \frac{j}{\epsilon} \qquad (15)$$

Expanding the complex valued cosine function into its real and imaginary parts yields the following,

$$\cos(nu) \cosh(nv) - j \sin(nu) \sinh(nv) = \pm \frac{j}{\epsilon} \qquad (16)$$

Therefore,

$$\cos(nu) \cosh(nv) = 0$$

Since,

$$\cosh(x) \leq 1.0 \quad \text{for any } x$$

Hence,

$$\cos(nu) = 0$$

$$nu = \frac{(2k+1)\pi}{2} \qquad \text{for } |k| = 0, 1, 2, \ldots$$

or

$$u_k = \frac{(2k+1)\pi}{2n} \qquad \text{for } k = 0, 1, 2, \ldots, 2n-1 \qquad (17)$$

Considering the imaginary part,

$$\sin(nu) \sinh(nv) = \pm \frac{j}{\epsilon} \qquad (18)$$

since

$$\sin(n \, u_k) = \pm 1.0 \qquad \forall \text{-} k$$

therefore,

$$\sinh(nv) = \pm \frac{1}{\epsilon}$$

or

$$v_k = \frac{\sinh^{-1}\left(\epsilon^{-1}\right)}{n} \qquad \forall \, k \tag{19}$$

Therefore eqns.(11) and (13) leads to the following:

$$\frac{4 - z_1^{-1} - z_1 - z_2 - z_2^{-1}}{4 \sin^2\left[\frac{\gamma\pi}{2}\right]} = \cos\left[w_k\right] \tag{20}$$

and $\qquad w_k = u_k + jv_k \qquad k = 0, 1, 2, \ldots, 2n-1$

From these roots the transfer function denominator polynomial can be developed. A direct solution for the z-domain variables in (19), to yield the polynomial factors, leads to an unrealizable filter. Although this filter is unrealizable, its responses can be generated. The magnitude response of an example filter is provided in Figure 1. Therefore an approximate solution is necessary if a realizable filter is to be attained.

Let,

$$z_i = e^{\xi_i}$$

hence,

$$(z_1 + z_1^{-1}) + (z_2 + z_2^{-1}) = e^{\xi_1} + e^{-\xi_1} + e^{\xi_2} + e^{-\xi_2} \tag{21}$$

by letting,

$$\xi_1 = \xi_2 = \xi$$

ensures that the solution is exact along $w_1T = w_2T$.

Using this relation in (21) yields the following:

$$(z_1 + z_1^{-1}) + (z_2 + z_2^{-1}) = 2(e^{\xi} + e^{-\xi}) \tag{22}$$

Using eqn. (22) to solve for the other terms in eqn. (19) yields the following values for ξ.

$$\xi_k = \cosh^{-1}\left[1 - \sin^2\left(\frac{\gamma\pi}{2}\right)\cos(w_k)\right] \tag{23}$$

It is necessary to represent (22) in the form of realizable polynomial factors. A factorization that satisfies eqn.(22) along $w_1T = w_2T$ is given by: (See Appendix)

$$\left[z_1^{-1} + z_2^{-1} - 2e^{\xi_k}\right]\left[z_1 + z_2 - 2e^{\xi_k}\right] \tag{24}$$

Since, the first factor represents the denominator of a stable filter and the second an unstable one, the filter is constructed as follows,

$$H(z_1, z_2) = \cfrac{1}{\displaystyle\prod_{k=0}^{2n-1} \left[(z_1^{-1} + z_2^{-1} - 2e^{\xi_k}) \right]} \tag{25}$$

Figures 2 through 5 offer the responses of an example filter along with its' root trajectories. The input parameters to this design technique are Amax, n and γ.

Conclusion: It can be stated that a viable method for the design of inherently stable, first quadrant, 2-D recursive digital filters has been developed. In addition to the technique presented, it is also possible to derive alternate arguments or implement the Butterworth approximation to develop additional design techniques, as illustrated in [13]. The stability of any filter designed is guaranteed through the filters development and the transfer function polynomial is easily studied due to its factorability. In addition, the filters lend themselves to efficient implementation due to their relative low order polynomials and minimal design computations.

References:

[1] D. E. Dudgeon and R. M. Mersereau, "Multidimensional Digital Signal Processing", Prentice-Hall, Inc., Englewood Cliffs, N.J., (1984).

[2] T. S. Huang, "Topics in Applied Physics: Two-Dimensional Digital Signal Processing I." Vol. 42, pp. 85-153, Springer-Verlag, Berlin, Germany (1981).

[3] G. Vachtsevanos, N. Papamarkos and B. Mertzios, "On the Approximation of the Magnitude Response of Two-Dimensional IIR Digital Filters Using Linear Programming", Circuits Systems and Signal Processing, Vol. 6, pp. 45-60, (1987).

[4] J. H. McClellan, "The Design of Two-Dimensional Digital Filters by Transformations", Proc. 7th Annual Princeton Conf. Inform. Sci. Syst., pp. 247-251, (1973).

[5] G. V. Mendonca, A. Antoniou and A. N. Venetsanopoulos, "Design of 2-Dimensional Pseudorated Digital Filters Satisfying Prescribed Specifications", IEEE Trans. CAS-31, No. 1, pp. 1-10, (Jan. 1987).

[6] J. M. Costa and A. N. Venetsanopoulos, "Design of Circularly Symmetric 2-Dimensional Recursive Filters", IEEE Trans. ASSP-22, No. 6, pp. 432-443, (Dec. 1974).

[7] M. P. Ekstrom, R. E. Twogood and J. W. Woods, "Two-Dimensional Recursive Filter Design: A Spectral Factorization Approach", IEEE Trans. ASSP-28, No. 1, pp. 16-26 (Feb. 1980).

[8] J. J. Murray, "A Design Method for Two-Dimensional Recursive Digital Filters", IEEE Trans. ASSP-30, No. 1, pp. 45-51, (Feb. 1982).

[9] J. W. Woods, J. H. Lee and I. Paul, "Two-Dimensional IIR Filter Design With Magnitude and Phase Error Criteria", IEEE Trans. ASSP-31, No. 4, pp. 886-893, (August 1983).

[10] G. A. Lampropoulos and M. A. Dahmy, "A New Technique for the Design of Two-Dimensional FIR and IIR Filters", IEEE Trans. ASSP-33, No. 1, pp. 268-279, (Feb. 1985).

[11] A. Chottera and G. A. Jullien, "Design of Two-Dimensional Recursive Digital Filters Using Linear Programming", IEEE Trans. Circuits and Systems, 29, pp. 817-826, (1982).

[12] T. S. Huang, J. W. Burnett and A. G. Deczky, "The Importance of Phase in Image Processing Filters", IEEE Trans. ASSP-23, No. 6, (Dec. 1975).

[13] T. J. Kent, "A Z-Domain Design Technique for Inherently Stable, Causal, Recursive 2-Dimensional Digital Filters", M.A.Sc. Thesis, University of Windsor, Windsor, Ontario, Canada, (1987).

APPENDIX

The factorization of eqn.(22) is approximately given by

$$\left(z_1^{-1} + z_2^{-1} - 2e^{\xi_k} \right)\left(z_1 + z_2 - 2e^{\xi_k} \right) = 0$$

for $\omega_1 T \simeq \omega_2 T$.

Expanding the above factors we get

$$\left(z_1^{-1} + z_2^{-1} \right)\left(z_1 + z_2 \right) - 2\left(z_1^{-1} + z_2^{-1} \right)e^{\xi_k}$$

$$- 2\left(z_1 + z_2 \right)e^{\xi_k} + 4\,e^{2\xi_k} = 0$$

or

$$e^{-\xi_k}\left(2 + z_1^{-1}z_2 + z_2^{-1}z_1 \right) - 2\left(z_1^{-1} + z_2^{-1} \right)$$

$$- 2\left(z_1 + z_2 \right) + 4e^{\xi_k} = 0$$

for $\omega_1 \simeq \omega_2$ we can write

$$e^{-\xi_k}(4) - 2\left(z_1^{-1} + z_2^{-1} \right) - 2\left(z_1 + z_2 \right) + 4\,e^{\xi_k} = 0$$

or

$$\left(z_1^{-1} + z_2^{-1} \right) + \left(z_1 + z_2 \right) = -2\left(e^{\xi_k} + e^{-\xi_k} \right)$$

which is eqn.(21).

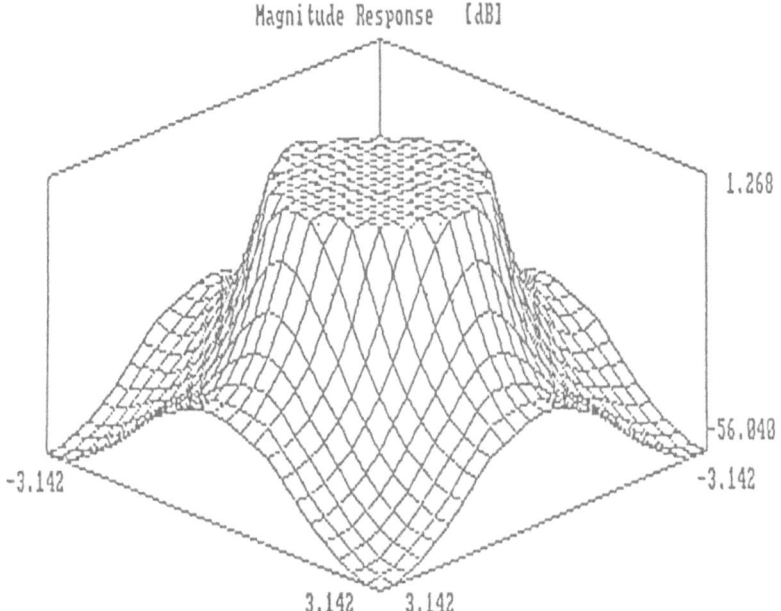

Type - Exact n = 4 Amax = 0.5 γ = 0.5
(0 dB at w_1T = w_2T = 0.01)

Figure 1: Magnitude Response for Exact Solution

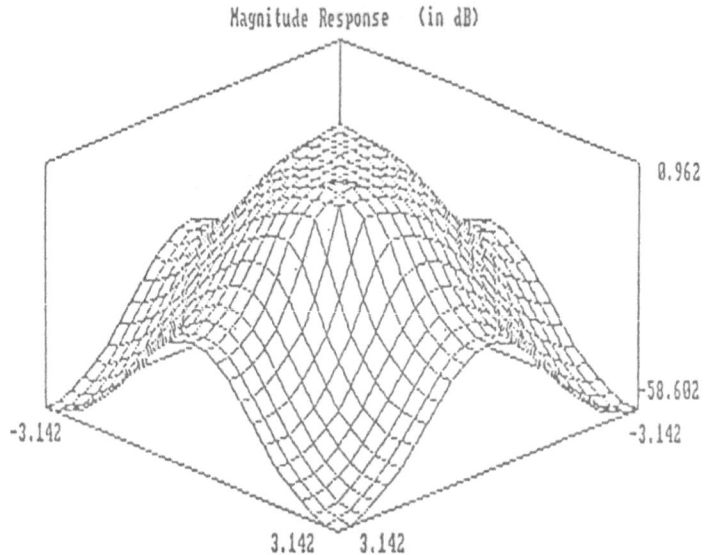

Magnitude Response (in dB)

Type - Cheb n = 4 Amax = 1.0 γ = 0.5
(0 dB at w_1T = w_2T = 0.0)
Figure 2: Example Chebyshev Low-Pass Filter Magnitude Response

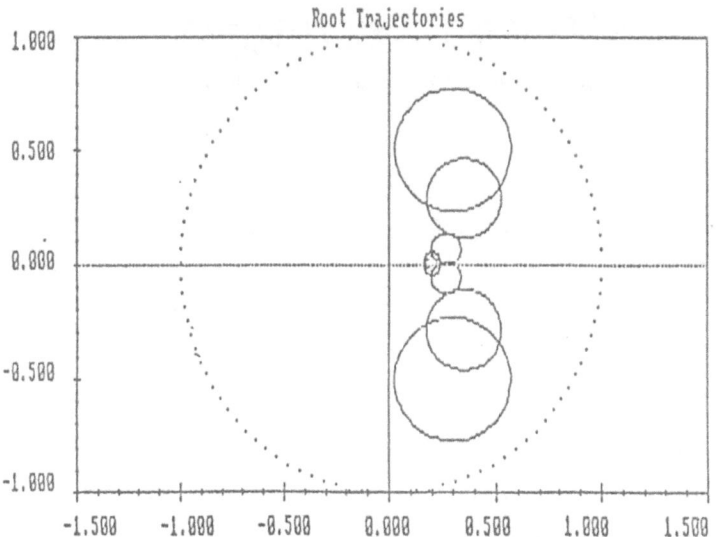

Root Trajectories

Type - Cheb n = 4 Amax = 1.0 γ = 0.5

Figure 3: Example Chebyshev Low-Pass Filter Root Trajectories

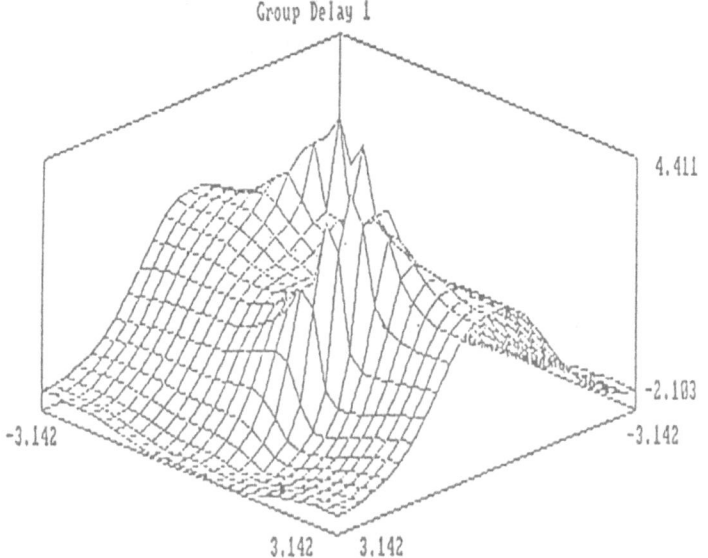

Type - Chep n = 4 Amax = 1.0 γ = 0.5

Figure 4: Example Chebyshev Low-Pass Filter Group Delay 1

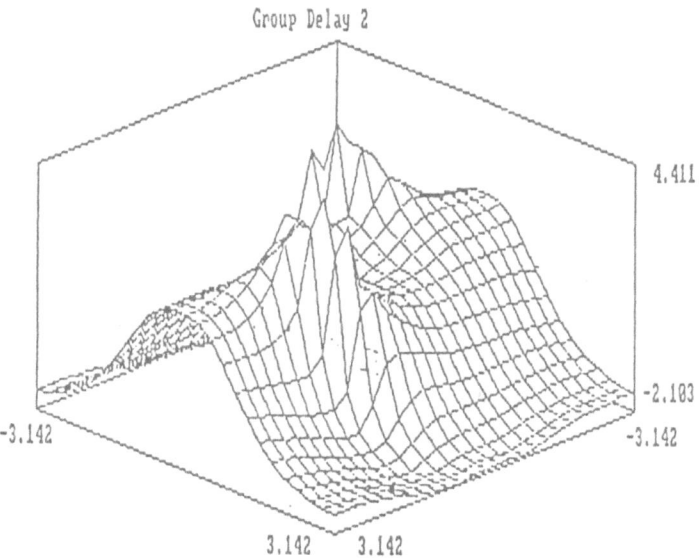

Type - Cheb n = 4 Amax = 1.0 γ = 0.5

Figure 5: Example Chebyshev Low-Pass Filter Group Delay 2

A DESIGN TECHNIQUE FOR VARIABLE 2-D
RECURSIVE DIGITAL FILTERS

Rachid Zarour and Moustafa M. Fahmy

Queen's University at Kingston

Introduction: The problem of designing 1-D recursive variable digital filters to have an arbitrarily desired frequency response has gained considerable attention in the past few years. Referred to as "variable" [1]-[3], [7], "tunable" [4] and "adjustable" [5],[6] the idea remains the same: designing a filter whose frequency response can be varied within a specified range, according to the need arising in real time processing. One of the important applications of such filters is the design of efficient time-varying filters (which can be seen as a special case of variable filters). Although practically all the applications of variable filters in the 1-D domain can be carried over to the 2-D domain, little attention has been given to the 2-D case. One of the first attempts in this direction, based on frequency transformation technique, is given in [8]. It describes two methods of realization. The first one is to make each coefficient in the realization adjustable and provide a mean of computing each from the tuning parameters. A simple computer program can be written to perform the polynomial multiplication involved in making the spectral transformation. This method is well-suited for first-order spectral transformation. However, second-order or two-variable transformations require more extensive revision of the realization because of the increased order [8]. The second approach for changing a realization in order to introduce a spectral transformation is to replace all z_1^{-1} and z_2^{-1} blocks by realizations of the all-pass functions representing the transformations. In realizations based on this technique, the tuning parameters are represented by directly adjustable elements. Some complications result, however. The replacement of z_1^{-1} and z_2^{-1} does not just increase the number of multipliers, but also requires additional memory. This is because the replacement of z_1^{-1} and z_2^{-1} blocks no longer allows the sharing of storage arrays. Also certain types of frequency transformation lead to free-delay loops which make the method unrealizable in such cases [1].

A more flexible and general approach is to treat the coefficients of the variable filter as functions of the spectral parameters defining the desired filter characteristics [5]. An implementation of such method for the 1-D case is given in [7]. Its main advantage is that the initial realization in not altered and that the number of multiplications needed to calculate the adjustable coefficients is generally small.

In this paper, the above method is extended to the 2-D case. Here the transfer function that models the two dimensional variable recursive filter assumes cascade realization. Instead of being invariant, the filter coefficients are represented by explicit analytic functions. The independent variables of these functions are the spectral parameters defining the desired magnitude response. An efficient technique for designing such filters is proposed. This technique optimizes several fixed filters having spectral parameters with reasonably spaced values and then uses a curve fitting technique in order to fit an analytic function to the coefficient values. This approach permits on-line variation of coefficients to meet the changing specifications. Examples to illustrate the proposed method are presented. For purpose of evaluation, the results are compared to optimally designed filters.

Variable filter derivation: The form of the digital filter transfer function is important for several reasons. It is known that, compared to the direct form, the cascade form has a lower sensitivity to quantization errors. Also, the stability of cascade filters is much easier to test than that of filters implemented in direct form. Thus, here we are going to consider 2-D recursive digital filters having a transfer function $H(z_1, z_2, \Psi)$ given by

$$H(z_1, z_2, \Psi) = a_0(\Psi) \frac{\prod_{i=1}^{M} H_i(z_1, z_2, \Psi)}{\prod_{i=M+1}^{N} H_i(z_1, z_2, \Psi)} \tag{1}$$

with

$$H_i(z_1, z_2, \Psi) = \sum_{k=0}^{M_i} \sum_{l=0}^{N_i} a_{k,l}^i(\Psi) z_1^{-k} z_2^{-l} \tag{2}$$

where

$\Psi = [\psi_1, \psi_2, ..., \psi_K]$ is the spectral vector,
$\psi_k; k = 1, ..., K$ are the spectral parameters of interest such as the cutoff frequency, the transition bandwidth, the bandpass mid-frequency, etc, and
$a_{k,l}^i(\Psi)$ are real functions.

By letting $z_1 = e^{j\omega_1}$ and $z_2 = e^{j\omega_2}$ one can compute the magnitude response of the filter using equation (3) given by:

$$|H(z_1, z_2, \Psi)| = |a_0(\Psi)| \frac{\prod_{i=1}^{M} |H_i(z_1, z_2, \Psi)|}{\prod_{i=M+1}^{N} |H_i(z_1, z_2, \Psi)|} \tag{3}$$

Here the filter coefficients are represented by the easily computable analytic functions $a_{k,l}^i(\Psi)$. The main assumption here is that if the frequency response changes slightly, then the corresponding change in the filter coefficients should also be small. Thus there exists a smooth relationship between the coefficient values and the frequency response. This relation may be exactly obtained for cases where the frequency transformation of [8] can be applied.

The form that $a_k^i(\Psi)$. should have is of considerable importance. Since any analytic function can be reasonably approximated over a closed range by multivariable polynomials, such polynomials will be used here to express $a_k^i(\Psi)$. More specifically, $a_k^i(\Psi)$ will have the form

$$\sum_{i_1=0}^{L_1} \sum_{i_2=0}^{L_2} \cdots \sum_{i_K=0}^{L_K} c_{i_1,i_2,\ldots,i_K} \prod_{m=1}^{K} \psi_m^{i_m} \qquad (4)$$

where

$c_{i_1,i_2,\ldots i_K}$ are the coefficients of $a_k^i(\Psi)$,

$L_1, L_2 \ldots, L_K$ are the chosen order of the polynomials, and

K is the number of spectral parameters of interest.

Example: if $K = 2$, $L_1 = 2$, $L_2 = 1$, then $a_k^i(\Psi)$ can be expressed as :

$$a_k^i(\Psi) = c_{0,0} + c_{0,1}\psi_2 + c_{1,0}\psi_1$$
$$+ c_{1,1}\psi_1\psi_2 + c_{2,0}\psi_1^2 + c_{2,1}\psi_1^2\psi_2$$

The technique proposed here for designing the variable filter is as follows:

Step (1): select a set of points $\{\Psi\}$, in the parameters range of interest, for which the corresponding optimal filters will be designed. It is clear that the computer time taken to perform this first step is proportional to the number of selected points. On the other hand, taking more points results in increased accuracy in determining the suitable coefficient functions. Thus, a compromise should be made when deciding on the number of these discrete points.

Step (2): Use a curve fitting technique (with the computed coefficients of the optimal filters as the specified points) to compute the values of the polynomials coefficients, $c_{i_1,i_2,\ldots,i_K}$ of equation (4). Once again, the higher the degrees of the polynomials chosen, the more accurate is the approximation of the behavior of the coefficient functions. On the other hand a large degree will slow down the design process and the on-line computation of the coefficient changes. Thus a trade-off must be considered in order to have an appropriate choice for every particular case. It should be noted that the different polynomials need not have the same degree, since, in general, the variations of the different coefficients are not the same.

Step (3): The variable filter coefficients can then be calculated for any value of Ψ in the specified range, using equation (4).

The proposed optimization technique: The ℓ^p nonlinear optimization technique has been employed successfully in the design of 2-D digital filters. Here a modification of the Fletcher-Powell algorithm proposed by Wan and Fahmy [9] has been used to perform the first step. This approach preserves the stability of the transfer function in each iteration. As result, a stability test in each iteration is not needed and thus considerable savings in the design time is achieved. Also the technique does not spend time refining the exact optimal step size. Instead, it takes a suboptimal one and proceeds to the next iteration. This results in further reduction of the overall computing time [9].

Consider the transfer function $H(z_1, z_2, \Psi)$ of 2-D digital filter given by (1). In order to obtain the values of $a_{k,r}^i(\Psi)$ for specific values of Ψ, the ℓ^p algorithm will minimize the performance index

$$L_p(A(\Psi)) = \sum_{q=1}^{Q} \sum_{r=1}^{R} \left[\, |H(z_q, z_r, \Psi)| - H_d(\omega_q, \omega_r, \Psi) \right]^p W(\omega_q, \omega_r, \Psi) \tag{5}$$

where

L_p is the weighted ℓ^p criterion defined at $A(\Psi)$,

p is an even positive integer,

$|H(z_1, z_2, \Psi)|$ is the designed magnitude response at $z_q = e^{j\omega_q}$, $z_r = e^{j\omega_r}$ defined at Ψ,

$H_d(\omega_q, \omega_r, \Psi)$ is the desired magnitude response at ω_q, ω_r defined at Ψ,

$\omega_q; q = 1, \ldots, Q$, $\omega_r; r = 1, \ldots, R$, are the the chosen normalized digital frequencies in $[0, \pi]$ defined in the frequency bands of interest, and

$A(\Psi)$ is the parameter vector of the designed filter coefficients defined at Ψ.

In applying the above algorithm to design the optimal filters for different Ψ, it is proposed that when moving from one point to the next, the initial filter coefficient values associated with Ψ_{i+1} are taken as those which have been already determined for Ψ_i. This is based on the assumption of coefficient continuity with the spectral parameters changes and results in great reduction of the computation effort.

In the second step, the coefficients of the filters designed above are then used to calculate the polynomials coefficients using a nonlinear least-squares algorithm [10]. For every coefficient function we minimize:

$$L_2(\Psi) = \sum_{\Psi} [a_{k,r}^i(\Psi) - \sum_{i_1=0}^{L_1} \sum_{i_2=0}^{L_2} \cdots \sum_{i_K=0}^{L_K} c_{i_1, i_2, \ldots, i_K} \prod_{m=1}^{K} \psi_m^{i_m}]^2 \tag{6}$$

It is worth noting that other techniques such as Chebyshev polynomials could have been used to achieve this second step.

<u>Illustrative examples:</u>   The proposed design method was implemented on an IBM 3081 machine running VM/CMS. The programs were written in standard FORTRAN-77.

In all the examples, the spectral parameters were chosen to be uniformly spaced in their specified ranges. The weights have been set to zero in the transition bands and one elsewhere. When referring to optimally designed filters we mean filters designed by the same optimization technique used for the fixed filters in step one.

<u>Example 1:</u>   A tenth order filter was designed to fit the magnitude characteristics of a family of circular symmetric lowpass filters specified by:

$$H_d(\omega_1, \omega_2, \psi) = \begin{cases} 1; & 0 \leq \sqrt{\omega_1^2 + \omega_2^2} \leq 0.34\pi + \psi \\ 0; & 0.58 + \psi \leq \sqrt{\omega_1^2 + \omega_2^2} \leq \pi \end{cases}$$

where $\psi \in [\,-0.08\pi, 0.08\pi\,]$.

This is a class of lowpass filters with a variable position of the transition band specified by the single spectral parameter ψ. Fourth order polynomials were used for the coefficient functions, i.e.,L = 4. For the optimization, 9 ψ samples were chosen. The total computer time taken was 20 seconds. The maximum absolute deviations of the designed responses from the desired one are compared to those of the optimally designed filters and grouped in Table 1. The magnitude response corresponding to the $\psi = -0.07\pi$ is plotted in Figure 1.

ψ/π values	optimal filter deviation	variable filter deviation
-0.07	0.93437E-02	0.18787E-01
-0.05	0.64695E-02	0.10362E-01
-0.03	0.78557E-02	0.11086E-01
-0.01	0.80101E-02	0.10008E-01
0.01	0.82382E-02	0.31706E-01
0.03	0.55672E-02	0.33743E-01
0.05	0.86603E-02	0.43474E-01
0.07	0.10966E-01	0.77292E-01

Table 1.   **Comparisons between variable and optimal lowpass filters**

In Table 1 the ψ values are the intermediate values between those used for the curve-fitting of the second step of the algorithm. This applies to all the examples of this paper.

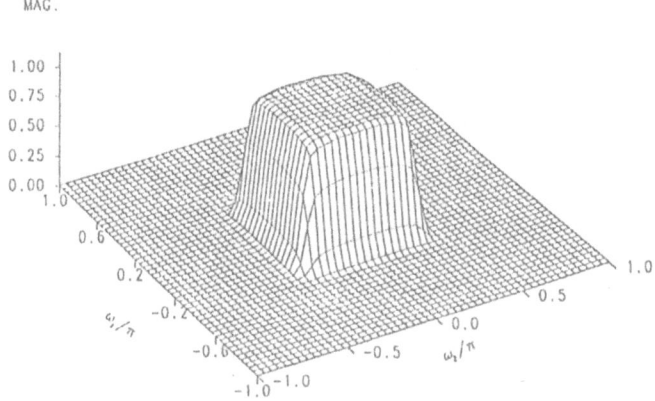

Fig. 1. Variable lowpass filter $\psi = -0.07\pi$

Example 2: A second order 2-D variable filter was designed to meet the following desired magnitude specifications of a family of fan filters:

$$H_d(\omega_1, \omega_2, \psi) = \begin{cases} 0; & |\omega_1| \geq |\omega_2| + 0.4\pi + \psi \\ 1; & |\omega_1| \leq |\omega_2| - 0.4\pi + \psi \end{cases}$$

where $\psi \in [-0.08\pi, 0.08\pi]$.

This is a class of fan filters with variable position of the transition band specified by the parameter ψ

Nine uniformly distributed samples were selected and Fourth order polynomials were used. The computer time taken was 35 seconds. The maximum absolute deviation of the variable and the optimally designed filter for different values of ψ are compared in Table 2. The magnitude response of the variable filter for $\psi = -0.07\pi$ is plotted in Figure 2.

ψ/π values	optimal filter deviation	variable filter deviation
-0.07	0.10506	0.10519
-0.05	0.94084E-01	0.94239E-01
-0.03	0.92301E-01	0.95796E-01
-0.01	0.12535	0.16786
0.01	0.73691E-01	0.77194E-01
0.03	0.67821E-01	0.90645E-01
0.05	0.70237E-01	0.81185E-01
0.07	0.10012	0.12249

Table 2. Comparisons between variable and optimal fan filters

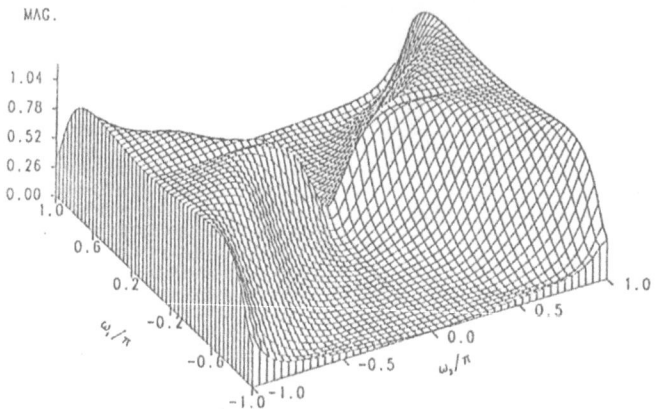

Fig. 2. Variable fan filter $\psi = -0.07\pi$

Example 3: A tenth order variable filter was designed to meet the following desired lowpass magnitude characteristics:

$$H_d(\omega_1, \omega_2, \Psi) = \begin{cases} 1; & 0 \leq \sqrt{\omega_1^2 + \omega_2^2} \leq 0.24\pi + \psi_1 \\ 0; & 0.50 + \psi_2 \leq \sqrt{\omega_1^2 + \omega_2^2} \leq \pi \end{cases}$$

where $\psi_1, \psi_2 \in [-0.08\pi, 0.08\pi]$.

The family of lowpass filters is similar to that of Example 1 except that it is now doubly parameterized by ψ_1 and ψ_2 to let the passband and stopband cutoff frequencies vary independent of each other. Third order polynomials, nine ψ_1 and nine ψ_2 samples were used. The total computer time was 157 seconds. The absolute maximum deviation of optimal and variable filters for different values of ψ_1 and ψ_2 are shown in Table 3. The magnitude response of the variable filter for $\psi_1 = -0.07\pi$ and $\psi_2 = 0.07\pi$ is shown in Figure 3. The speed at which the filter coefficients are obtained greatly overweights the fact that the deviation of these filters is larger than their optimal counterparts.

ψ_1/π values	ψ_2/π values	optimal filter deviation	variable filter deviation
-0.07	-0.07	0.30121E-01	0.30409E-01
-0.07	0.07	0.30002E-01	0.30550E-01
0.07	-0.07	0.86954E-01	0.88687E-01
0.07	0.07	0.57120E-01	0.97743E-01

Table 3.   Comparisons between variable and optimal lowpass filters

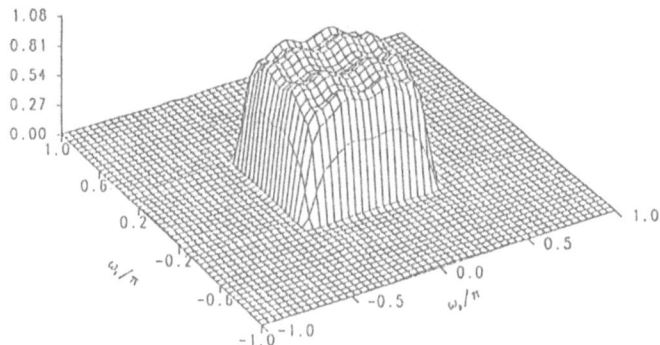

Fig. 3. Variable lowpass filler $\psi_1 = -0.07\pi$, $\psi_2 = 0.07\pi$

Conclusion: In this work, a technique for the design of 2-D variable recursive digital filters is presented. By using multivariable polynomials to approximate the behavior of the filter coefficients against a variation in the spectral parameters, a variable filter transfer function is established. An efficient algorithm to determine the multivariable polynomials is described. The design examples show that the variable filters match reasonably well their specifications. The fact that this technique provides us with the means of computing almost instantaneously the variations in the filter coefficients makes it very attractive for applications that need on-line filter variations.

References:

[1] W. Schüssler and W. Winkelnkemper, "Variable digital filter," *Arch. Elek. Ubertragung,* vol. 24, no. 1, pp. 524-525, 1970, also in *Digital signal processing,* L. R. Rabiner and C. M. Rader, Eds.

[2] D. H. Johnson, "Variable digital filter having a recursive structure," *IEEE Trans. Acoust., Speech, signal processing,* vol. ASSP-27, pp. 98-99, Feb. 1979.

[3] K. Steiglitz, "A note on variable recursive digital filters," *IEEE Trans. Acoust., Speech, signal processing,* vol. ASSP-28, pp. 111-112, Feb. 1980.

[4] S. K. Mitra, Y. Neuvo, and H. Roivainen, "Variable cutoff frequency digital IIR filters," in *Proc. IASTED int. symp. Appl. Signal Processing Digital Filtering,* Paris, France, pp. 5-8, Jun. 1985.

[5] H. E. Mutluay, "N-dimensional digital filters: analysis and design," Ph.D. Thesis, Queen's University at Kingston, 1983.

[6] K. Y. Sia and M. M. Fahmy, "Design of adjustable-coefficients recursive digital filters," *Proc. 30 symposium on circuits and systems.* Syracuse, New York, pp. 250-53, Aug. 1987.

[7] R. Zarour and M. M. Fahmy, "A design technique for variable digital filters," submitted for publication to *IEEE trans. on circuits and systems.*

[8] N. A. Pendergrass, S. K. Mitra and E. I. Jury, "spectral transformations for Two-Dimensional digital filters," *IEEE trans. on circuits and systems,* vol. 23, no. 1, pp. 26-35, Jan. 1976.

[9] Y. Wan and M. M. Fahmy, "Design of N-D digital filters with finite wordlength coefficients," submitted for publication to *IEEE trans. on circuits and systems.*

[10] P. E. Gill and W. Murray, "algorithms for the solution of the nonlinear least-squares problem," *SIAM J. Numer. Anal.,* vol. 15, no. 5, pp. 977-92, Oct. 1978.

Acknowledgements: This work has been supported in part by The National Science and Engineering Research Council of Canada (NSERC) under grant A4149

The authors are with the Department of Electrical Engineering, Queen's University, Kingston, Ontario, Canada, K7L 3N6

Analytical Methods for the Design of 2-D Elliptically Symmetric Digital Filters of Arbitrary Orientation Using Generalized McClellan Transformation

N. Nagamuthu and M.N.S. Swamy
Concordia University

I. Introduction

The current research interests in the area of **two-dimensional (2-D)** digital filters are towards the development of efficient methods for the design and implementation of the filters. In applications like *automatic quality control, computer vision*, and *robotics* there is need for *real-time image processing* due to large volume of data. The 2-D image signal, to be processed, may have a *spatial variation along a certain direction* and it may not coincide with the spatial coordinates along which the sampling is performed. To process such signals, **2-D elliptically symmetric digital filters of arbitrary orientation** are employed.

A simple and fast method of designing 2-D zero-phase **finite/infinite impulse response (FIR/IIR)** filters is to transform a one-dimensional (1-D) zero-phase finite/infinite impulse response filter by using McClellan transformation [1]. This method is capable for designing 2-D zero-phase filters of even *higher order* in a *modest amount of computer time*, and there are *efficient implementations* for the filters [2]. Since the 2-D zero-phase IIR filter obtained by using the transformation is unstable, it is **spectrally factorized** [3] into 4 **quarter-plane** [4, 5] or 2 **half-plane** [6] filters which are stable.

The **original McClellan transformation** [1] has only cosine terms; therefore, it is only suitable for designing *2-D quadrantally symmetric filters* [7]. When a horizontal or vertical elliptical cutoff contour is rotated, it no longer possesses quadrantal symmetry, but possesses only centro-symmetry [8]. So, in order to have a better approximation for this type of contours, we should use the **generalized McClellan transformation** which has sine terms also [7, 9]. It is given by

$$\cos(\omega) = t_{00} + t_{01}\cos(\omega_2) + t_{10}\cos(\omega_1) + t_{11}\cos(\omega_1)\cos(\omega_2) + s_{11}\sin(\omega_1)\sin(\omega_2)$$

$$= F(\omega_1,\omega_2) \tag{1}$$

For $0 \leq \omega \leq \pi$, the left hand side of (1) *i.e.* $\cos(\omega)$ varies from 1 to -1. So, the right hand side of (1) should satisfy the constraint

$$|F(\omega_1,\omega_2)| \leq 1, \qquad 0 \leq \omega_1 \leq \pi, \qquad 0 \leq \omega_2 \leq \pi. \tag{2}$$

The McClellan transformation coefficients $[t_{ij}$'s, $(i,j) \in (0,1)$ and $s_{11}]$ in (1) are generally found using an *optimization method* [10] so that the 1-D **pass-band cutoff frequency** ω_0 is mapped onto the desired 2-D pass-band cutoff contour. If both 1-D and 2-D filters are low-pass, one of the desirable requirement is that the 1-D origin be mapped onto the 2-D origin *i.e.* $0 \rightarrow (0,0)$. This gives the constraint equation

$$t_{00} + t_{01} + t_{10} + t_{11} = 1. \tag{3}$$

Using (3), any one of the t_{ij}'s can be expressed as a function of the other three. Thus, the total number of independent coefficients in (1) to be optimized reduces from 5 to 4.

Even though the **optimization method** [10] gives the *best result*, it is computationally very expensive. The amount of computation *exceeds* the arithmetic and/or speed capability of low cost, stand-alone 2-D signal processors and therefore, it is *not suitable* for *real-time applications*. Also, the *nonlinear optimization method* may be *unreliable* because of the *local minima*. To overcome these problems, recently, Nguyen and Swamy [11], and Reddy and Hazra [12] have proposed **analytical methods** to find the coefficients. The method [11] uses *second-order* sine terms and gives *good approximation for lower values* of frequency specifications. It is *much simpler* than the method [12]. A disadvantage of the method is that its scaled coefficients produce *two extra pass-bands* around (π, π) and $(-\pi, \pi)$, in addition to the normal pass-band around $(0, 0)$. The method [12] uses *first-order* sine terms and gives *reasonably good approximation for higher values* of frequency specifications as well. It is *more complex* than the method [11], and it *does not* give good results for lower values of the ratio of minor-axis $(2\omega_a)$ to major-axis $(2\omega_b)$. The *analytical methods* presented here, *do not have* the disadvantages of [10]-[12], and *have* the advantages of [11]. First, we consider rotated versions of **vertical** [*i.e.* type-(a)] elliptical contour. In Sections II and III, formulas are presented for unscaled and scaling-free coefficients. Then, we consider rotated versions of **horizontal** [*i.e.* type-(b)] elliptical contour. The corresponding formulas are given in Sections IV and V. In Section VI, we compare the results of several analytical methods and optimization method. Section VII is the summary.

II. APPROXIMATION FOR COEFFICIENTS FOR TYPE-(A) CONTOURS

Consider a *type-(a) elliptical cutoff contour* rotated about the 2-D origin by an angle θ in the anticlockwise direction. It is depicted in Fig. 1. The equation of the rotated elliptical contour is

$$\frac{\omega_1^2}{\omega_A^2} + \frac{\omega_2^2}{\omega_B^2} + \frac{\omega_1\omega_2}{\omega_C} = 1 \tag{4a}$$

where

$$\omega_A^2 = \omega_a^2\left[1 - \left\{1 - \frac{\omega_a^2}{\omega_b^2}\right\}\sin^2(\theta)\right]^{-1} \tag{4b}$$

$$\omega_B^2 = \omega_b^2\left[1 - \left\{1 - \frac{\omega_b^2}{\omega_a^2}\right\}\sin^2(\theta)\right]^{-1} \tag{4c}$$

and

$$\omega_C = \omega_a\omega_b\left[\left\{\frac{\omega_a}{\omega_b} - \frac{\omega_b}{\omega_a}\right\}\sin(2\theta)\right]^{-1}. \tag{4d}$$

To have a *better fit* for the nonquadrantal contour, and to have a 2-D filter with *minimum filter order*, we use the *first-order* generalized McClellan transformation *i.e.* (1). It should be noted that in (1), only first-order cosine terms and sine term are used to provide the *quadrantally symmetric* and *centro-symmetric component* respectively. Fig. 2 shows the 2-D unit impulse response of the transformation in (1).

'$\cos(x)$' and '$\sin(x)$' can be approximated by their *truncated power series* as

$$\cos(x) \approx 1 - \frac{x^2}{2} + \frac{x^4}{24} \quad \text{and} \quad \sin(x) \approx x - \frac{x^3}{6} + \frac{x^5}{120}. \tag{5a,5b}$$

Using these in (1), and neglecting sixth- and higher- order powers, we get,

$$1 - \frac{\omega^2}{2} + \frac{\omega^4}{24} = (t_{00} + t_{01} + t_{10} + t_{11}) - \frac{1}{2}(p_1\omega_1^2 + p_2\omega_2^2)$$

$$+ \frac{1}{24}(p_1\omega_1^4 + p_2\omega_2^4 + 6\ell_{11}\omega_1^2\omega_2^2)$$

$$+ s_{11}\left\{\omega_1\omega_2 - \frac{\omega_1^3\omega_2}{6} - \frac{\omega_1\omega_2^3}{6}\right\} \qquad (6a)$$

where

$$p_1 = \ell_{10} + \ell_{11} \quad \text{and} \quad p_2 = \ell_{01} + \ell_{11}. \qquad (6b,6c)$$

Let us assume that ω_0 is exactly mapped onto the rotated contour (4a). Substituting $\omega = \omega_0$, ω_2^2 value from (4a), and (3) in (6a), the *unconstrained linear error function* {defined similar to (16) in [10]} is written as

$$\frac{1}{2}\left[-\omega_0^2\left\{1 - \frac{\omega_0^2}{12}\right\} + \omega_B^2\left\{1 - \frac{\omega_B^2}{12}\right\}p_2\right]$$

$$+ \frac{\omega_1^2}{2}\left[p_1 + \frac{\omega_B^2}{\omega_A^2}\left\{\frac{\omega_B^2}{6} - 1\right\}p_2 - \frac{R_a}{12}\right]$$

$$+ \frac{\omega_1\omega_2}{2}\left[\left\{\frac{\omega_B^2}{\omega_C}p_2 + 2s_{11}\right\}\left\{\frac{\omega_B^2}{6} - 1\right\}\right]$$

$$+ \frac{\omega_1^3\omega_2}{6}\left[s_{11} + \frac{R_a}{4\omega_C} - \frac{\omega_B^2}{\omega_A^2}\left\{\frac{\omega_B^2}{2\omega_C}p_2 + s_{11}\right\}\right]$$

$$- \frac{\omega_1^4}{24}\left[p_1 + \frac{\omega_B^4}{\omega_A^4}p_2 - \frac{R_a}{\omega_A^2}\right] = 0. \qquad (7a)$$

where

$$R_a = \omega_B^2\left[\frac{\omega_B^2}{\omega_C}\left\{\frac{\omega_B^2}{\omega_C}p_2 + 4s_{11}\right\} + 6\ell_{11}\right]. \qquad (7b)$$

Since the error must be zero, the right hand side of (7a) is taken to be zero. Equating the constant, ω_1^2, $\omega_1\omega_2$, and $\omega_1^3\omega_2$ term error components to zero, and solving the resulting equations, we get the following *simple formulas*:

$$p_2 = \omega_0^2\left\{1 - \frac{\omega_0^2}{12}\right\}\left[\omega_B^2\left\{1 - \frac{\omega_B^2}{12}\right\}\right]^{-1} \qquad (8a)$$

$$p_1 = p_2\frac{\omega_B^2}{6}\left\{1 + \frac{6 - \omega_B^2}{\omega_A^2}\right\} \qquad (8b)$$

$$l_{11} = \frac{p_2}{3}\left\{1 + \frac{\omega_B^4}{2\omega_C^2}\right\} \tag{8c}$$

$$s_{11} = -\frac{\omega_B^2}{2\omega_C}p_2 \tag{8d}$$

From this derivation, we conclude that the **analytical formulas** (8a)-(8d) are *approximate solutions* to the *linear optimization problem* itself. Also, it is *more straight forward, easy to understand, and useful in deriving* the **scaling-free formulas** given in the next Section. Since on a rotated ellipse, the ratios ω_1^2/ω_A^2 and ω_2^2/ω_B^2 are *not necessarily* less than one, the analysis above is *not as accurate* as for the case of *type-(a)* or *type-(b)* elliptical contours. Therefore, the formulas given in this paper are *valid only if* ω_0, ω_a, and ω_b are *less than* 1 rad/sec (*i.e.* $<0.3183\pi$).

A. Finding the Coefficients:

To find the coefficients, first ω_A^2, ω_B^2, and ω_C are calculated using (4b)-(4d) from the given **frequency specifications** (ω_0, ω_a, ω_b, and θ). Then p_2 is calculated from (8a). Using these values in (8b)-(8d), p_1, l_{11}, and s_{11} are calculated. Other l_{ij} values are found by using (6b), (6c), and (3).

B. Special Cases:

When $\theta = 0$, $\omega_A = \omega_a$, $\omega_B = \omega_b$, and $\omega_C = \infty$. In the above formulas, (8d) *i.e.* s_{11} becomes zero, and out of the remaining formulas (8b) and (8c) do not converge to the formulas for *type-(a) contours* in [13]. The reason is that in this derivation ω_1^4 term is not forced to zero. Similarly, when $\theta = 0.5\pi$, $\omega_A = \omega_b$, $\omega_B = \omega_a$, $\omega_C = \infty$, and two of the above formulas *i.e.* (8b) and (8c) do not converge to the corresponding formulas for *type-(b) contours* in [13] for the same reason. When $\theta = 0.25\pi$, we get the following *simplified formulas*:

$$\omega_A^2 = \omega_B^2 = (2\omega_a^2\omega_b^2) / (\omega_a^2 + \omega_b^2) \tag{9a}$$

$$\omega_C = (\omega_a^2\omega_b^2) / (\omega_a^2 - \omega_b^2) \tag{9b}$$

$$p_1 = p_2 = \omega_0^2\left\{1 - \frac{\omega_0^2}{12}\right\}\left[\omega_A^2\left\{1 - \frac{\omega_A^2}{12}\right\}\right]^{-1} \tag{10}$$

(8c) and (8d).

Further, when $\omega_a = \omega_b = \omega_c$ (say), $\omega_A = \omega_B = \omega_c$ and $\omega_C = \infty$, and the formulas (10) and (8c) converge to the formulas for **type-(c) (circular) contours** in [14]. Since $p_1 = p_2$ in (10), we get $l_{10} = l_{01}$. These simplified formulas *reduce the computations* for those specifications with $\theta = 0.25\pi$ and different values of ω_0, ω_a, and ω_b.

Example 1: The specifications for a *rotated type-(a) elliptical contour* are $\omega_0 = \omega_b = 0.5\pi$, $\omega_a = 0.25\pi$, and $\theta = \pi/6$.

The **unscaled coefficients** obtained by using our *simple formulas* (8a)-(8d) are given below:

$l_{00} = -2.07969140$	$l_{01} = 0.47150514$
$l_{10} = 1.50443560$	$l_{11} = 1.10375070$
$s_{11} = 1.16932420$	
$E_{2\,MSE} = 0.47201728 \times 10^{-2}$	$E_{2\,max} = 0.12511143$
$E_{1\,MSE} = 0.10775088 \times 10^{-1}$	$E_{1\,max} = 0.42234110$

Using (16) in [15], we get the following **scaled coefficients**:

$$T_{00} = -0.18077894 \qquad T_{01} = 0.18077894$$
$$T_{10} = 0.57681294 \qquad T_{11} = 0.42318706$$
$$S_{11} = 0.44832848 \qquad \omega_0' = 0.28851292\pi$$
$$E_{2\,MSE} = 0.69519661 \times 10^{-3} \qquad E_{2\,max} = 0.47968740 \times 10^{-1}$$
$$E_{1\,MSE} = 0.10775088 \times 10^{-1} \qquad E_{1\,max} = 0.42234115$$

It should be noted that in this example, ω_0 and ω_b are *not less* than 1 rad/sec. Even then, the mean square error (MSE) values are *reasonably good.* For lower frequency specifications, the MSE will be still less. To reinforce this point, in the next example, we halve all the frequency specifications and keep the same rotation angle.

Example 2: We use the same design example in [11]. The specifications for a *rotated type-(a) elliptical contour* are $\omega_0 = \omega_b = 0.25\pi$, $\omega_a = 0.125\pi$, and $\theta = \pi/6$.

The *unscaled coefficients* obtained by using our *simple formulas* (8a)-(8d) are given below:

$$l_{00} = -2.60203370 \qquad l_{01} = 0.51192058$$
$$l_{10} = 1.89175360 \qquad l_{11} = 1.19835950$$
$$s_{11} = 1.26955380$$
$$E_{2\,MSE} = 0.16556856 \times 10^{-4} \qquad E_{2\,max} = 0.82058487 \times 10^{-2}$$
$$E_{1\,MSE} = 0.26051774 \times 10^{-3} \qquad E_{1\,max} = 0.99643828 \times 10^{-1}$$

Using (16) in [15], we get the scaled coefficients. These are given in column 1 of Table I. By comparing the error values of the above two examples, we observe that the approximation method gives *better results for lower values of frequency specifications* than for higher values. It is the *simplest* of all the existing methods [10-12].

III. APPROXIMATION FOR SCALING-FREE COEFFICIENTS FOR TYPE-(A) CONTOURS

We define P_1 and P_2 similar to p_1 and p_2 by replacing l_{ij}'s in (6b) and (6c) by T_{ij}'s which are the *scaled coefficients*. Let S_{11} be the *scaled coefficient* of s_{11}. The sine term in (1) *does not affect* the derivations in [15]. Therefore the maximum and minimum values of the function (F_{max} and F_{min}) and their locations do not change. They are affected only by the values of l_{ij}'s. S_{11} is obtained from the formula $S_{11} = C_1 s_{11}$ where C_1 is a **scaling factor** defined in [15]. S_{11} can also be expressed from the definition of U_{11} $[T_{ij} = f^{\dagger}(U_{ij})]$ by replacing l_{11} or l in its numerator by s_{11}.

The scaling-free generalized McClellan transformations are obtained from the scaling-free original transformations in Section V of [15] by adding the sine term '$S_{11}\sin(\omega_1)\sin(\omega_2)$'. The number of independent coefficients in them varies from 0 to 3 compared to 4 in the generalized original transformation. Also, the value of the coefficients can be expressed in **sum-of-powers-of-two** (SOPOT) as far as possible. The 2-D filters obtained by the transformations can be implemented by using the schemes described in [2]. The *number of multiplications* required per output sample will be *small* because, the multiplications by powers of two are done by bit shifting. Totally multiplierless 2-D filter structures are possible by quantizing the coefficients to their nearest SOPOT value. These implementations are *very attractive* for *low cost real-time signal processing applications.*

Using the *coefficients* (U_{ij}'s and S_{11}) of the *type-(a) scaling-free generalized transformations* in the equations derived from (7a), we get the following *simple scaling-free formulas.*

$^{\dagger} f \triangleq$ function of

For all the four cases the *common formula* for P_2 is

$$P_2 = \omega_0^2 \left\{ 1 - \frac{\omega_0^2}{12} \right\} \left[\omega_B^2 \left\{ 1 - \frac{\omega_B^2}{12} \right\} \right]^{-1} \tag{11}$$

The individual formulas for each case are as follows:

(i)
$$S_{11} = \omega_C \left[\frac{3}{\omega_B^2} \left\{ \frac{1}{\omega_B^2} - \frac{1}{4} - P_2 \left\{ \frac{1}{\omega_A^2} + \frac{1}{4} \right\} \right\} + \frac{P_2}{2} \left\{ \frac{1}{\omega_A^2} - \frac{\omega_B^2}{2\omega_C^2} \right\} \right] \tag{12a}$$

$$U_{01} = - U_{10} = (P_2 - 1) / 2 \tag{12b}$$

(ii),(iii)
$$S_{11} = - \frac{P_2 \omega_B^2}{2\omega_C} \tag{13a}$$

$$U_{01} = P_2 \left[1 + \frac{2}{\omega_A^2} - \frac{\omega_B^2}{3} \left\{ \frac{\omega_B^2}{2\omega_C^2} + \frac{1}{\omega_A^2} \right\} \right] - \frac{2}{\omega_B^2} \tag{13b}$$

$$U_{10} = 1 + U_{01} - P_2 \tag{13c}$$

(iv)
$$P_1 = \omega_B^2 \left[P_2 \left\{ \frac{1}{\omega_A^2} + \frac{1}{4} - \frac{\omega_B^2}{6} \left\{ \frac{\omega_B^2}{2\omega_C^2} + \frac{1}{\omega_A^2} \right\} \right\} - \frac{1}{4} \right] \left[1 - \frac{\omega_B^2}{4} \right]^{-1} \tag{14a}$$

$$U_{11} = (P_1 + P_2 - 1) / 2 \tag{14b}$$
$$U_{10} = 1 - U_{01} = P_1 - U_{11} \tag{14c}$$

The *advantage* of the above formulas is that they give the scaling-free coefficients *directly*. Also, the 1-D cutoff frequency ω_0 remains *the same*. On the other hand, the coefficients calculated by using (8a)-(8d) *require scaling* and ω_0 changes to ω_0' after scaling.

When $\theta = 0$, $\omega_A = \omega_a$, $\omega_B = \omega_b$, $\omega_C = \infty$, and $S_{11} = 0$. The above formulas for each case converge to the *scaling-free formulas* of the corresponding case for *type-(a) contours* in [15].

IV. APPROXIMATION FOR COEFFICIENTS FOR TYPE-(B) CONTOURS

In this Section, we follow a similar derivation as in the Section II. The equation of a *type-(b) elliptical cutoff contour* of arbitrary orientation is

$$\frac{\omega_1^2}{\omega_B^2} + \frac{\omega_2^2}{\omega_A^2} - \frac{\omega_1 \omega_2}{\omega_C} = 1 \tag{15}$$

where ω_A^2, ω_B^2, and ω_C are as defined before [(4b)-(4d)]. Substituting $\omega = \omega_0$, ω_2^2 value from (15), and (3) in (6a), the *unconstrained linear error function* is written as

$$\frac{1}{2} \left[- \omega_0^2 \left\{ 1 - \frac{\omega_0^2}{12} \right\} + \omega_A^2 \left\{ 1 - \frac{\omega_A^2}{12} \right\} P_2 \right]$$

$$+ \frac{\omega_1^2}{2} \left[p_1 + \frac{\omega_A^2}{\omega_B^2} \left\{ \frac{\omega_A^2}{6} - 1 \right\} p_2 - \frac{R_b}{12} \right]$$

$$+ \frac{\omega_1 \omega_2}{2} \left[\left\{ - \frac{\omega_A^2}{\omega_C} p_2 + 2 s_{11} \right\} \left\{ \frac{\omega_A^2}{6} - 1 \right\} \right]$$

$$+ \frac{\omega_1^3 \omega_2}{6} \left[s_{11} - \frac{R_b}{4\omega_C} - \frac{\omega_A^2}{\omega_B^2} \left\{ s_{11} - \frac{\omega_A^2}{2\omega_C} p_2 \right\} \right]$$

$$- \frac{\omega_1^4}{24} \left[p_1 + \frac{\omega_A^4}{\omega_B^4} p_2 - \frac{R_b}{\omega_B^2} \right] = 0. \qquad (16a)$$

where

$$R_b = \omega_A^2 \left[- \frac{\omega_A^2}{\omega_C} \left\{ - \frac{\omega_A^2}{\omega_C} p_2 + 4 s_{11} \right\} + 6 l_{11} \right]. \qquad (16b)$$

Equating the constant, ω_1^2, $\omega_1 \omega_2$, and $\omega_1^3 \omega_2$ term error components to zero, and solving the resulting equations, we get the following *simple formulas*:

$$p_2 = \omega_0^2 \left\{ 1 - \frac{\omega_0^2}{12} \right\} \left[\omega_A^2 \left\{ 1 - \frac{\omega_A^2}{12} \right\} \right]^{-1} \qquad (17a)$$

$$p_1 = p_2 \frac{\omega_A^2}{6} \left\{ 1 + \frac{6 - \omega_A^2}{\omega_B^2} \right\} \qquad (17b)$$

$$l_{11} = \frac{p_2}{3} \left\{ 1 + \frac{\omega_A^4}{2\omega_C^2} \right\} \qquad (17c)$$

$$s_{11} = \frac{\omega_A^2}{2\omega_C} p_2 \qquad (17d)$$

A.   Finding the Coefficients:

To find the coefficients, first ω_A^2, ω_B^2, and ω_C are calculated using (4b)-(4d) from the given *frequency specifications* (ω_0, ω_a, ω_b, and 0). Then p_2 is calculated from (17a). Using these values in (17b)-(17d), p_1, l_{11}, and s_{11} are calculated. Other l_{ij} values are found by using (6b), (6c), and (3).

B. Special Cases:

When $\theta = 0$, or $\theta = 0.5\pi$, in the above formulas, (17d) $i.e.$ s_{11} becomes zero, and the formula for p_2 $i.e.$ (17a) converges to the that for $type$-(b), or $type$-(a) $contours$ in [13], respectively. The remaining formulas $i.e.$ (17b) and (17c) do not converge to the corresponding formulas in [13] because, in this derivation ω_1^4 term is not forced to zero. When $\theta = 0.25\pi$, we get the same $simplified$ $formulas$ (9a)-(10), and (17c) and (17d). Further, when $\omega_a = \omega_b = \omega_c$ (say), $\omega_A = \omega_B = \omega_C$, $\omega_C = \infty$, and the formulas (10) and (17c) converge to the formulas for $type$-(c) $(circular)$ $contours$ in [14].

V. APPROXIMATION FOR SCALING-FREE COEFFICIENTS FOR TYPE-(B) CONTOURS

Using the $coefficients$ (U_{ij}'s and S_{11}) of the $type$-(b) $scaling$-$free$ $generalized$ $transformations$ in the equations derived from (16a), we get the following $simple$ $scaling$-$free$ $formulas$.

For all the four cases the $common$ $formula$ for P_2 is

$$P_2 = \omega_0^2 \left\{ 1 - \frac{\omega_0^2}{12} \right\} \left[\omega_A^2 \left\{ 1 - \frac{\omega_A^2}{12} \right\} \right]^{-1} \tag{18}$$

Additionally, the $condition$ for the cases (i)-(iii) is $\omega_0 = \omega_A$. The individual formulas for each case are as follows:

(i)
$$S_{11} = \omega_A^2 / (2\omega_C) \tag{19a}$$

$$P_1 = \omega_A^2 \left[\frac{\omega_A^4}{6\omega_C} \left\{ \frac{1}{2} - \frac{1}{\omega_C} \right\} + \frac{1}{\omega_B^2} \left\{ 1 - \frac{\omega_A^2}{6} \right\} + \frac{1}{4} \right] \left[1 - \frac{\omega_A^2}{4} \right]^{-1} \tag{19b}$$

$$U_{01} = -U_{10} = (1 - P_1) / 2 \tag{19c}$$

(ii),(iii)
$$U_{01} = \frac{2}{3} \left\{ 1 - \frac{\omega_A^4}{4\omega_C} \right\} \tag{20a}$$

$$P_1 = \omega_A^2 \left[\frac{1}{\omega_B^2} + \frac{1}{6} \left\{ 1 - \frac{\omega_A^2}{\omega_B^2} \right\} \right] \tag{20b}$$

$$U_{10} = P_1 + U_{01} - 1 \tag{20c}$$

(iv)
$$S_{11} = \frac{\omega_A^2}{2\omega_C} P_2 \tag{21a}$$

$$P_1 = \omega_A^2 \left[P_2 \left\{ \frac{1}{\omega_B^2} + \frac{1}{4} - \frac{\omega_A^2}{6} \left\{ \frac{\omega_A^2}{2\omega_C^2} + \frac{1}{\omega_B^2} \right\} \right\} - \frac{1}{4} \right] \left[1 - \frac{\omega_A^2}{4} \right]^{-1} \tag{21b}$$

$$U_{11} = (P_1 + P_2 - 1) / 2 \tag{21c}$$

$$U_{10} = 1 - U_{01} = P_1 - U_{11} \tag{21d}$$

204

When $\theta = 0$, the formulas for the cases (i) and (iv) converge to the *scaling-free formulas* of the corresponding case for *type-(b) contours* in [15]. But, the formulas for the cases (ii) and (iii) do not converge similarly because, in this derivation ω_1^4 term is not forced to zero; they converge only when the value of U_{01} for the type-(b) contour is 2/3.

VI. Comparison of Several Analytical Methods and Optimization Method

For the purpose of comparing several *analytical methods* and *optimization method*, let us choose the same contour approximation problem in Example 2. We have used the **mean square error (MSE)** and the **maximum absolute error (max)** of the **linear and nonlinear error functions** $E_2(\omega_1)$ and $E_1(\omega_1)$ {defined similar to (16) and (14) in [10]} to compare different methods. The results obtained by using several methods are given in Table I.

The scaled optimum solution (column 3) gives the *lowest* $E_{2\,MSE}$ and is of the order of 10^{-7}. The scaled analytical solution [12] (column 2) gives the *next lowest* $E_{2\,MSE}$ and is of the order of 10^{-6}. The scaled (column 1), and scaling-free (column 4) approximate solutions give $E_{2\,MSE}$s *close* to this value, and is of the order of 10^{-5}. In all the solutions the $E_{1\,MSE}$ is *higher* than the $E_{2\,MSE}$.

The $E_{2\,MSE}$ values of the scaled optimum and the scaled approximate solutions in [11] are of the order of 10^{-10} and 10^{-6} respectively. These are *slightly lower* than those in our method because, the method [11] uses a transformation with *two sine terms* and hence, one more independent coefficient and increased degree of freedom. However, the method [11] has a *serious drawback* as discussed in Section I, and *increases* the order of the designed filters. We have observed by several examples that for both the scaled and scaling-free solutions, the MSE *decreases* in our method, while it *increases* in the method [11], with *increase* in θ $(0 \leq \theta \leq \pi/4)$. These solutions for two other values of θ (i.e. $\pi/12$ and $\pi/4$) are given in Table II. Even though the method [12] gives *slightly lower* error, it is *more complex* than our method and the method [11]. Also, it has a *problem* that when ω_a/ω_b is very low, F_{max} may not occur at (0,0). The described methods *do not* have these disadvantages.

VII. Summary

Analytical methods are developed for choosing the coefficients of the *first-order generalized McClellan transformation* for the design of *2-D elliptically symmetric digital filters of arbitrary orientation* from 1-D filters. Formulas are derived for calculating the coefficients by forcing the lower order significant terms in the power series expansion of the linear error function to zero. Formulas are also presented for calculating the coefficients of the *first-order scaling-free generalized McClellan transformation*. We consider *all the cases* of the rotated vertical and horizontal elliptical cutoff contours. The results obtained by using the derived formulas *agree very well* with the results of the optimization method and other analytical methods, and their accuracy is *more than sufficient* for many engineering applications. Since all the formulas are *extremely simple*, they are very useful for *real-time adaptive design* of 2-D digital filters. Also, the *number of multiplications* required to implement the designed filters is *small* and therefore, they are very attractive for *real-time digital filtering*. The described methods *do not* have the disadvantages of [10]-[12], and *have* many advantages.

Acknowledgment

This work was supported in part by the *Natural Sciences and Engineering Research Council of Canada* under Grant A7739.

The authors' address is *Department of Electrical and Computer Engineering, Concordia University, 1455 De Maisonneuve Blvd. W., Montreal, Quebec H3G 1M8, Canada.*

REFERENCES

[1] J.II. McClellan, "The Design of Two-Dimensional Digital Filters by Transformation", *Proc. 7th Ann. Princeton Conf. Inform. Sci. and Syst.*, pp. 247-251, March 1973.

[2] R.M. Mersereau, "Two-Dimensional Nonrecursive Filter Design", in *Two-Dimensional Digital Signal Processing I*, T.S. Huang, Ed. Berlin: Springer-Verlag, 1981.

[3] N. Nagamuthu and M.N.S. Swamy, "Transformation Design of N-D Recursive Digital Filters Using Spectral Factorization", *Proc. Int. Conf. Computers, Syst., and Signal Process.*, Bangalore, India, pp. 1632-1635, Dec. 10-12, 1984.

[4] N. Nagamuthu, M.A. Sid-Ahmad, and M. Ahmadi, "Design of 2-D Recursive Digital Filters with Constant Group-Delay Characteristics", *J. Franklin Inst.*, Vol. 320, Nos. 3/4, pp. 191-200, Sep./Oct. 1985.

[5] N. Nagamuthu and M.N.S. Swamy, "Spectral Transformation Design of 2-D Recursive Digital Filters - A Review and New Results", *Proc. 26th Midwest Symp. Circuits Syst.*, Puebla, Mexico, pp. 270-274, Aug. 15-16, 1983.

[6] N. Nagamuthu and M.N.S. Swamy, "On the Transformation Design of Half-Plane 2-D Recursive Digital Filters Using Spectral Factorization", *Proc. IEEE Int. Elect. Electronics Conf. and Exposition (ELECTRONICOM'83)*, Toronto, Ont., pp. 520-524, Sep. 26-28, 1983.

[7] P.K. Rajan and M.N.S. Swamy, "Quadrantal Symmetry Associated with Two-Dimensional Digital Transfer Functions", *IEEE Trans. Circuits Syst.*, Vol. CAS-25, No. 6, pp. 340-343, June 1978.

[8] M.N.S. Swamy and P.K. Rajan, "Symmetry in Two-Dimensional Digital Filters and Its Application", Chapter 9 in *Multidimensional Systems: Techniques and Applications*, S.G. Tzafestas, Ed., New York: Marcel Dekker Inc., pp. 527-584, 1986.

[9] R.M. Mersereau, "The Design of Arbitrary 2-D Zero-phase FIR Filters Using Transformations", *IEEE Trans. Circuits Syst.*, Vol. CAS-27, No. 2, pp. 142-144, Feb. 1980.

[10] R.M. Mersereau, W.F.G. Mecklenbrauker, and T.F. Quatieri, "McClellan Transformations for Two-Dimensional Digital Filtering: I-Design", *IEEE Trans. Circuits Syst.*, Vol. CAS-23, No. 7, pp. 405-414, July 1976.

[11] D.T. Nguyen and M.N.S. Swamy, "Approximation Design of 2-D Digital Filters with Elliptical Magnitude Response of Arbitrary Orientation", *IEEE Trans. Circuits Syst.*, Vol. CAS-33, No. 6, pp. 597-603, June 1986.

[12] M.S. Reddy and S.N. Hazra, "Design of Elliptically Symmetric Two-Dimensional FIR Filters with Arbitrary Orientation", *IEEE Trans. Circuits Syst.*, (Submitted for publication.)

[13] N. Nagamuthu and M.N.S. Swamy, "Analytical Methods for the Design of 2-D Elliptically Symmetric Digital Filters Using McClellan Transformation", *Proc. IEEE Pacific Rim Conf. Commu., Comp., Signal Process.*, Vancouver, BC, June 1 & 2, 1989. (Accepted for publication.)

[14] N. Nagamuthu and M.N.S. Swamy, "Analytical Methods for the Design of 2-D Circularly Symmetric Digital Filters Using McClellan Transformation", *Proc. IEEE Int. Symp. Circuits Syst.*, Portland, Oregon, May 9-11, 1989. (Accepted for publication.)

[15] N. Nagamuthu and M.N.S. Swamy, "Scaled and Scaling-free McClellan Transformations for the Design of 2-D Low-Pass Digital Filters", *Proc. Canadian Conf. Elect. Comp. Eng.*, (CCECE'88), Vancouver, BC, pp. 633-636, Nov. 3-4, 1988.

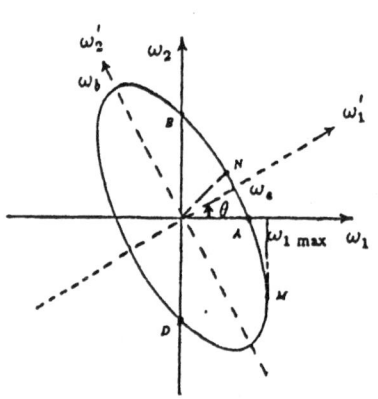

Fig. 1 Geometry of a Rotated Type-(a) Elliptical Cutoff Contour.

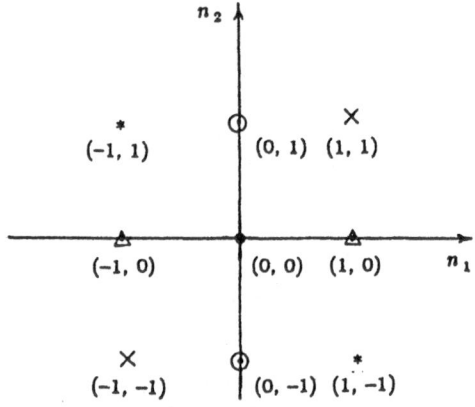

Fig. 2 Two-dimensional Unit Impulse Response $[f(n_1, n_2)]$ of the First-order Generalized McClellan Transformation.
$f(0, 0) = t_{00}$,
$f(0, 1) = f(0, -1) = t_{01}/2$,
$f(1, 0) = f(-1, 0) = t_{10}/2$,
$f(1, 1) = f(-1, -1) = (t_{11} - s_{11})/4$,
$f(1, -1) = f(-1, 1) = (t_{11} + s_{11})/4$.

Para-meter	Scaled Approx. Solution* (8a)-(8d) & (16) in [15] $\omega_0 = 0.25\pi$ †	Scaled Analytical Solution* [12] $\omega_0 = 0.25\pi$ †	Scaled Optimum Solution* [12] $\omega_0 = 0.25\pi$ †	Scaling-free Approx. Solution (11), (13a)-(13c) $\omega_0 = \omega_0$ †
t_{00}	-0.16566402	-0.10120310	-0.14574660	-0.17148958
t_{01}	0.16566402	0.10120310	0.14574660	0.17148958
t_{10}	0.61219558	0.55844520	0.60295680	0.61746900
t_{11}	0.38780442	0.44155480	0.39704320	0.38253100
s_{11}	0.41084377	0.39968230	0.40195920	0.41125363
ω_0'	0.13970893π	0.13843520π	0.13850030π	0.13970893π
$E_{2\,MSE}$	$0.17339359 \times 10^{-5}$	$0.10034494 \times 10^{-6}$	$0.87546093 \times 10^{-7}$	$0.16212458 \times 10^{-5}$
$E_{2\,max}$	$0.26846471 \times 10^{-2}$	$0.32057939 \times 10^{-3}$	$0.30466212 \times 10^{-3}$	$0.26605428 \times 10^{-2}$
$E_{1\,MSE}$	$0.26051869 \times 10^{-3}$	$0.11799751 \times 10^{-4}$	$0.13896269 \times 10^{-4}$	$0.25461967 \times 10^{-3}$
$E_{1\,max}$	$0.99643893 \times 10^{-1}$	$0.31863704 \times 10^{-1}$	$0.34061167 \times 10^{-1}$	$0.99600123 \times 10^{-1}$

The coefficients are actually T_{ij}'s and S_{11}. † One of the specification.

Table I Comparison of Results by Several Analytical Methods and Optimization Method for Type-(a) Elliptical Cutoff Contour with $\omega_a = 0.125\pi$, $\omega_b = 0.25\pi$, and $\theta = \pi/6$.

Para-meter	Scaled Approximate Solution [*] $\omega_0 = 0.25\pi$ [†]		Scaling-free Approximate Solution $\omega_0 = \omega_0'$ [†]	
	$\theta = \pi/12$ (8a)-(8d) & (16) in [15]	$\theta = \pi/4$ (9a)-(10), (8c) & (8d)	$\theta = \pi/12$ (11) & (13a)-(13c)	$\theta = \pi/4$ (11) & (13a)-(13c)
t_{00}	-0.13655077	-0.42666667	-0.14067431	-0.43459899
t_{01}	0.13655077	0.42666667	0.14067431	0.43459899
t_{10}	0.80077045	0.42666667	0.80455045	0.43366931
t_{11}	0.19922955	0.57333333	0.19544955	0.56633069
s_{11}	0.20960462	0.60000000	0.20990916	0.60055780
ω_0'	0.13031066π	0.15818894π	0.13031066π	0.15818894π
$E_{2\,MSE}$	$0.25224430 \times 10^{-6}$	$0.26017336 \times 10^{-7}$	$0.24241316 \times 10^{-6}$	$0.52116429 \times 10^{-8}$
$E_{2\,max}$	$0.32798648 \times 10^{-2}$	$0.13260160 \times 10^{-3}$	$0.32789260 \times 10^{-2}$	$0.13273189 \times 10^{-3}$
$E_{1\,MSE}$	$0.60322057 \times 10^{-3}$	$0.27795555 \times 10^{-6}$	$0.59526261 \times 10^{-3}$	$0.18192724 \times 10^{-6}$
$E_{1\,max}$	0.14124343	$0.15587450 \times 10^{-1}$	0.14133315	$0.13344013 \times 10^{-1}$

[*] The coefficients are actually T_{ij}'s and S_{11}. [†] One of the specification.

Table II Comparison of Scaled and Scaling-free Approximate Solutions for Different Values of θ for Type-(a) Elliptical Cutoff Contour with $\omega_a = 0.125\pi$ and $\omega_b = 0.25\pi$.

AN EFFICIENT ALGORITHM FOR THE DESIGN OF CIRCULAR SYMMETRIC LINEAR PHASE RECURSIVE DIGITAL FILTERS WITH SEPARABLE DENOMINATOR TRANSFER FUNCTION

M. Ahmadi
Univeristy of Windsor

Henry J. J. Lee
University of Windsor

M. Shridhar
University of Michigan-Dearborn

V. Ramachandran
Concordia University

Introduction. There are many problems that call for digital filtering of sampled 2-D data. For example, the geophysical industry uses 2-D filtering for processing seismic records and magnetic data. 2-D digital filters are also used for enhancement of low-quality images, for example, X-ray images, and compensation of linear optical degradation.

It has been reported [1] that these filters can be of better value in image processing application if they have linear phase characteristics. Unfortunately, of the number of existing methods for the design of 2-D recursive digital filters, very few have addressed design of 2-D filters with linear phase characteristics [2-5]. These design techniques were generally for the class of non-separable numerator and denominator 2-D transfer function. It has been shown recently [6-8] that most conventional 2-D filters with circular symmetric frequency response can be designed using a separable denominator, non-separable numerator transfer function of the form shown below.

$$H\left(z_1, z_2\right) = \frac{N\left(z_1, z_2\right)}{D_1\left(z_1\right) \cdot D_2\left(z_2\right)} = \frac{\sum\limits_{i=0}^{M_1} \sum\limits_{j=0}^{N_1} n(i, j) \, z_1^{-i} \, z_2^{-j}}{\left(\sum\limits_{i=0}^{M_2} d_{1j} \, z_1^{-i}\right)\left(\sum\limits_{j=0}^{N_2} d_{2j} \, z_2^{-j}\right)} \qquad (1)$$

where $Z = \exp j\omega T$.

To ensure the stability of the designed filter

$$D_i\left(z_i\right) \neq 0 \qquad \text{for} \quad \left|z_i\right| \geq 1 \qquad i = 1, 2 \qquad (2)$$

In this paper without loss of generality we assume that $M_1 = N_1 = M_2 = N_2 = N$. It has also been shown [9] that if the following conditions are imposed on the coefficients of the numerator of the 2-D transfer function (1), a quadrantal symmetry magnitude response can be obtained.

$$n(i, j) = n(N-i, j) = n(i, N-j) = n(N-i, N-j)$$

with $D_1(z_1) = D(z_1)$ and $D_2(z_2) = D(z_2)$

In this case the transfer function of eqn.(1) can be written as

$$H\left(z_1, z_2\right) = \frac{\sum_{i=0}^{M} \sum_{j=0}^{N} n'(i, j) \cos iw_1 \cos jw_2}{\left[\sum_{i=0}^{M} d_i z_1^{-i}\right]\left[\sum_{i=0}^{M} d_i z_2^{-i}\right]} \qquad (3)$$

From the above expression an octagonal symmetry filter can be obtained if

$$n'(i, j) = n'(j, i) \qquad (4)$$

Different types of symmetry are depicted in Fig. (1).

By designing a 2-D filter we mean calculation of the coefficients of the filter transfer function eqn.(1) or (3) in such a way so that the magnitude response and/or phase response of the designed filter approximates a desired one beside maintaining the stability designed filter. The latter requires that the roots of the 1-D poly-nomials of eqn.(1) or (3) be calculated at the end of each design process. If any of the roots 1-D polynomials and found to be outside the unit circle in z_1 and z_2 plane hence instability, they should be replaced by their mirror image with respect to their corresponding unit circle in the z_1 and z_2 plane to stabilize the filter.

This stabilization procedure unfortunately changes the group delay character-istics of the designed filter, i.e., if the designed filter has constant group delay responses (linear-phase), at the end of this process the group delay responses will no longer remain constant.

In this paper based on the properties of the positive definite and semi-definite matrices we generate a polynomial which has all its zeros inside the unit circle and assign it to the denominator of eqn.(3). The new coefficients of the desired trans-fer function are used as the parameters of optimization so that a desired magnitude and phase responses are obtained.

Generation of 1-D Hurwitz Polynomial. Any positive definite matrix can be decomposed as

$$\Delta_1 = A \Gamma A^T s + G \qquad (5)$$

where s is complex variable. A is an upper triangular matrix with unity elements in its diagonal, Γ is a diagonal matrix with non-negative elements, and G is a skew symmetric matrix. It is known that Δ_1 is always physically realizable [10]. Therefore D = det Δ_1 constitutes the even or odd part of a Hurwitz polynomial in s. In this case

$$B(s) = D + k_1 \frac{\partial D}{\partial s} \qquad (6)$$

is a Hurwitz polynomial (HP) in s where k is a positive number. Using the above technique a 2nd order HP can be obtained as follows

$$\Delta_1 = \begin{bmatrix} 1 & a \\ 0 & 1 \end{bmatrix} \begin{bmatrix} \gamma_1^2 & 0 \\ 0 & \gamma_2^2 \end{bmatrix} \begin{bmatrix} 1 & 0 \\ s & 1 \end{bmatrix} s + \begin{bmatrix} 0 & g \\ -g & 0 \end{bmatrix} \qquad (7)$$

$$D = \det \Delta_1 = \gamma_1^2 \, \gamma_2^2 \, s^2 + g^2 \qquad (8)$$

$$B(s) = \gamma_1^2 \, \gamma_2^2 \, s^2 + 2k \, \gamma_1^2 \, \gamma_2^2 \, s + g^2 \qquad (9)$$

Assuming k = 2 gives

$$B(s) = \gamma_1^2 \, \gamma_2^2 \, s^2 + 4\gamma_1^2 \, \gamma_2^2 \, s + g^2 \qquad (10)$$

To obtain the discrete version of the above polynomial bilinear transformation will be utilized.

Formulation of the Design Problem. In the design method to be presented here a 1-D HP is generated using eqns.(5) or (6). The discrete version of this polynomial is obtained by the application of the bilinear transformation. It should be noted that this would yield a rational function with a numerator having all its zeros inside the unit circle while the denominator is a polynomial in z having all its zeros at -1. The numerator of this rational function which is a stable 1-D poly-nomial in z is then assigned to the denominator of eqn.(3). Now the coefficients of this 2-D transfer function can be used as the parameter of optimization so that desired magnitude and phase responses are met.

A look at the transfer function in eqn.(3) would reveal that the numerator is basically a zero phase polynomial in variable z_1 and z_2 and has no effect on the phase response of the transfer function. In fact it is easy to show that the phase response is generated through the two 1-D denominator polynomials. The obvious approach to calculate the parameters of the filter transfer function is to separately use the coefficients of the denominator polynomials for phase only specification and the numerator's coefficients for achieving the overall magnitude response of the 2-D filter. This approach however, has a drawback, that is through the phase approximation step the magnitude of the two 1-D all pole transfer function may generate spikes which cannot be compensated by the numerator is 2-variable zero-phase polynomial. Therefore the following modification to the above approach has been added.

Assuming that the two 1-D polynomial in eqn.(3) are identical except for the variable. We calculate the error between the magnitude response of the ideal and the designed 1-D filter as follows.

$$E_{mag}\left(j\omega_{in}, \psi \right) = \left| H_I \left(e^{j\omega_{in}T} \right) \right| - \left| H_D \left(e^{j\omega_{in}T} \right) \right| \qquad i = 1, 2 \qquad (11)$$

ψ is the coefficient vector,

where $\left| H_I \left(j\omega_i \right) \right|$ is the ideal magnitude response of the 2-D filter $\left(\left| H_I \left(j\omega_1, j\omega_2 \right) \right| \right)$ along ω_i axis for i = 1, 2 and $\left| H_D \right|$ is the magnitude response of the designed 1-D all pole filter and defined as

$$\left| H_D\left(e^{jw_{in}T}\right) \right| = \left| \frac{1}{\sum\limits_{j=0}^{N} d_{ij} z_i^{-j}} \right| \qquad \text{for } i = 1, 2 \qquad (12)$$

The error between the group–delay response of the ideal and the designed filter is defined as

$$E_\tau\left(w_{in}, \psi\right) = \tau_I T - \tau\left(w_{in}\right) \qquad i = 1, 2 \qquad (13)$$

where τ_I is the ideal group–delay response of the 2-D filter $\left(\tau_I(w_1, w_2)\right)$ along $w_i = i = 1, 2$ axis where $\tau(w_i)$ is the group–delay response of the designed filter. Now the general least mean square error is calculated using the following relationship

$$E_g\left(w_{in}, \psi\right) = \sum_{n \in I_{ps}} E^2_{Mag}\left(jw_{in}\right) + \sum_{n \in I_p} E^2_\tau\left(jw_{in}\right) \qquad i = 1, 2 \qquad (14)$$

where I_{ps} is a set of all discrete frequency points along w_i, $i = 1, 2$ axis in the passband and the stopband of the filter while I_p is a set of all discrete frequency points along w_i, $i = 1, 2$ axis in the passband only.

Now the coefficient vector ψ can be calculated by minimizing E_g in eqn.(14). This is a simple non-linear optimization problem and can be solved by using any suitable unconstrained non-linear optimization technique. In this paper we used Fletcher and Powell technique [11]. Now the coefficients of the numerator can be used as the parameters of the optimization to achieve the overall magnitude response. In this phase of the design approach the error of the magnitude response is calculated using the relationship

$$E_{Mag}\left(jw_{1m}, jw_{2n}, \psi\right) = \left| H_I\left(e^{jw_{1m}T}, e^{jw_{2n}T}\right) \right| - \left| H_D\left(e^{jw_{1m}T}, e^{jw_{2n}T}\right) \right| \qquad (15)$$

where ψ is the coefficient vector (the coefficients of the numerator polynomials in Eqn.(3)) and $|H_I|$ is the magnitude response of the ideal 2-D filter while $|H_D|$ is the magnitude response of the designed filter. The least mean square error is calculated using the relationship

$$E\left(jw_{1m}, jw_{2n}, \psi\right) = \sum_{m, n \in I_{ps}} E^2_{Mag}\left(jw_{2m}, jw_{2n}, \psi\right) \qquad (16)$$

where I_{ps} is a set of all discrete frequency points along w_1, w_2 axis covering both the passband and the stopband of the filter. By minimizing E in eqn.(16) using any of the non-linear or linear unconstraint optimization technique coefficients of the numerator can be calculated to obtain overall magnitude response.

This technique is extremely fast and efficient and a considerable reduction in computation cost can be achieved over the method reported in [5]-[7].

Design Example. In this section we design an octagonal symmetry 2-D filter with the followingtions using the method discussed earlier.

$$\left| H_I \left(e^{j\omega_{1m}T}, e^{j\omega_{2n}T} \right) \right| = \begin{cases} 1 & 0 \leq \sqrt{\omega_{1m}^2 + \omega_{2m}^2} \leq 1. \text{ rad/sec} \\ \\ 0 & 2.5 \leq \sqrt{\omega_{1m}^2 + \omega_{2n}^2} \leq 5 \text{ rad/sec} \end{cases}$$

with constant group-delay response. The order of the filter chosen is two. Table (1) shows the coefficients of the designed filter while Fig. (2a-d) shows the magnitude and group-delay responses with respect to ω_1 and ω_2 of the designed filter.

Conclusions. In this paper a method is presented for the generation of 1-variable HP using the properties of the positive definite and positive semi-definite matrices. We have also presented a method for the design of 2-D recursive digital filter with separable denominator transfer function and octagonal symmetry satisfying a prescribed magnitude and group-delay specifications. The method consists of two steps. In the first step the two 1-D all pole filter is used to approximate the group-delay and magnitude responses along $\omega_1 - \omega_2$ axis while in the second step the numerator coefficients are used to approximate the overall magnitude response.

This method is very efficient and is very easy to implement. The usefullness of the technique is illustrated by example.

Acknowledgements. Dr. M. Ahmadi is Professor of Electrical Engineering at the University of Windsor, N9B 3P4, Canada. Mr. H. J. J. Lee is a graduate student in the Department of Electrical Engineering, University of Windsor. Dr. M. Shridhar is Professor and Head of the Department of Electrical and Computer Engineering at the University of Michigan-Dearborn, Dearborn, 48128, U.S.A. Dr. V. Ramachandran is Professor of Electrical Engineering at Concordia University, Montreal, Quebec, Canada, H3G 1M8. Both Dr. Shridhar and Dr. Ramachandran are also Adjunct Professors in the Department of Electrical Engineering, University of Windsor.

The authors wish to thank N.S.E.R.C. of Canada for its financial support of this research.

Table (1)   Value of the Coefficients of the Designed 2-D Filter

The Denominator Coefficients	The Numerator Coefficients
$\alpha_{11} = \alpha_{21} = 2.5092$	$n'_{00} = -0.1757$
$\alpha_{21} = \alpha_{22} = 0.1831$	$n'_{01} = n'_{10} = 0.4312$
$g_1 = g_2 = 1.0153$	$n'_{11} = 0.4427$

References.

[1] T. S. Huang, J. W. Burnett, A. G. Deczky, "The Importance of Phase in Image Processing Filters". IEEE Trans., 1975, ASSP-23, pp. 529-542.

[2] A. Chottera, G. A. Jullien, "Designing Near Linear Phase Recursive Filters Using Linear Programming". Proc. IEEE International Conference on Acoustic, Speech and Signal Processing, 1977, pp. 82-92.

[3] S.A.M. Aly, M. M. Fahmy, "Design of Two-Dimensional Recursive Digital Filters
     With Specified Magnitude and Group Delay Characteristics". IEEE Trans., 1978,
     CAS-25, pp. 908-915.

[4] S. Golikeri, M. Ahmadi, V. Ramachandran, "Design of 2-D Recursive Digital
     Filters Satisfying Prescribed Magnitude and Constant Group Delay Response".
     Electron. Lett., 19, pp. 9-11.

[5] M. Ahmadi, V. Ramachandran, "New Method for Generating Two-Variable VSHPs And
     Its Application In The Design of Two-Dimensional Recursive Digital Filters With
     Prescribed Magnitude and Constant Group Delay Responses". IEE Proc. Vol. 131,
     Pt. G, No. 4, August 1984, pp. 151-155.

[6] K. Rajan, M.N.S. Swamy, "Design of Separable Denominator 2-Dimensional Digital
     Filters Possessing Real Circular Symmetric Frequency Responses". Proc. IEE,
     Vol. 129, Pt. G, No. 5, October 1982, pp. 235-240.

[7] M. Ahmadi, M. T. Boraie, V. Ramachandran, C. S. Gargour, "Design of 2-D
     Recursive Digital Filters With Constant Group Delay Characteristics Using
     Separable Denominator Transfer Function anA New Stability Test". IEEE Trans.
     on Acoustics, Speech and Signal Processing, Vol. ASSP-33, No. 4, October 1985,
     pp. 1316-1318.

[8] C. L. Chan, H. K. Kwan, "Simple and Computationally Efficient Design of Two-
     Dimensional Circularly Symmetric IIR Digital Filters".   Int. J. Electronics,
     Vol. 64, No. 2, 1988, pp. 229-237.

[9] P. Karivaratharajan, M.N.S. Swamy, "Quadrantal Symmetry Associated With Two-
     Dimensional Digital Transfer Functions".   IEEE Trans. on Circuits and Systems,
     Vol. CAS-25, June 1978, pp. 340-343.

[10] H.J.J. Lee, "Generation of 1-D and 2-D Stable Polynomials And Its Application
     In The 1-D and 2-D Recursive Filter Design".   M.A.Sc. thesis, University of
     Windsor, 1988.

[11] R. Fletcher, M.J.D. Powell, "A Rapidly Convergent Descent Method for
     Minimization". Computer Journal, Vol. 6, 1964, pp. 163-168.

214

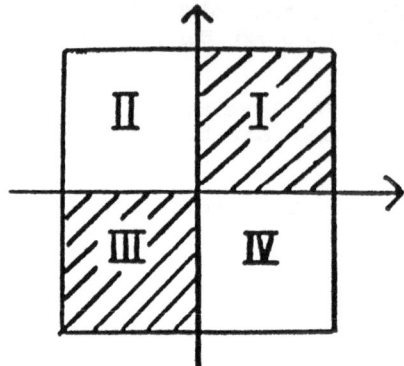

a: Frequency characteristic of·general z-transfer
   function with real coefficients

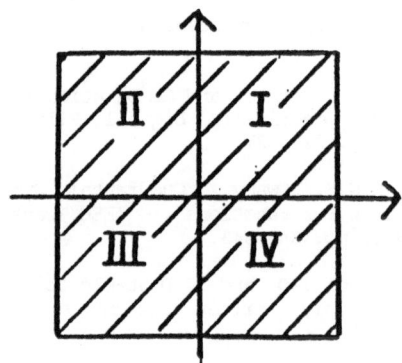

b: Quadrantal symmetric freqency characteristics

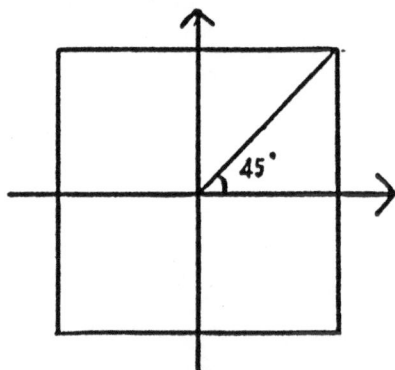

c: Octagonal symmetric frequency characteristics

Figure 1

DESIGN OF OCTAGONAL SYMM IIR FILTER
LOW-PASS

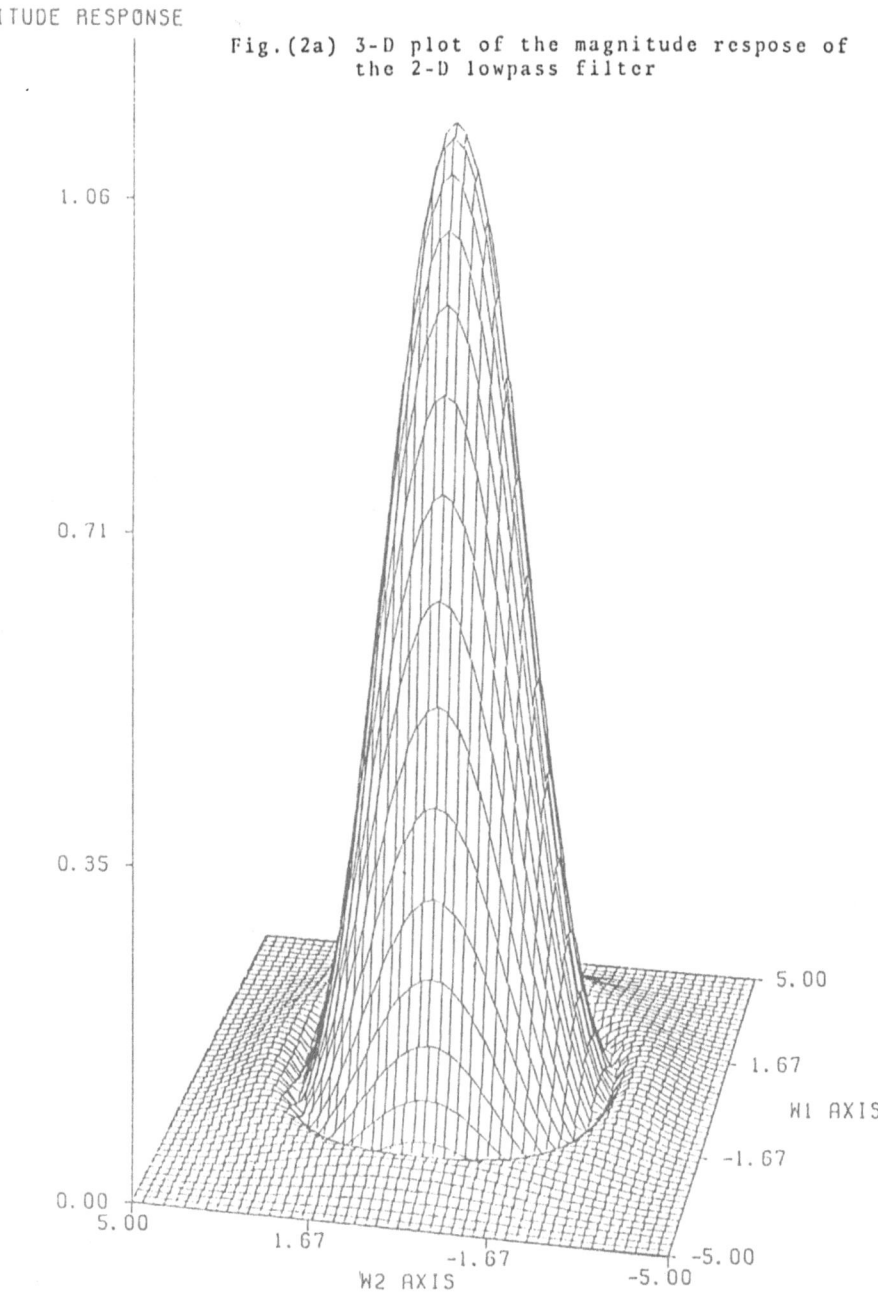

MAGNITUDE RESPONSE

Fig.(2a) 3-D plot of the magnitude respose of
the 2-D lowpass filter

1.06

0.71

0.35

0.00

5.00
1.67
-1.67
-5.00

W2 AXIS

5.00
1.67
-1.67
-5.00

W1 AXIS

OCTAGONAL SYMMETRIC
LOW-PASS

Fig(2b) Contour plot of the 2-D lowpass filter

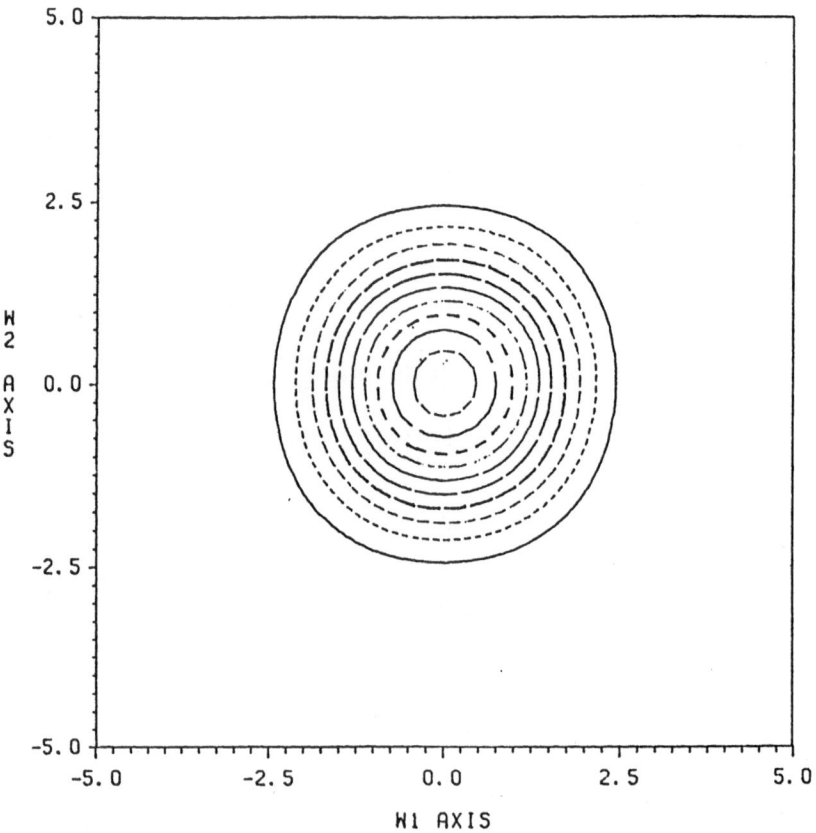

MAGNITUDE RESPONSE

——— 0.1	········ 0.2	··—·· 0.3
——— 0.4	——— 0.5	——— 0.6
·—··· 0.7	———— 0.8	—·—— 0.9
——— 1.0		

Fig.(2c) 3-D plot of the group-delay response
of the 2-D lowpass filter(w_1)

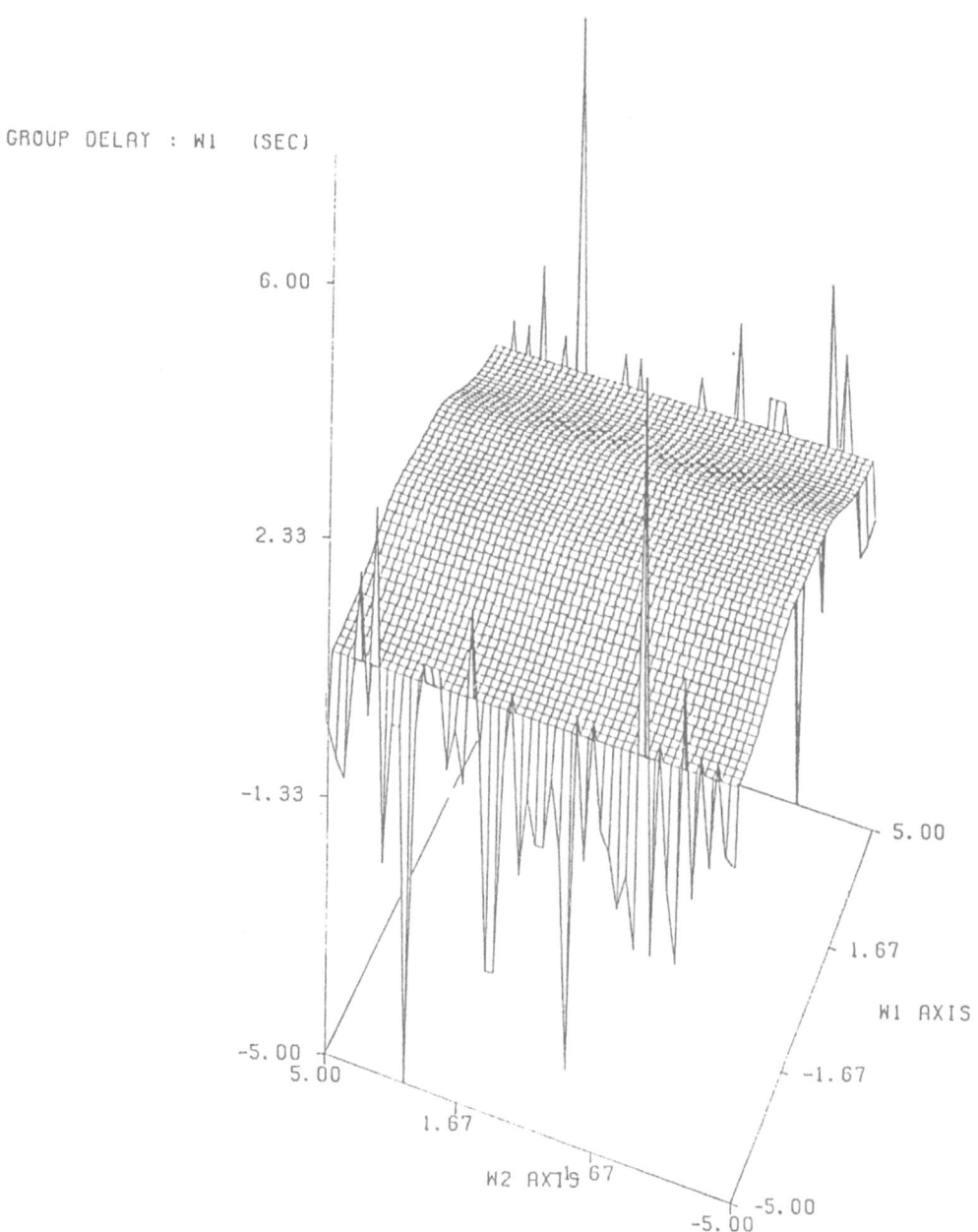

GROUP DELAY : W1   (SEC)

Fig.(2d) 3-D plot of the group-delay response
of the 2-D lowpass filter (w_2)

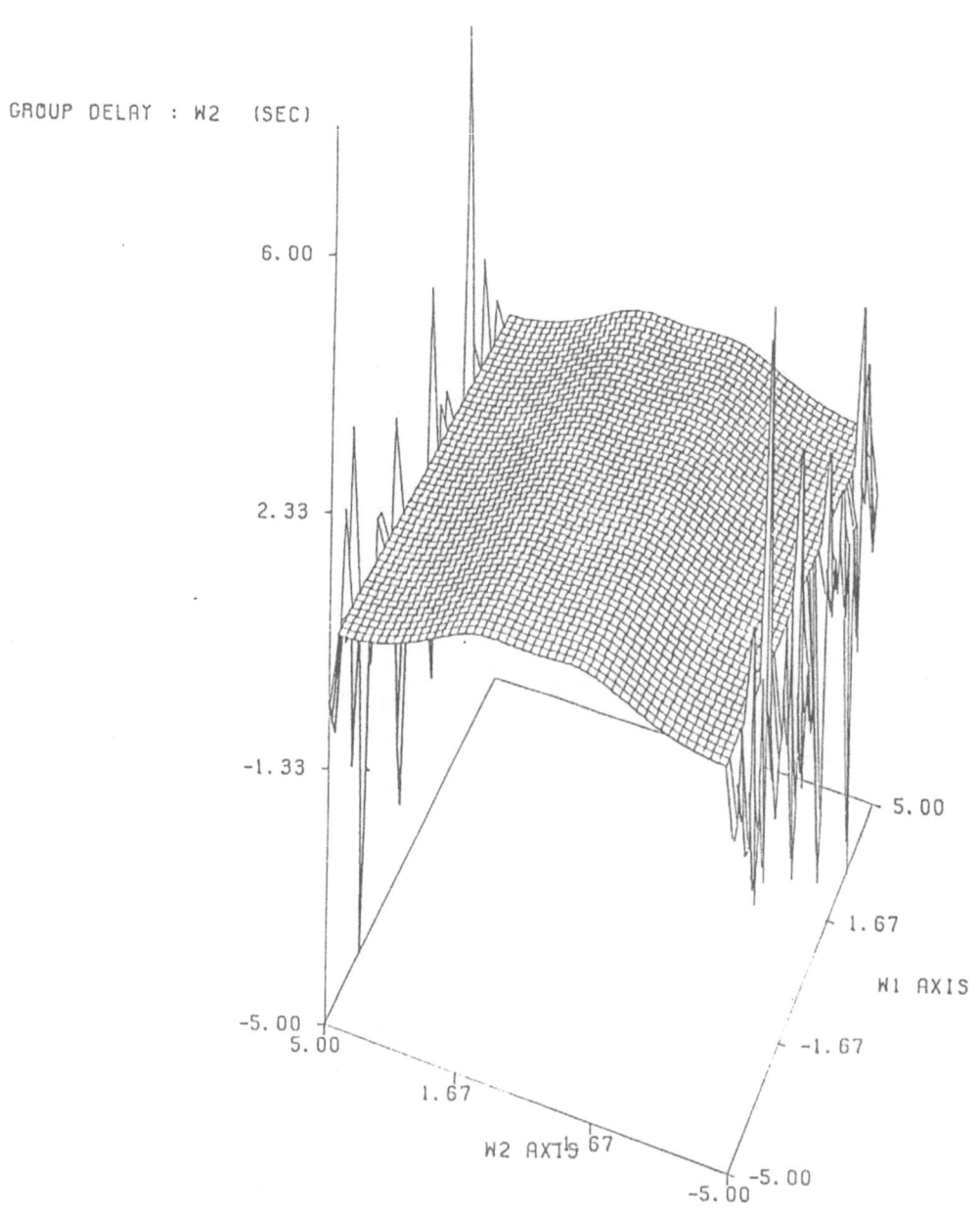

A SIMULATION STUDY OF NEAR- AND FAR-END CROSSTALK CANCELLATION FOR MULTI-CHANNEL DATA TRANSMISSION

Pedro Crespo and Michael L. Honig

Bell Communications Research

1. INTRODUCTION

Due to the impending digitization of the public switched telephone network, there is currently considerable interest in techniques for high-speed data transmission over twisted pair copper wires in the local subscriber loop. One of the major impairments associated with these channels is cross-coupling, or crosstalk, between neighboring twisted pairs within a binder group [1]. The two types of crosstalk, near-end and far-end, are illustrated in Figure 1, which shows a full-duplex multi-channel communications system. Near-end crosstalk (NEXT) refers to the interference into a specific receiver from a transmitter on the same side of the channel, whereas far-end crosstalk (FEXT) refers to the interference into a receiver from a transmitter on the far side of the channel.

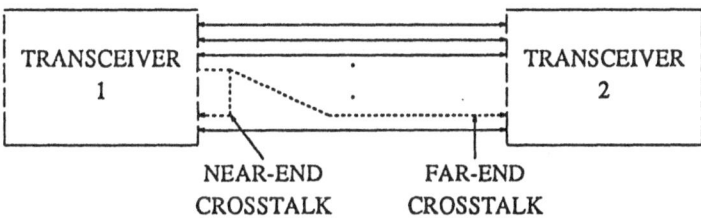

Figure 1. Near- and Far-end Crosstalk

NEXT power is much greater than FEXT power since FEXT suffers the attenuation caused by the channel, whereas NEXT is coupled directly from the offending transmitter into the receiver.

Crosstalk in subscriber loops is typically caused by capacitive and inductive imbalance between neighboring twisted pairs [2] and is therefore a linear phenomenon. Consequently, it has been pointed out in [3] that the deleterious effects of crosstalk can be substantially reduced by the appropriate use of adaptive linear Finite Impulse Response (FIR) filters. Here we present a series of simulation results that illustrate the benefit derived from such an approach. It will be assumed that each end of the multi-input/multi-output channel is terminated at a single physical location. While this does not apply to the subscriber loop in general, it does apply to many situations. For example, a business customer occupying a single building may utilize at least one cable between the business premises and the central office. Also, trunks between central offices satisfy this assumption. If only one end of the bundle is terminated at a single location such as the central office, then the transceiver simulated here can be used to increase the data rate from the central office to customer premises.

A channel consisting of four 24-gauge twisted pairs 12 kft in length was simulated from a standard transmission line model [5]. This model was used to generate the impulse response of each individual twisted pair and the near- and far-end crosstalk impulse responses between twisted pairs.

Results indicate that by using the techniques proposed in [3], a bit rate of 3.2 Mbps can be communicated over the *aggregate* channel simultaneously in each direction (full-duplex transmission). This rate is five times the current basic ISDN access rate (144 Kbps, full-duplex).

2. SYSTEM MODEL

A simplified discrete-time model of the multi-channel communications system is shown in Figure 2a. The received signal vector at time T is

$$r(T) = H * s^f(T) + G * s^n(T) + n(T) \qquad (1)$$

where $s^f(T)$ is the vector of far-end input symbols at time T, $s^n(T)$ is the vector of near-end input symbols at time T (assuming full-duplex transmission), $n(T)$ is a noise vector, "$*$" denotes convolution, and T is discrete. $H(T)$ and $G(T)$ are the far- and near-end matrix impulse responses, respectively. In particular, the ikth component of H, $h_{ik}(T)$, is the far-end response on channel i at time T to a unit pulse on channel k, and similarly, $g_{ik}(T)$ is the analogous near-end response. Note that $h_{ii}(T)$ and $g_{ii}(T)$ are, respectively, the self-impulse response and echo response for channel i. The first term on the right of (1) therefore represents intersymbol interference (ISI) and FEXT, and the second term on the right represents NEXT and echo.

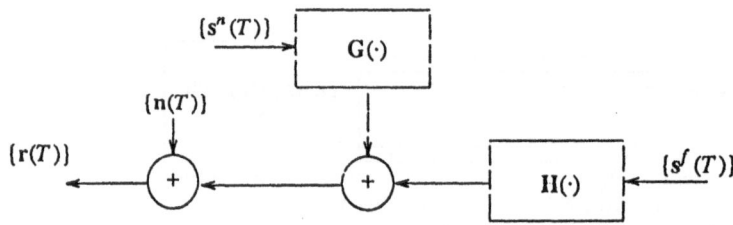

Figure 2a. Simplified Channel Model.

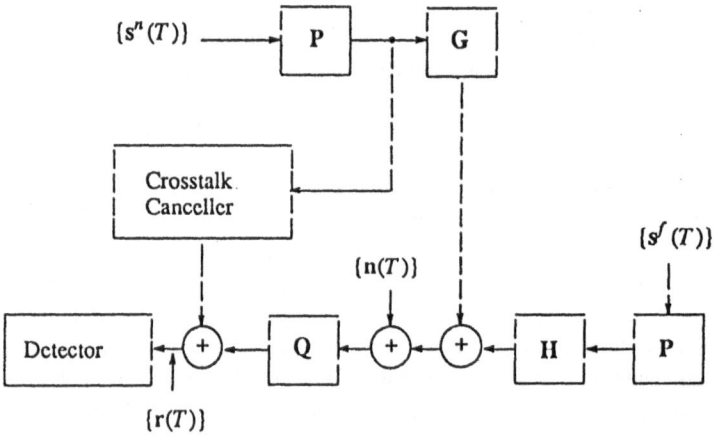

Figure 2b. Block diagram showing pre- and post-processors and NEXT canceller.

Since the sequence $\{s^n(T)\}$ is known to the transmitter, an adaptive multi-channel FIR filter can be used to cancel the NEXT term $\mathbf{G} * \mathbf{s}^n(T)$. This is illustrated in Figure 2b, which shows a crosstalk canceller (CTC), i.e., an adaptive multi-channel FIR filter, connected between the near-end transmitter and receiver. Similarly, a multi-channel decision feedback equalizer (DFE) (not shown in Figure 2b) can be used to remove ISI and FEXT [3]. Also shown in Figure 2b are linear pre- and post-processors $\mathbf{P}$ and $\mathbf{Q}$ which will be discussed shortly.

Figure 3 shows a block diagram of the simulated receiver which includes NEXT and FEXT cancellers. Each canceller is a single-input/single-output adaptive FIR filter. In particular, CTC_{ik} and DFE_{ik} cancel NEXT and FEXT, respectively, from channel i to k. The input to the FEXT cancellers is the (vector) sequence of detected symbols. The FEXT cancellers therefore can cancel FEXT due only to the postcursors of the impulse response, i.e., $\mathbf{H}(T)$ for $T > \tau$, where τ is the sampling delay. In general, a linear multi-channel pre-equalizer is needed to remove ISI and FEXT caused by precursors, i.e., $\mathbf{H}(T)$ for $T < \tau$. Because of the sharp rise of the self impulse response $h_{ii}(t)$ used to generate the simulation results in the next Section (see Figure 4a), equalization of the received signal can be adequately accomplished by cancelling the postcursors with a feedback filter. Consequently, the pre-equalizer has been ommitted in Figure 3. The boxes labelled with "H" shown in Figure 3 are hybrids, which separate the incoming signal from the outgoing signal, again assuming full-duplex transmission. The box labeled "EC" is the echo canceller for channel 1, and the boxes labelled $\mathbf{E}_R + \mathbf{E}_I$ are the pre/post-processors, which will be explained shortly.

Assuming that the near-end data is uncorrelated with the far-end data, then the minimum mean square error solution for the impulse response of the filter CTC_{ik} is $g_{ik}(0), g_{ik}(1), \ldots, g_{ik}(M_{ik})$, where M_{ik} is the number of taps in CTC_{ik}. Similarly, the MMSE solution for the impulse response of the filter DFE_{ik} is $h_{ik}(\tau), h_{ik}(\tau+1), \ldots, h_{ik}(\tau+L_{ik})$, where τ is the sampling delay, and L_{ik} is the corresponding number of taps. Note that M_{ii} and L_{ii} are the number of taps in the echo canceller and DFE, respectively, for channel i.

For simplicity we use the transversal gradient algorithm to update the taps of the filters shown in Figure 3. Let $d_l^{ik}(T)$ denote the lth coefficient of CTC_{ik} at time T, and $c_l^{ik}(T)$ denote the lth coefficient of DFE_{ik} at time T. The input to the slicer for channel k is

$$x_k(T) = r_k(T) - \sum_{i=1}^{N} \sum_{l=0}^{M_{ik}-1} d_l^{ik}(T-1)s_i^n(T-l) - \sum_{i=1}^{N} \sum_{l=0}^{L_{ik}-1} c_l^{ik}(T-1)s_i^f(T-l) \qquad (2)$$

where N is the number of channels. If pre-equalization is necessary, then $r_k(T)$ in (2) is the output of the linear equalizer for channel k. Let $s_k^f(T)$ denote the Tth detected bit at the receiver on channel k. The filter coefficients are updated according to

$$e_k(T) = \gamma_k(T-1)s_k^f(T) - x_k(T) \qquad (3)$$

$$d_l^{ik}(T) = d_l^{ik}(T-1) + \beta_1 s_i^n(T-l)e_k(T) \qquad (4a)$$

$$c_l^{ik}(T) = c_l^{ik}(T-1) + \beta_2 s_i^f(T-l)e_k(T) \qquad (4b)$$

where β_1 and β_2 are small step-sizes and $\gamma_k(T)$ is adapted to minimize mean square error. Observe that the single error e_1 is used to adapt the coefficients of all of the filters shown in Figure 3.

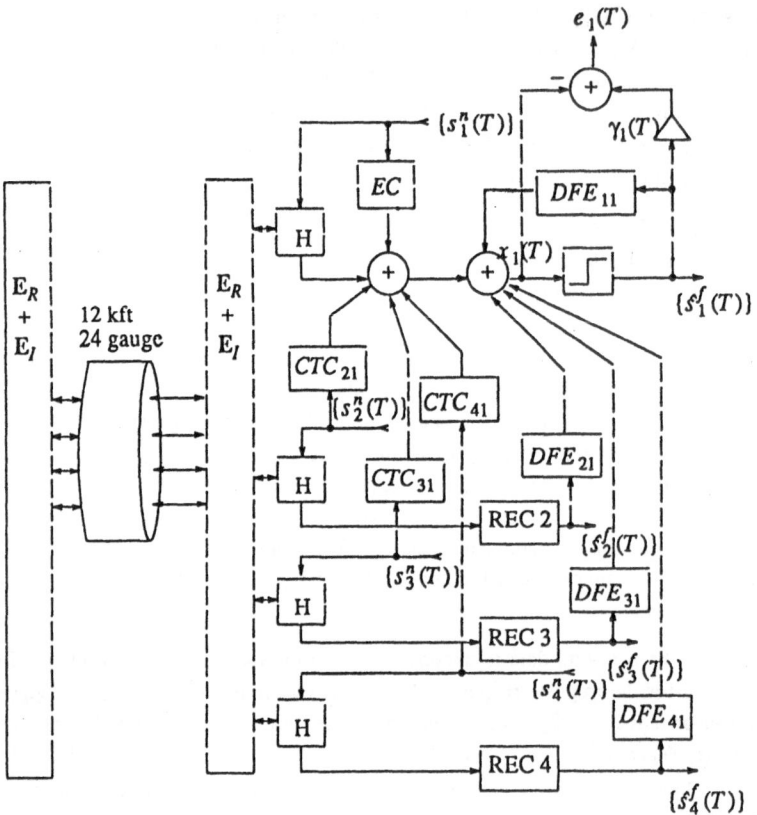

Figure 3. Block diagram of simulated receiver.

2.1 Pre- and Post-Processors

According to Figure 2b, the received signal vector is

$$r(T) = Q*H*P*s^f(T) + Q*G*P*s^n(T) + Q*n(T) \qquad (5)$$

where $P(T)$ and $Q(T)$ are the impulse responses associated with the pre- and post-processors, respectively. It was pointed out in [3] that if H is circulant, corresponding to an aggregate channel with circular symmetry, then H can be diagonalized by taking P and Q to be the FFT and inverse FFT (IFFT) operators, respectively. In this case, $P(T) = Q^*(T) = \delta_{0,T} E$, where $[E]_{lm} = \frac{1}{N} e^{2\pi j(l-m)/N}$, and $\delta_{0,T}$ is the Kronecker delta. Note that the corresponding transfer functions $\hat{P}(\omega)$ and $\hat{Q}(\omega)$ are independent of ω. It is clear from (5) that if both H and G are circularly symmetric, then this choice for P and Q diagonalizes both H and G, eliminating the need for crosstalk cancellation. Of course, this situation is never quite true in practice, although adding FFT and IFFT pre- and post-processors may shorten the crosstalk impulse response, thereby reducing the complexity of the crosstalk cancellers.

To derive the form of the pre- and post-processors shown in Figure 3, we note that for circulant $H(T)$,

$$E \, H(T) \, E^{-1} = \Lambda(T) \tag{6}$$

where $\Lambda(T)$ is diagonal and $E^{-1} = E^*$ where "*" denotes complex conjugate transpose. Let

$$E = E_R + jE_I \tag{7}$$

where $[E_R]_{ik} = \frac{1}{N}\cos\,[2\pi(i-1)(k-1)/N]$ and $[E_I]_{ik} = \frac{1}{N}\sin\,[2\pi(i-1)(k-1)/N]$. Then E_R and E_I satisfy the following properties:

$$E_R' = E_R, \quad E_I' = E_I \tag{8}$$

where " ' " denotes transpose, and

$$E_R \, E_I = E_I \, E_R = 0. \tag{9}$$

Because H is real and symmetric, (6) implies that

$$E_R \, HE_R + E_I \, HE_I = \Lambda \tag{10}$$

and similarly,

$$E_R \, \Lambda E_R + E_I \, \Lambda E_I = H. \tag{11}$$

Consequently,

$$\Lambda = (E_R + E_I)H(E_R + E_I) - 2E_I \, HE_R = (E_R + E_I)H(E_R + E_I) \tag{12}$$

since (9) and (11) imply that $E_I \, HE_R = 0$. The pre- and post-processors can therefore be realized as identical resistive networks given by $E_R + E_I$. For the case $N = 4$ we have that

$$E_R + E_I = \frac{1}{4}\begin{bmatrix} 1 & 1 & 1 & 1 \\ 1 & 1 & -1 & -1 \\ 1 & -1 & 1 & -1 \\ 1 & -1 & -1 & 1 \end{bmatrix}. \tag{13}$$

3. SIMULATION RESULTS

The simulated channel consists of a bundle of four identical 24-gauge twisted pairs 12 kft in length. The impulse response of each twisted pair was computed from a standard transmission line model [5], and is shown in Figure 4a. The NEXT impulse response from pair i to pair k ($g_{ik}(T)$), shown in Figure 4b, was computed from the NEXT transfer function [4]

$$g_{ik}(\omega) = j\omega\int_0^l e^{-2\Gamma(\omega)x} c_{ik}(x)dx \tag{14}$$

where l is the length of the cable, $\Gamma(\omega) = \sqrt{(R + j\omega L)(G + j\omega C)}$ is the propagation constant where R, L, G, and C are the wire resistance, inductance, conductance, and capacitance per unit length, respectively, and $c_{ik}(x)$ is the capacitive coupling between pairs i and k as a function of length. In general, the coupling capacitance can be a very complicated function of distance [4]. However, as a first-order approximation we assume that $c_{ik}(x) = c_{ik}$ is constant and l is very large, so that (14) becomes

$$g_{ik}(\omega) = \frac{j\omega c_{ik}}{2\Gamma(\omega)}. \tag{15}$$

Because the crosstalk cancellers are adaptive, their performance should depend primarily on the length of the NEXT impulse response, rather than on its detailed behavior. Consequently, the preceding NEXT model is assumed adequate for our purposes. The FEXT impulse response between pairs i and k, shown in Figure 4c, was computed from the FEXT transfer function [4]

$$\hat{h}_{ik}(\omega) = j\omega \int_0^l c_{ik}(x)dx.$$

(16)

Assuming again that $c_{ik}(x)$ is constant, then $\hat{h}_{ik}(\omega) = j\omega c_{ik}l$.

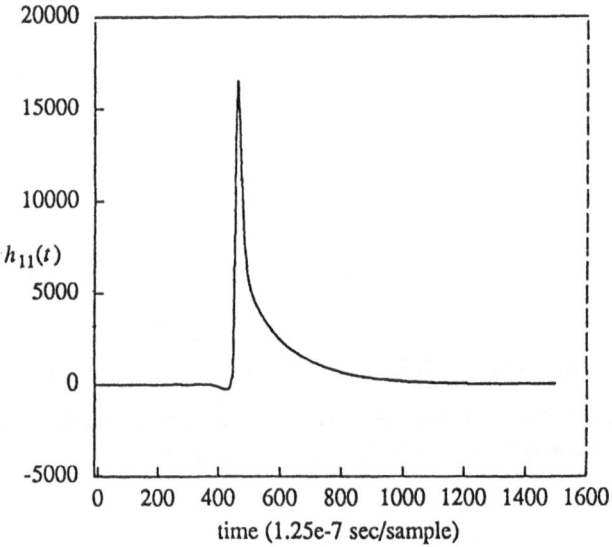

Figure 4a. Self Impulse Response

The transmitted signal for each channel has the form

$$s(t) = \sum_k a_k p(t-kT)$$

(17)

where $a_k \in \{\pm 1\}$ is an element of a pseudorandom sequence, and

$$p(t) = \begin{cases} 1, & 0 \le t \le T \\ 0 & \text{elsewhere} \end{cases}$$

(18)

The data rate on each scalar channel is $1/T = 800$ kbps. Since no linear pre-equalizer is included in the receiver, a timing recovery circuit is required to select a sampling phase that ensures the precursors are much smaller than the cursor $h_{11}(\tau)$, where τ is the sampling delay. For the following simulation results the sampling phase is chosen so that the cursor is at least 25 times larger than the precursors.

The constants R, C, L, and G used to simulate the NEXT and FEXT impulse response are given in Table 1. Let C be the capacitive coupling matrix, i.e., $[C]_{ik} = c_{ik}$ and $c_{ii} = C$. To assess any advantage that might be gained by using the FFT and IFFT operators as pre- and post-processors, two different capacitive coupling matrices were simulated. The first case represents a small perturbation from the ideal circulant case, whereas the second case represents a larger deviation from the circulant case.

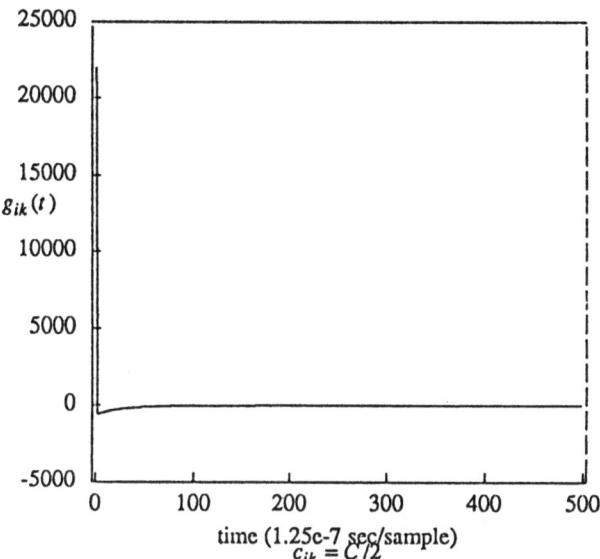

$$c_{ik} = C/2$$

Figure 4b. NEXT Impulse Response.

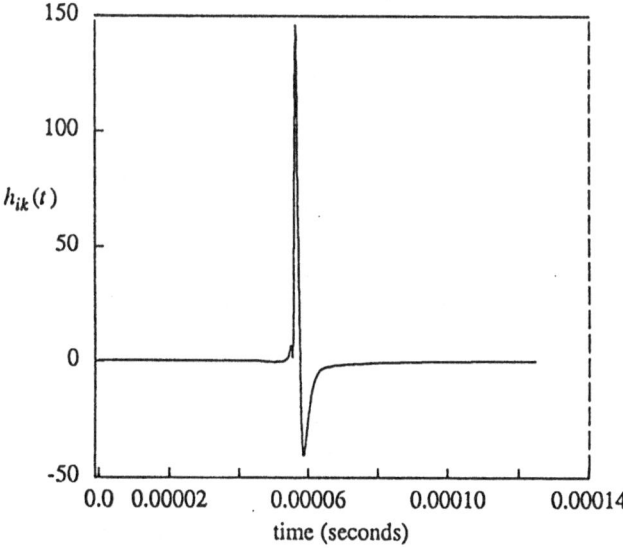

Figure 4c. FEXT impulse response.

Case (i) Small Perturbation

$$C = \begin{bmatrix} C & 0.9c & 0.85c & 1.1c \\ 0.9c & C & 1.05c & 0.9c \\ 0.85c & 1.05c & C & 1.15c \\ 1.1c & 0.9c & 1.15c & C \end{bmatrix} \qquad (19a)$$

Case (ii) Large Perturbation

$$C = \begin{bmatrix} C & 0.9c & 3c & 2c \\ 0.9c & C & c & 1.5c \\ 3c & c & C & 4c \\ 2c & 1.5c & 4c & C \end{bmatrix} \qquad (19b)$$

Two values for the nominal coupling capacitance c were simulated, $c = C/10$ and $c = C/2$.

The first set of simulation results corresponding to the case $c = C/10$ is shown in Tables 2a and 2b. The performance criterion used is Signal-to-Noise Ratio (SNR) defined as

$$SNR = 10 \log \left[\frac{\frac{1}{A} \sum_{T=0}^{A-1} [\hat{a}_T \gamma_1(T)]^2}{\frac{1}{A} \sum_{T=0}^{A-1} e_1^2(T)} \right] = 10 \log \frac{\sum_{T=0}^{A-1} \gamma_1^2(T)}{\sum_{T=0}^{A-1} e_1^2(T)} \qquad (20)$$

where $e_1(T)$ is given by (3), and $\hat{a}_T$ and $\gamma_1(T)$ are the receiver estimates of the transmitted symbol a_T and the cursor of the impulse response, $h_{11}(\tau)$, respectively. The averages were computed for $A = 120$ samples, and after a delay of 2300 samples to ensure convergence of the adaptive filters. Tables 2 and 3 show SNR computed for various combinations of taps in the NEXT and FEXT cancellers. The computed values of SNR both with and without pre- and post-processors are shown. In all cases the number of taps in the echo canceller and in the feedback filter (DFE_{11}) are 10 and 70, respectively. Also, the step-size β_2 used to adjust the taps of DFE_{11} was 0.003, the step-size used to adjust the taps of the echo canceller and NEXT cancellers was 0.002, and the step-size used to adjust the taps of the FEXT cancellers ($DFE_{i1}, i \neq 1$) was 0.001. The results in Table 2a (2b) are for the small (large) perturbation case. Tables 3a and 3b show the analogous results to those shown in Tables 2a and 2b for the case $c = C/2$.

Figures 5a-c show eye diagrams corresponding to three of the entries in Table 3b. Figure 5a assumes three taps per NEXT canceller, five taps per FEXT cancellers, and that the pre- and post-processors are present. Figure 5b assumes no crosstalk cancellation, but with pre- and post-processing, and Figure 5c assumes neither crosstalk cancellation nor pre- and post-processing. As the number of taps in the NEXT cancellers are reduced, crosstalk interference increases, and the eye opening narrows. Note that the eye in Figure 5c is closed.

4. CONCLUSIONS

Tables 2 and 3 indicate that for the channels considered, adding the FFT and IFFT pre- and post-processors increases the SNR in most cases. This improvement becomes more significant as the number of NEXT and FEXT canceller taps is reduced. In particular, if the number of taps in the NEXT or FEXT canceller spans the length of the corresponding impulse response, then there is nothing to be gained by adding the pre- and post-processors. On the other hand, if NEXT and FEXT cancellation is ommitted, then the results indicate that the pre- and post-processors reduce the amount of crosstalk distortion present.

Since the NEXT impulse response shown in Figure 4b is quite short, only a few NEXT canceller taps (i.e., less than four) are needed to adequately cancel NEXT generated from our model. The FEXT impulse response shown in Figure 4c is longer than the NEXT impulse response. Consequently, the SNR shown in the right most column of Table 3 drops significantly when the number of taps per FEXT canceller is reduced to five. Tables 2a and 2b show the SNR as being an insensitive function of the number of taps in the NEXT and FEXT cancellers. The large coupling capacitance used to generate the results in Tables 3a and 3b was selected to increase this sensitivity in performance.

Note that the SNR sometimes increases when the number of taps per NEXT or FEXT canceller is reduced. This implies that the reduction in mean squared error obtained from adding extra taps to the cancellers is less than the increase in mean squared error due to tap fluctuations caused by the adaptive algorithm. Tables 2 and 3 indicate that the SNR is approximately maximized with three taps per NEXT canceller and five taps per FEXT canceller. The results also indicate that including the pre- and post-processors virtually eliminates any improvement in performance obtainable from FEXT cancellation. The modest increase in complexity associated with the addition of the proposed pre- and post-processors and NEXT cancellers appears to buy a significant improvement in performance for the channels simulated.

REFERENCES

1. D. G. Messerschmitt, "Design Issues in the ISDN U-Interface Transceiver," *IEEE Journal on Selected Areas in Communications,* Vol. SAC-4, No. 8, pp. 1281-1293, Nov. 1986.

2. G. A. Campbell, "Crosstalk Formulae for Non-Loaded Circuits", and "Crosstalk Formulae for Phantom Circuits", *B.S.T.J.,* vol. 14, no. 4, pp. 559-570, 1935.

3. M. L. Honig, K. Steiglitz, and B. Gopinath, "Multi-Channel Signal Processing for Data Communications in the Presence of Crosstalk", submitted to *IEEE Transactions on Communications.*

4. H. Cravis and T. V. Crater, "Engineering of T1 Carrier System Repeatered Lines," *B. S. T. J.,* Vol. 42, No. 2, pp. 431-486, March 1963.

5. Members of Technical Staff, AT&T Bell Laboratories, *Transmission Systems for Communications,* 5th ed. AT&T Bell Laboratories, 1982.

Cable Parameters
$R = 281.30\ \Omega/\text{mile}$
$L = 0.972e-3\ \text{H/mile}$
$G = 2.217e-6\ \Omega^{-1}/\text{mile}$
$C = 0.083e-6\ \text{F/mile}$

Table 1.

SNR vs. Number of NEXT and FEXT Canceller Taps (small perturbation case, c= C/10)			
# NEXT taps	# FEXT taps	With FFT/IFFT	Without FFT/IFFT
10	20	22.5	22.6
5	20	23.1	23.1
3	20	23.1	23.0
1	20	23.3	23.1
5	5	23.7	23.9
3	5	23.7	24.1
1	5	23.9	24.2
5	0	24.4	22.8
3	0	24.4	22.7
1	0	24.6	22.8
0	0	24.4	20.5

Table 2a.

SNR vs. Number of NEXT and FEXT Canceller Taps (large perturbation case, c= C/10)			
# NEXT taps	# FEXT taps	With FFT/IFFT	Without FFT/IFFT
10	20	22.4	22.4
5	20	22.9	22.9
3	20	23.0	22.9
1	20	23.0	22.5
5	5	23.4	23.6
3	5	23.5	23.7
1	5	23.5	23.2
5	0	23.2	18.6
3	0	23.2	18.5
1	0	23.3	18.4
0	0	20.2	14.8

Table 2b.

SNR vs. Number of NEXT and FEXT Canceller Taps (small perturbation case, c= C/2)			
# NEXT taps	# FEXT taps	With FFT/IFFT	Without FFT/IFFT
10	20	21.9	21.9
5	20	22.2	21.7
3	20	22.2	21.0
1	20	22.3	19.1
5	5	22.5	22.0
3	5	22.5	21.4
1	5	22.6	19.2
5	0	23.0	13.5
3	0	23.0	13.4
1	0	23.2	13.1
0	0	22.4	7.8

Table 3a.

SNR vs. Number of NEXT and FEXT Canceller Taps (large perturbation case, c= C/2)			
# NEXT taps	# FEXT taps	With FFT/IFFT	Without FFT/IFFT
10	20	20.9	19.9
5	20	21.3	19.0
3	20	21.3	17.7
1	20	20.7	14.4
5	5	21.4	18.5
3	5	21.5	17.2
1	5	20.8	13.8
5	0	17.8	6.8
3	0	17.7	6.6
1	0	17.5	6.3
0	0	11.5	1.1*

Table 3b.

*Eye is closed

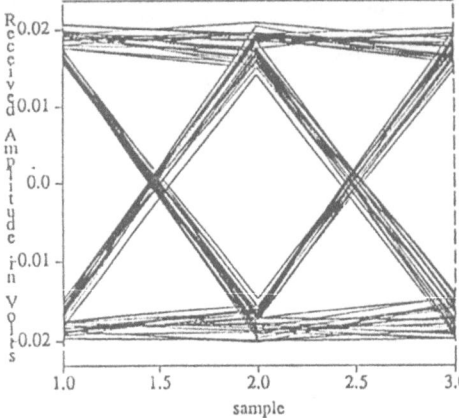

Figure 5a. Eye diagram corresponding to entry in Table 3b. The receiver contains five taps/NEXT canceller, three taps/FEXT canceller, and pre- and post-processing. Baud-rate sampling is assumed.

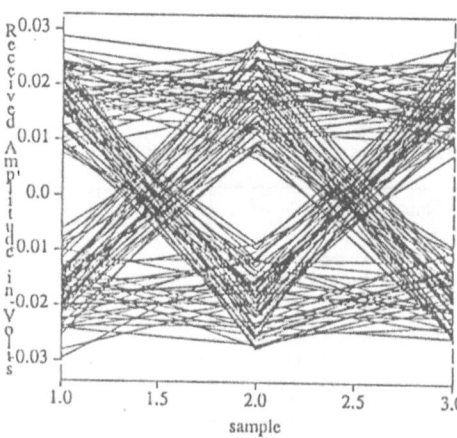

Figure 5b. Eye diagram corresponding to entry in Table 3b. The receiver contains pre- and post-processors without crosstalk cancellation.

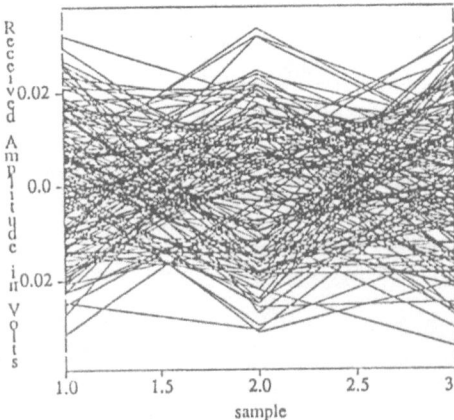

Figure 5c. Eye diagram corresponding to entry in Table 3b assuming no crosstalk cancellers and no pre- and post-processors.

ANALYSIS OF A CLASS OF ADAPTIVE NONLINEAR PREDICTORS

H. Vincent Poor

University of Illinois

Rajiv Vijayan

University of Illinois

1. Introduction. In this paper, we derive and analyze a nonlinear one-step predictor for a discrete-time random process which is the sum of a random binary sequence, a narrow-band (NB) Gaussian signal and white Gaussian noise. This study was prompted by the need to suppress NB interference in direct-sequence spread-spectrum communication systems, and thus we will discuss our results in this framework. However, the results will also apply to other situations where the above probabilistic model is valid.

In a direct-sequence spread-spectrum system [1], the message signal is modulated with a binary pseudo-noise (PN) signal before transmission, so that the transmission bandwidth is much greater than the message bandwidth. At the receiver, the incoming signal is despread by correlating it with the PN sequence. It has been shown in [2] - [10] that the NB interference rejection capability of spread-spectrum systems can be improved substantially by suitably processing the received signal prior to correlating it with the PN sequence. This processing is based on the following idea. Since the spread signal has a nearly flat spectrum, it cannot be predicted accurately from its past values unless, of course, we make use of our knowledge of the PN sequence. The interfering signal, being narrow-band, can be predicted accurately. Hence, a prediction of the received signal based on previously received values will, in effect, be an estimate of the interfering signal. By subtracting the predicted value obtained at each sampling instant from the signal received during the subsequent instant and using the resulting prediction error as the input to the correlator, the effect of the interfering signal can be reduced.

Previous work in this area has primarily involved the use of transversal filters to suppress the interference. Such a filter forms a linear prediction of the received signal based on previous samples. This estimate is subtracted from the delayed received signal to obtain an error signal which is used as the input to the correlator. The filter tap coefficients are updated using a suitable adaptive algorithm, such as the Widrow LMS algorithm. The performance of the interference suppression filter can be evaluated in terms of the improvement in signal-to-noise ratio (SNR) achieved. For the purpose of such analyses, the direct-sequence signal can be modeled as being an independent, identically distributed (i.i.d.) binary sequence. Note that such a sequence is highly non-Gaussian. Thus, the *optimum* filter for predicting a narrow-band process in the presence of such a sequence will, in general, be *nonlinear*. In this paper, our aim is to exploit the binary nature of the direct-sequence signals to obtain alternative nonlinear filters which perform better for cancellation of narrow-band interference than the linear filters hitherto employed for this purpose.

2. A Nonlinear Filter. In this section, we will describe the spread-spectrum communication signal that leads to the model of interest and evaluate the performance of interference rejection filters when the statistics of the NB interference are known to the receiver of such a signal.

2.1. A Model for the Received Signal. The low pass equivalent of a direct-sequence spread-spectrum modulation waveform is given by

$$m(t) = \sum_{k=0}^{L-1} c_k q(t - k\tau_c) ,$$ (2.1)

where L is the number of PN chips per message bit, τ_c is the chip interval, c_k is the kth chip of the PN sequence, and $q(t)$ is a rectangular pulse of duration τ_c .

The total transmitted signal may be expressed as

$$s(t) = \sum_k b_k m(t - kT_b) ,$$ (2.2)

where $\{ b_k \}$ is the binary information sequence and $T_b = L\tau_c$ is the bit duration.

The received signal has the form

$$z(t) = s(t) + n(t) + i(t) ,$$ (2.3)

where $n(t)$ is white Gaussian noise and $i(t)$ is narrow-band interference.

We assume that the received signal is chip-matched-filtered and sampled at the chip rate of the PN sequence to yield samples

$$z_k = s_k + n_k + i_k ,$$ (2.4)

where the discrete-time sequences $\{ s_k \}$, $\{ n_k \}$, and $\{ i_k \}$ are due to $\{ s(t) \}, \{ n(t) \}$, and $\{ i(t) \}$, respectively.

Under the assumption that the PN sequence is truly random, we can consider $\{ s_k \}$ to be a sequence of i.i.d. random variables taking on values +1 or -1 with equal probability. $\{ n_k \}$ is a sequence of i.i.d. zero mean Gaussian random variables with variance σ_n^2. The sequences $\{ n_k \}$, $\{ s_k \}$ and $\{ i_k \}$ are assumed to be mutually independent. This is the model for the received signal that will be used in the remainder of this paper.

2.2. Nonlinear Filters for Autoregressive Interference. We begin our analysis by modeling the narrow-band interference $\{ i_k \}$ as a Gaussian autoregressive process of order p, i.e., we assume a model of the form

$$i_k = \sum_{i=1}^{p} \phi_i i_{k-i} + e_k ,$$ (2.5)

where $\{ e_k \}$ is a white Gaussian process, and where the autoregressive parameters $\phi_1, \phi_2, \cdots, \phi_p$ are known to the receiver.

Under this model, the received signal has a state space representation as follows.

$$x_k = \Phi x_{k-1} + w_k ,$$ (2.6)

$$z_k = H x_k + v_k ,$$ (2.7)

where

$$x_k = [i_k \ i_{k-1} \ \cdots \ i_{k-p+1}]^T ,$$

$$
\Phi = \begin{bmatrix}
\phi_1 & \phi_2 & \cdots & \phi_{p-1} & \phi_p \\
1 & 0 & \cdots & 0 & 0 \\
0 & 1 & \cdots & 0 & 0 \\
\cdots & \cdots & \cdots & \cdots & \cdots \\
0 & 0 & \cdots & 1 & 0
\end{bmatrix} ,
$$

$$
w_k = [e_k \ 0 \ \cdots \ 0]^T ,
$$

$$
H = [1 \ 0 \ \cdots \ 0] ,
$$

and

$$
v_k = s_k + n_k .
$$

When $\{v_k\}$ and $\{w_k\}$ are Gaussian processes, the minimum-mean-squared error estimates of the state x_k, and hence of the interference i_k, are given recursively by the Kalman-Bucy filtering equations, which also yield the optimum linear estimates for this model [13]. However, since v_k is the sum of two independent random variables, one of which is Gaussian and the other takes on values +1 or -1 with equal probability, its density is the weighted sum of two Gaussian densities, i.e.,

$$
p_{v_k}(v)= \frac{1}{2} \left[N_{\sigma_n^2}(v - 1) + N_{\sigma_n^2}(v + 1) \right] , \tag{2.8}
$$

where the zero-mean Gaussian probability density $N_{\sigma^2}(\cdot)$ is defined by

$$
N_{\sigma^2}(x)= \frac{1}{\sqrt{2 \pi} \sigma} e^{-x^2 / 2\sigma^2} .
$$

For the case where the measurement noise density is a Gaussian sum, the exact conditional mean estimator has been shown by Sorenson and Alspach [11] to have exponentially increasing complexity which renders it unsuitable for practical implementation.

In [12], Masreliez has developed an approximate conditional mean (ACM) filter for estimating the state of a linear system with Gaussian state noise and non-Gaussian measurement noise. This filter is derived by making the approximating assumption that the state prediction density $p(x_k \mid Z^{k-1})$ is Gaussian with mean x_k and covariance matrix M_k. Under this assumption, the filtered estimate $\hat{x}_k$ and its conditional covariance P_k are obtained recursively through the update equations

$$
\hat{x}_k = x_k + M_k H^T g_k (z_k) \tag{2.9}
$$

$$
P_k = M_k - M_k H^T G_k (z_k)H M_k \tag{2.10}
$$

$$
M_{k+1} = \Phi P_k \Phi^T + Q_k \tag{2.11}
$$

$$
x_{k+1} = \Phi \hat{x}_k \tag{2.12}
$$

where

$$
g_k (z_k)= -\left| \frac{\partial p (z_k \mid Z^{k-1})}{\partial z_k} \right| \cdot \left| p (z_k \mid Z^{k-1}) \right|^{-1} \tag{2.13}
$$

$$G_k(z_k) = \frac{\partial g_k(z_k)}{\partial z_k} ,$$

(2.14)

and

$$Q_k = E\{w_k w_k^T\} .$$

Since $p(x_k \mid Z^{k-1})$ is assumed to be Gaussian, using (2.7) and the fact that v_k is independent of Z^{k-1}, we obtain the following expression for the observation prediction density.

$$p(z_k \mid Z^{k-1}) = \frac{1}{2}\left[N_{(HM_kH^T+\sigma_n^2)}(z_k - H\bar{x}_k - 1) + N_{(HM_kH^T+\sigma_n^2)}(z_k - H\bar{x}_k + 1)\right] .$$

On substituting this expression into (2.13) and (2.14), we then have

$$g_k(z_k) = \frac{1}{(HM_kH^T+\sigma_n^2)}\left[z_k - H\bar{x}_k - \tanh\left[\frac{z_k - H\bar{x}_k}{HM_kH^T+\sigma_n^2}\right]\right],$$

(2.15)

and

$$G_k(z_k) = \frac{1}{(HM_kH^T+\sigma_n^2)}\left[1 - \frac{1}{(HM_kH^T+\sigma_n^2)}\mathrm{sech}^2\left[\frac{z_k - H\bar{x}_k}{HM_kH^T+\sigma_n^2}\right]\right].$$

(2.16)

The ACM filter is thus seen to have a structure similar to that of the standard Kalman-Bucy filter. The time update equations (2.11) and (2.12) are identical to those in the Kalman-Bucy filter. The measurement update of (2.9) involves correcting the predicted value by a non-linear function of the prediction residual $z_k - H\bar{x}_k$. The nature of the nonlinearity is determined by the probability density of the observation noise. For our case of interest, this function is given by (2.15).

2.3. Computer Simulation Results.   Computer simulations have been carried out to evaluate the capability of the ACM filter to reject an autoregressive interferer. When the interference is modeled as a first-order autoregressive process, which does not have a very sharply peaked spectrum, the performance of the ACM filter does not seem to be appreciably better than that of the Kalman-Bucy filter. However, when the spectrum of the interference is made to be more sharply peaked by increasing the order of the autoregression, the ACM filter is found to give significant performance gains over the Kalman filter.

Sample simulation results are presented in Table 1. In this simulation, the interfering signal is obtained by passing white noise through a second-order IIR filter with both poles at $z = 0.99$; i.e.,

$$i_k = 1.98\ i_{k-1} - 0.9801\ i_{k-2} + e_k ,$$

where $\{e_k\}$ is white Gaussian noise. The power of the background thermal noise is kept constant at $\sigma_n^2 = 0.01$. The various signal-to-noise ratios involved are defined as follows.

$$\text{SNR at the filter input} \triangleq \frac{E(s_k^2)}{E(\mid z_k - s_k\mid^2)} .$$

SNR at the filter output $\triangleq \dfrac{E(s_k^2)}{E(|\epsilon_k - s_k|^2)}$

where ϵ_k is the output of the filter. Therefore,

SNR improvement $= \dfrac{E(|z_k - s_k|^2)}{E(|\epsilon_k - s_k|^2)}$.

In making these definitions, we have neglected any distortion of the direct-sequence signal caused by the suppression filter. This is a reasonable approximation when sufficiently long spreading sequences with good autocorrelation properties are used, especially in the case of predicted estimates, which are of more interest since they will be used in the adaptive filters to be discussed later.

The SNR at the input is varied by changing the power of the interfering signal. The table lists the steady-state performance of both the ACM and Kalman-Bucy estimators, using filtered as well as predicted estimates. The figures shown are the average performance figures over ten trials. On each trial, the filter was allowed to run until its performance attained a steady state. Then, the average SNR improvement over 500 data points was calculated. This is the method used for all the simulations in this paper.

From these results, it is seen that, when the statistics of the interfering signal are known to the receiver, nonlinear filtering techniques offer considerably better interference rejection properties than do linear filters. This indicates that when the interfering signal has unknown parameters, adaptive nonlinear filters based on the techniques considered here could give better performance than conventional adaptive linear filters. This issue is addressed in the following section.

3. Adaptive Nonlinear Filters. In Section 2, we considered nonlinear filters for suppressing interfering signals whose parameters are constant and known to the receiver. In practice, the parameters of the interference are rarely known to the receiver. Also, the interference can have statistics that vary with time. Therefore, an effective suppression filter should be able to adapt itself to variations in the interference characteristics. In this section, we will consider adaptive nonlinear filters which can track the interfering signal and reject it.

Table 1. Simulation Results for AR Interferer

Input SNR (dB)	SNR Improvement (dB)			
	Kalman Filter		ACM Filter	
	Filtered Estimates	Predicted Estimates	Filtered Estimates	Predicted Estimates
-20.0	27.6	26.7	42.9	37.5
-15.0	24.5	23.9	37.6	33.7
-10.0	19.8	19.4	31.3	28.3
-5.0	16.5	16.2	26.1	24.0
0.0	12.5	12.3	20.9	19.3
5.0	9.6	9.5	15.5	14.3

3.1. Adaptive ACM Filters. In the previous section, it was seen that for rejecting an autoregressive interference with known parameters, the ACM filter performs appreciably better than the optimal linear filter. Therefore, in order to suppress an interference with unknown or varying parameters, an obvious method of implementing an adaptive filter would be to identify the AR parameters of the interference recursively and to carry out the ACM filtering algorithm using the estimates obtained at each instant.

For estimating the AR parameters of the interference, we note that the received signal, being the sum of an autoregressive process and a white process, is an autoregressive moving-average (ARMA) process whose AR parameters are the same as those of the interfering signal. Thus, the parameter estimation problem of interest is reduced to that of estimating the AR parameters of an ARMA process. Several algorithms for estimating these parameters recursively are discussed by Ljung and Söderström in [14]. One possible approach is the Recursive Maximum Likelihood (RML) algorithm which has been used in [15] and [16] for adaptive signal processing problems similar to the problem of interest. However, when the model for the interfering signal has poles close to the unit circle, which is the case for narrow-band signals, the parameter estimates obtained by the RML algorithm converge rather slowly. Our simulations have shown the ACM filter to be sensitive to variations in the parameter values. Therefore, the obvious approach of directly adapting the ACM filter using estimates of AR parameters is not promising. Hence other approaches are necessary to obtain effective adaptive nonlinear filtering algorithms.

3.2. Adaptive Filters Based on the LMS Algorithm. In [5] and [9], Iltis, Li, and Milstein have applied linear transversal filters to the problem of narrow-band interference rejection. In these studies, the tap weights $a_1, a_2, \cdots, a_L$ of an Lth order filter are adjusted using the Widrow LMS algorithm as follows :

$$\Theta_k = \Theta_{k-1} + \mu \epsilon_k X_k \qquad (3.1)$$

where Θ_k is the tap-weight vector obtained at time k ; i.e.,

$$\Theta_k \triangleq [\, a_1(k), a_2(k), \cdots, a_L(k) \,]^T \;\; ,$$

$$X_k \triangleq [\, z_{k-1}, z_{k-2}, \cdots, z_{k-L} \,]^T \;\; ,$$

μ is a tuning constant controlling the stability and rate of convergence of the algorithm, and ϵ_k is the prediction error; i.e.,

$$\epsilon_k = z_k - \hat{z}_k \;\; , \qquad (3.2)$$

where

$$\hat{z}_k \triangleq X_k^T \Theta_{k-1} \qquad (3.3)$$

is the predicted value of the received signal z_k based on the L immediate past values. The prediction error ϵ_k is used as the input to the PN correlator.

It has been shown in [4] and [5] that better interference rejection can be obtained by using a two-sided interpolation filter rather than a one-sided prediction error filter.

In Section 2, it was seen that in the ACM filter, the predicted value of the state was obtained as a linear function of the previous estimate modified by a nonlinear function of the prediction error. We now use the same approach to modify the adaptive linear prediction filter described above. In order to show the influence of the prediction error explicitly, the equation for the linear one-sided prediction filter can be written as

$$\hat{z}_k = \sum_{i=1}^{L} a_i(k-1)[\hat{z}_{k-i} + \epsilon_{k-i}] .$$ (3.5)

We make the assumption, similar to that made in the derivation of the ACM filter, that the prediction residual ϵ_k is the sum of a Gaussian random variable and a binary random variable. If the variance of the Gaussian random variable is σ_k^2, then the nonlinear transformation appearing in the ACM filter can be written as

$$\rho_k(\epsilon_k) = \epsilon_k - \tanh\left|\frac{\epsilon_k}{\sigma_k^2}\right| .$$ (3.6)

By transforming the prediction error in equation (3.5) using the above nonlinearity, we get a nonlinear transversal filter for the prediction of z_k, namely,

$$\hat{z}_k = \sum_{i=1}^{L} a_i(k-1)[\hat{z}_{k-i} + \rho_{k-i}(\epsilon_{k-i})] .$$ (3.7)

The structure of this filter is shown in Figure 1.

In order to implement the filter of (3.7), an estimate of the parameter σ_k^2 and an algorithm for updating the tap weights must be obtained. A useful estimate for σ_k^2 is $\hat{\sigma}_k^2 = \Delta_k - 1$ where Δ_k is a sample estimate of the prediction error variance. The LMS algorithm (3.1) - (3.2) can still be used for the tap-weight updates. Note, however, that the LMS algorithm is a gradient algorithm based on the *linear* prediction of (3.3). An alternative gradient algorithm was derived in [17] based on the nonlinear prediction of (3.7). However, due to the increased complexity and unsatisfactory performance of that algorithm, we will restrict our attention to nonlinear filters using the LMS algorithm.

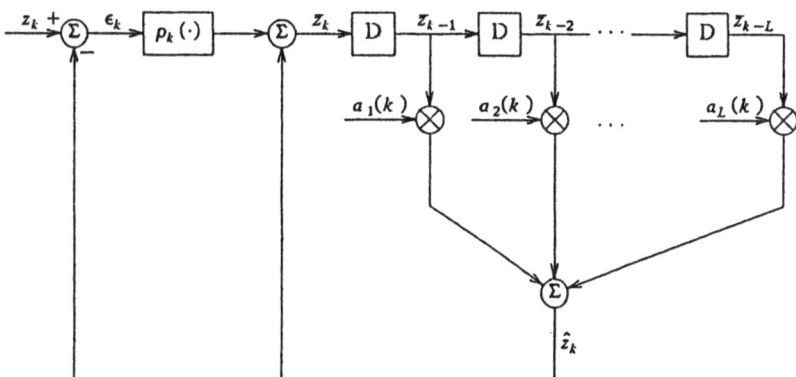

Figure 1. The Adaptive Nonlinear Prediction Filter

We now analyze the performance of the adaptive algorithms discussed in this section.

3.3. Computer Simulation Results. Computer simulations have been carried out to compare the performance of the two adaptive interference rejection filters described above. Filter A is the linear two-sided interpolation filter using the LMS algorithm. This is the best of the adaptive linear filters studied previously. Filter B is the nonlinear prediction filter of (3.7) with the coefficients being updated using the LMS algorithm.

Table 2 presents sample simulation results for a single-tone sinusoidal interferer. The performance figures were obtained by averaging over ten runs, each consisting of 500 data points. In this table, the frequency of the interfering tone is kept constant at 0.15 radian and its power is varied. The background thermal noise power is kept constant at $\sigma_n^2 = 0.01$. The number of filter taps is $L = 10$ for each filter. The bound in the last column refers to the optimum SNR improvement that can be obtained using a fixed two-sided linear interpolation filter with L taps and optimum tap weights. This quantity can be calculated using an expression derived in [4] for this case of a single-tone interferer. Table 3 presents analogous results for the case in which the interference is the narrow-band Gaussian signal considered in Section 2.

These simulation results indicate that filter B performs significantly better than filter A for both types of interference, and that its performance is substantially better than that of the theoretically optimum linear filter with the same number of taps.

3.4. Connection to Decision Feedback. Equation (3.7) for the prediction can be rewritten as

Table 2. Adaptive Filter Performance for Sinusoidal Interferer

Input SNR (dB)	SNR Improvement (dB)		
	Filter A	Filter B	Bound
-20	27.8	33.3	28.6
-15	23.2	28.3	23.6
-10	18.4	23.3	18.6
-5	13.4	18.6	13.7

Table 3. Adaptive Filter Performance for AR Interferer

Input SNR (dB)	SNR Improvement (dB)	
	Filter A	Filter B
-20	28.4	33.1
-15	23.7	28.6
-10	19.2	24.9
-5	15.0	21.7

$$\hat{z}_k = \sum_{i=1}^{L} a_i \left| z_{k-i} - \beta_{k-i}(\epsilon_{k-i}) \right| \tag{3.8}$$

where

$$\beta_k(\epsilon_k) \triangleq \epsilon_k - \rho_k(\epsilon_k) = \tanh\left[\frac{\epsilon_k}{\sigma_k^2}\right] \tag{3.9}$$

We make the assumption, analogous to that made in deriving the ACM filter, that the prediction error ϵ_k can be decomposed as

$$\epsilon_k = s_k + y_k \tag{3.10}$$

where y_k is independent of s_k and

$$p(y_k \mid Z^k) = N_{\sigma_k^2}(y_k) \tag{3.11}$$

Under this assumption, $\beta_k(\epsilon_k)$ is the minimum-mean-squared-error estimate of s_k based on the prediction error. This estimate is subtracted from the received signal to obtain the input to the linear part of the predictor.

In the limit as $\sigma_k^2 \rightarrow 0$, the function $\beta_k(\cdot)$ converges pointwise to the function

$$\text{sgn}(x) = \begin{cases} 1 & \text{if } x > 0 \\ 0 & \text{if } x = 0 \\ -1 & \text{if } x < 0 \end{cases}$$

In this limiting case, we are able to estimate the direct-sequence signal perfectly so that the input to the linear part of the predictor consists of only the narrow-band interference and white noise. Hence, the linear part is just a finite-length predictor for the sum of white noise and interference, whose optimum tap weights are given by the Wiener-Hopf equations. Also, the LMS algorithm with sufficiently small step size will converge to the optimum tap weights.

If $\sigma_k^2 > 0$, using the nonlinearity $\text{sgn}(\epsilon_k)$ instead of $\beta_k(\epsilon_k)$ corresponds to subtracting the MAP estimate of ϵ_k from the received signal rather than the MMSE estimate. Alternatively, it can be viewed as a choice between making hard and soft decisions about the signal, which are then fed back to the predictor. The filter with the $\text{sgn}(\cdot)$ nonlinearity is widely known as the decision-feedback filter which has been used to suppress inter-symbol interference in communication channels and narrow-band interference in QPSK systems [18].

3.5. The Case of Unknown Amplitude. In the above discussion, we had assumed that the amplitude of the direct-sequence signal is known and that it is equal to 1. For an arbitrary amplitude A, the nonlinearity becomes

$$\beta_k(\epsilon_k) = A \tanh\left[\frac{A \epsilon_k}{\sigma_k^2}\right] .$$

Therefore, if the filter is to be implemented in a situation where the amplitude of the signal is not known a priori, we have to incorporate an estimator of the signal amplitude into the predictor. For this, we again assume the density for ϵ_k given by equations (3.10) and (3.11). Under this assumption,

$$E(|\epsilon_k| \mid Z^k) = A + (2/\pi)^{\frac{1}{2}}\sigma_k e^{-A^2/2\sigma_k^2} - 2AQ\left[\frac{A}{\sigma_k}\right]$$

(3.12)

where

$$Q(x) \triangleq \frac{1}{\sqrt{2\pi}} \int_x^\infty e^{-t^2/2}dt \ .$$

If $A/\sigma_k \gg 1$, i.e., if the signal-to-noise ratio at the filter output is large, the right-hand side of equation (3.12) will be dominated by the first term. Hence, a reasonably good estimate of the signal amplitude can be obtained by averaging the absolute values of the prediction errors. This estimate can then be used in place of A in equation (3.12). One such estimator is given by

$$\hat{A}_{k+1} = \hat{A}_k + \gamma[\ |\epsilon_{k+1}| - \hat{A}_k\ ] \ ; \ \hat{A}_0 = 0 \ .$$

(3.13)

Asymptotically, $E(\hat{A}_k) \to E(|\epsilon_k|)$ and $\hat{A}_k$ has the form

$$\hat{A}_k = \gamma \sum_{i=0}^\infty (1-\gamma)^i |\epsilon_{k-i}| \ .$$

(3.14)

An advantage of this estimator over the arithmetic mean is that it can track time-varying amplitudes.

Setting the initial estimate to zero corresponds to starting with a linear predictor. This is a reasonable approach if we do not have any knowledge of the signal amplitude to start with. As more samples are received, the acquired information about the amplitude is used via equation (3.14) to introduce a nonlinearity in the prediction.

Simulations indicate that the nonlinear predictor of (3.8) performs well when used in conjunction with the amplitude estimator of (3.13). Also, under these conditions, the soft-decision-feedback predictor of (3.8) performs better than the hard-decision-feedback filter.

4. Conclusions. In this paper, we have considered nonlinear prediction filters for the rejection of narrow-band interference in spread-spectrum systems. The motivation for using prediction filters is that the signal of interest, being broadband, cannot be predicted well whereas the narrow-band interference can be predicted very well. Though previous approaches to this problem involved the use of linear prediction filters, the fact that the received signal is highly non-Gaussian indicated that using nonlinear filters could result in better performance.

In Section 2, for the case of an autoregressive interferer, the problem was cast as a state estimation problem with Gaussian state noise and non-Gaussian observation noise. An approximately optimum nonlinear filter based on the Masreliez approximation [12] was used in this case. Simulations indicate that this filter performs considerably better than the optimum linear estimator. Based on the structure of this nonlinear estimator, an adaptive nonlinear filtering algorithm was derived in Section 3. Computer simulations indicate that this filter is more effective than the conventional adaptive linear filter for rejecting interferers with unknown parameters. It was also shown that this filter can be viewed as a generalization of both linear and decision-feedback filters, in which we utilize our knowledge of the prediction error statistics to make a soft decision about the binary signal, which is then fed back to the predictor.

5. Acknowledgement. This research was supported by the U.S. Army Research Office under Contract DAAL 03-86-K-0093.

H. Vincent Poor	Rajiv Vijayan
Coordinated Science Laboratory	Coordinated Science Laboratory
University of Illinois	University of Illinois
1101 W. Springfield Avenue	1101 W. Springfield Avenue
Urbana, IL 61801	Urbana, IL 61801

REFERENCES

[1] G. R. Cooper and C. D. McGillem, *Modern Communications and Spread Spectrum*. New York: McGraw-Hill, 1986.

[2] F. M. Hsu and A. A. Giordano, "Digital whitening techniques for improving spread-spectrum communications performance in the presence of narrow-band jamming and interference," *IEEE Trans. Comm.*, vol. COM-26, pp. 209-216, Feb. 1978.

[3] J. W. Ketchum and J. G. Proakis, "Adaptive algorithms for estimating and suppressing narrow-band interference in PN spread-spectrum systems," *IEEE Trans. Comm.*, vol. COM-30, pp. 913-924, May 1982.

[4] L-M. Li and L. B. Milstein, "Rejection of narrow-band interference in PN spread-spectrum systems using transversal filters," *IEEE Trans. Comm.*, vol. COM-30, pp. 925-928, May 1982.

[5] L-M. Li and L. B. Milstein, "Rejection of pulsed CW interference in PN spread-spectrum systems using complex adaptive filters," *IEEE Trans. Comm.*, vol. COM-31, pp. 10-20, Jan. 1983.

[6] E. Masry, "Closed-form analytical results for the rejection of narrow-band interference in PN spread-spectrum systems - Part I: Linear prediction filters," *IEEE Trans. Comm.*, vol. COM-32, pp. 888-896, Aug. 1984.

[7] R. A. Iltis and L. B. Milstein, "Performance analysis of narrow-band interference rejection techniques in DS spread-spectrum systems," *IEEE Trans. Comm.*, vol. COM-32, pp. 1169-1177, Nov. 1984.

[8] E. Masry, "Closed-form analytical results for the rejection of narrow-band interference in PN spread-spectrum systems - Part II: Linear interpolation filters," *IEEE Trans. Comm.*, vol. COM-33, pp. 10-19, Jan. 1985.

[9] R. A. Iltis and L. B. Milstein, "An approximate statistical analysis of the Widrow LMS algorithm with application to narrow-band interference rejection," *IEEE Trans. Comm.*, vol. COM-33, pp. 121-130, Feb. 1985.

[10] E. Masry and L. B. Milstein, "Performance of DS spread-spectrum receiver employing interference-suppression filters under a worst-case jamming condition," *IEEE Trans. Comm.*, vol. COM-34, pp. 13-21, Jan. 1986.

[11] H. W. Sorenson and D. L. Alspach, "Recursive Bayesian estimation using Gaussian sums," *Automatica*, vol. 7, pp. 465-479, 1971.

[12] C. J. Masreliez, "Approximate non-Gaussian filtering with linear state and observation relations," *IEEE Trans. Automat. Contr.*, vol. AC-20, pp. 107-110, Feb. 1975.

[13] H. V. Poor, *An Introduction to Signal Detection and Estimation*. New York: Springer-Verlag, 1988.

[14] L. Ljung and T. Söderström, *Theory and Practice of Recursive Identification*. Cambridge, MA: MIT Press, 1983.

[15] B. Friedlander, "System identification techniques for adaptive signal processing," *IEEE Trans. Acoust., Speech, Signal Processing*. vol. ASSP-30, pp. 240-246, Apr. 1982.

[16] B. Friedlander, "A recursive maximum likelihood algorithm for ARMA line enhancement," *IEEE Trans. Acoust., Speech, Signal Processing*, vol. ASSP-30, pp. 651-657, Aug. 1982.

[17] R. Vijayan and H. V. Poor, "Improved algorithms for the rejection of narrow-band interferers from direct-sequence signals," in *Proc. 22nd Annual Conference on Information Sciences and Systems*, Princeton, NJ, pp. 851-856, 1988.

[18] L-M. Li and L. B. Milstein, "Rejection of CW interference in QPSK systems using decision-feedback filters," *IEEE Trans. Comm.*, vol. COM-31, pp. 473-483, Apr. 1983.

RECURSIVE ARMA PARAMETER ESTIMATION WITH A DISCERNING UPDATE STRATEGY- FINITE PRECISION EFFECTS

Ashok K. Rao   Yih-Fang Huang

University of Notre Dame   University of Notre Dame

Introduction.   The performance of adaptive filter algorithms in finite precision environments has received a lot of attention in the past few years. The problem is important because a practical implementation of these algorithms will impose constraints on the word-length, which may cause significant degradation in the performance. For example, some of the fast least-squares algorithms, though appealing in theory, have been found to be unstable in finite word-length implementations [1]. Round-off and quantization errors affect different adaptive algorithms in different ways. The accumulation of round-off errors in the recursive least-squares (RLS) algorithm can cause the inverse of the associated estimated covariance matrix to become indefinite and the algorithm to diverge fairly early, especially if the order of the filter is large[2,3]. This effect is pronounced if the data is ill conditioned, i.e., the data autocorrelation matrix has a large eigenvalue spread. On the other hand, it can take millions of iterations before the effect of quantization errors becomes noticeable in the widely used LMS algorithm[4].

In this paper we first study the effects of roundoff errors in a fixed point implementation of the so-called Optimal Bounding Ellipsoid (OBE) algorithm[5]. This algorithm estimates recursively the coefficients of autoregressive with exogenous inputs (ARX) processes. One of the main features of this algorithm is a discerning update strategy. This feature, obtained by the introduction of an information dependent updating/forgetting factor, yields a modular structure thereby increasing the potential for concurrent and pipelined processing of signals. The presence of such a forgetting factor also gives the algorithm the ability to track time varying parameters.

The OBE algorithm belongs to a broad family of algorithms known as membership set parameter estimation algorithms [6],[7],[8]. These algorithms are particularly useful when the statistical properties of the noise sequence {v(t)} are unknown, but instantaneous bounds on its magnitude are available. In the past few years, there has been a resurgence of interest in these algorithms. However, the key issue of finite precision effects has not received much attention. We have found that in small word-length situations, the performance of the OBE algorithm is superior to that of the RLS algorithm (with and without forgetting factor). The EOBE algorithm, which is an extension of the OBE algorithm to ARMA models[9] is studied next and simulation results also indicate that the EOBE algorithm has better numerical properties than the extended least-squares (ELS) algorithm(see[3]for details of the RLS and ELS algorithms).

This paper is organised as follows: The first section introduces the concept of membership-set parameter estimation and describes in fuller detail the OBE algorithm. The next section presents the extension of the OBE algorithm for parameter estimation of ARMAX processes. The simulation procedure and the simulation results are presented in the following two sections. The paper concludes with discussions of the simulation results.

The OBE Algorithm. Membership-set parameter estimation is concerned with the determination of sets of parameters which are consistent with the measurements, model structure, and noise constraints. The model and noise representations commonly used are

$$y(t) = \theta^{*T} \Phi(t) + v(t), \quad |v(t)| \le r^{1/2}(t) \tag{1}$$

where $\{y\}$ is a sequence of scalar observations, θ^* is the parameter vector to be identified, $\Phi(t)$ is a n-vector of variables known at time t, $\{v\}$ is the noise sequence and $\pm r^{1/2}(t)$ are the time varying noise bounds. Given a sequence $\{y(i), \Phi(i)\}$, $i = 1..k$, the optimal membership set

$$\psi^0{}_k = \cap^k{}_{i=1} S_i$$

where

$$S_i = \{ \theta : (y(i) - \theta^T \Phi(i))^2 \le r(i), \ \theta \in R^n \}$$

From a geometrical viewpoint, S_i is a convex polytope in R^n and contains the true parameter vector. Finding $\psi^0{}_k$ is often computationally intractable and it is therefore necessary to approximate $\psi^0{}_k$ by some set which approximates it closely and which can be described and updated economically. The different membership-set algorithms differ in the way the optimal membership set is approximated and in the method used to obtain an optimum(in some sense) set.

The OBE algorithm estimates the coefficients of ARX processes described by

$$y(t) = a_1 y(t-1) +..+ a_n y(t-n) + b_0 u(t) + b_1 u(t-1) +..+ b_m u(t-m) + v(t)$$

where y(t) is the output, u(t) is the input and v(t) is the noise contaminating the observations. The above equation can be recast as :

$$y(t) = \theta^{*T} \Phi(t) + v(t)$$

where

$$\theta^* = [a_1, a_2 ... a_n, b_0, b_1, ..., b_m]^T$$

is the vector of true parameters, and

$$\Phi(t) = [y(t-1), y(t-2), .. y(t-n), u(t), u(t-1), .. u(t-m)]^T$$

is the regressor vector. It is assumed that the noise is uniformly bounded in magnitude, i.e., there exists $\gamma_0 \ge 0$, such that

$$v^2(t) \le \gamma_0^2 \qquad \text{for all t, hence}$$

$$(y(t) - \theta^{*T} \Phi(t))^2 < \gamma_0^2$$

Let S_t be a subset of the euclidean space R^{n+m+1}, defined by

$$S_t = \{ \theta : (y(t) - \theta^T \Phi(t))^2 \le \gamma_0^2, \ \theta \in R^{n+m+1} \}$$

The OBE algorithm starts off with a large ellipsoid, E_0, in R^{n+m+1} which contains all possible values of the modelled parameter θ^*. After the first observation $y(1)$ is acquired, an ellipsoid is found which bounds the intersection of E_0 and the convex polytope S_1. To hasten convergence, this ellipsoid must be optimized in some sense, say minimum volume[7] or by any other criterion[5,10]. Denoting the optimal ellipsoid by E_1, one can proceed exactly as before with the future observations and obtain a sequence of optimal bounding ellipsoids(OBE) { E_t }.

The center of the ellipsoid E_t can be taken as the parameter estimate at the t-th instant and is denoted by $\theta(t)$. If at a particular time instant i, the resulting optimal bounding ellipsoid would be of a " smaller size ", thereby implying that the data point $y(i)$ contains some " information " regarding the parameter estimates, then the parameter estimates are updated. Otherwise E_i is set equal to E_{i-1}, and the estimates are not updated. In essence, the recursive estimator consists of two modules, an information evaluator followed by an updating processor. At each data point, the received data proceed to the updating processor only if the information evaluator indicates that some fresh information is contained in the data.

Specifically, let the ellipsoid E_{t-1} at the (t-1)-th instant be formulated by

$$E_{t-1} = \{ \theta: (\theta - \theta(t-1))^T P^{-1}(t-1) (\theta - \theta(t-1)) \leq \sigma^2(t-1) \}$$

for some positive definite matrix $P(t-1)$ and a non-negative scalar $\sigma^2(t-1)$. Then, given $y(t)$, an ellipsoid which bounds $E_{t-1} \cap S_t$ "tightly" is

$$\{ \theta: (1 - \lambda_t)(\theta - \theta(t-1))^T P^{-1}(t-1)(\theta - \theta(t-1)) + \lambda_t (y(t) - \theta^T \Phi(t))^2$$

$$\leq (1 - \lambda_t) \sigma^2(t-1) + \lambda_t \gamma_0^2 \}$$

where the forgetting factor λ_t satisfies $0 \leq \lambda_t \leq \alpha < 1$, with α being a user chosen upper bound on the forgetting factor. The size of the bounding ellipsoid is related to the scalar $\sigma^2(t-1)$ and the eigenvalues of $P(t-1)$. The update equations for $\theta(t)$, $P(t)$ and $\sigma^2(t)$, derived in [5], are as follows

$$\theta(t) = \theta(t-1) + K(t)\delta(t) \tag{2a}$$

$$\delta(t) = y(t) - \theta^T(t-1) \Phi(t) \tag{2b}$$

$$K(t) = \frac{\lambda_t P(t-1)\Phi(t)}{1-\lambda_t + \lambda_t G(t)} \tag{2c}$$

$$G(t) = \Phi^T(t) P(t-1) \Phi(t) \tag{2d}$$

$$P(t) = \frac{1}{1-\lambda_t}[I - K(t)\Phi^T(t)] P(t-1) \tag{2e}$$

where $\Phi(t)$ is the regressor vector which contains present and previous input and output samples.

The optimal ellipsoid which bounds the intersection of E_{t-1} and S_t is defined in terms of an optimal value of λ_t. For the OBE algorithm of [5], the optimum value λ_t^* is determined by minimization of $\sigma^2(t)$ with respect to λ_t at every time instant. The minimization procedure results in a discerning update procedure. In particular, λ^*_t is set equal to zero (no update) if

$$\sigma^2(t) + \delta^2(t) \leq \gamma_0^2(t) \tag{3}$$

On the other hand, if (3) is not satisfied, then the optimal value of λ_t is computed as follows:

$$\lambda^*_t = \min(\alpha, v_t) \tag{4}$$

where

$$v_t = \begin{cases} \alpha & \text{if } \delta^2(t) = 0 \\[2ex] \dfrac{1-\beta(t)}{2} & \text{if } G(t) = 1 \\[2ex] \dfrac{1}{1-G(t)}\left[1 - \sqrt{\dfrac{G(t)}{1 + \beta(t)\,(G(t) - 1)}}\,\right] & \text{if } \beta(t)\,(G(t)-1) + 1 > 0 \\[2ex] \alpha & \text{if } \beta(t)\,(G(t) - 1) + 1 \leq 0 \end{cases}$$

and

$$\beta(t) \triangleq \frac{\gamma^2 - \sigma^2(t-1)}{\delta^2(t)}$$

The above recursions(2), and the selective update criterion (3,4), along with the initial values

$$P^{-1}(0) = I, \quad \theta(0) = 0 \text{ and } \sigma^2(0) = 1/\varepsilon \text{ with } \varepsilon \ll 1$$

form the basis of the OBE estimation algorithm. Note that the OBE algorithm is similar in form to the weighted recursive least-squares (WRLS) algorithm, with the information dependent updating factor acting as a weighting factor on the observations. Note also that the complexity of the information evaluation procedure (3) is much less than that of the updating procedure (2).

The Extended OBE algorithm. An ARMA (n,r) process is of the form

$$y(t) = a_1 y(t-1) + .. + a_n y(t-n) + w(t) + c_1 w(t-1) + .. + c_r w(t-r) \tag{5}$$

where $y(t)$ is the output and $w(t)$ is the input noise which is assumed to be uncorrelated and unknown. If $w(t)$ is assumed to be bounded in magnitude by γ_0, then the OBE algorithm can be extended to this ARMA parameter estimation problem, if estimates of $w(t-1), w(t-2).., w(t-r)$ are available[9]. The algorithm is essentially the same except for the following changes:

(i) The regressor vector is now given by

$$\Phi(t) = [\, y(t-1), \,..\, y(t-n), \varepsilon(t-1), \,...\, \varepsilon(t-r)\,]^T$$

where

$$\varepsilon(t) = y(t) - \theta^T(t)\Phi(t)$$

is the *a posteriori* prediction error and the parameter estimate $\theta(t)$ is the estimate of the a_i, $i = 1,2,..,n$ and c_j, $j = 1,2,..,r$.

(ii) The γ_0 in (3), which is the upper bound on the noise, is replaced by γ, an upper bound on the magnitude of the output $y(t)$.

It is easily shown that, for the EOBE algorithm, minimizing $\sigma^2(t)$ with respect to λ_t at every time instant yields the same updating criterion (3) and the same algorithm for determining the optimum value of the forgetting / updating factor λ^*_t, as in [5]. The algorithm thus retains the discerning update strategy and the modular adaptive filter structure.

Simulation Setup. A fixed point implementation of the OBE algorithm was simulated by performing the operations in integer arithmetic. The input and output observations, which are generated as floating point numbers, are converted to integers by the formula

$$x_{quant} = \begin{cases} \text{INT}(\ x.\ 2^{ibit} + 0.5)\ ,\ x > 0 \\ \text{INT}(\ x.\ 2^{ibit} - 0.5)\ ,\ x \leq 0\ . \end{cases}$$

where *ibit* is the number of bits assigned for the integer representation of the fractional part of the real number x. In the simulations, since an integer is stored in 32 bits, all registers and word sizes are 32 bits. Multiplication is performed by forming the product in a 48-bit word, scaling down by 2^{-ibit}, and then rounding off to the nearest integer. Inner products are formed similarly by accumulating the products in a 48-bit word, scaling down and then rounding off.

The upper bound α on the forgetting factor, has to be chosen with care in the fixed point implementation of the OBE and EOBE algorithms. If α is chosen greater than 0.1, then the elements of the matrix P often increase rapidly in magnitude and overflows can occur. The reason for this is that in the initial stages, the optimum value of the forgetting factor λ equals α fairly often. Consequently, since $1- \lambda$ appears in the denominator of (2e), the magnitude of the elements of P can increase and cause overflows. On the other hand, if α is chosen too small then the algorithm takes more iterations to converge and the number of updates increases. A value of $\alpha =0.1$ was found to yield a satisfactory convergence rate and inhibit overflows in the update equation for P(t).

In addition to α, the initial value $\sigma^2(0)$ has to be chosen small enough to prevent overflows in the subsequent calculations of λ^*. This is because if, at any time t, $\sigma^2(t-1)$ is large and $\delta^2(t)$ is small then $\beta = (\gamma^2 - \sigma^2(t-1))/\delta^2(t)$ can become a very large negative number and the product $\beta(G-1)$ can overflow. However, if overflows can be detected and a saturation value is used for β, then the calculation of λ^* will not be affected. Since β is negative and large in magnitude, $1+\beta(G-1)$ is a large positive or negative number, depending on whether G is greater than or less than unity. In case $1+\beta(G-1)$ is positive, then it can be seen from (4) that v_t is greater than unity, and consequently $\lambda^* = \alpha$. On the other hand if $1+\beta(G-1)$ is negative then $\lambda^* = \alpha$ from (4). Thus large values of $\sigma^2(0)$ can be used if care is taken to account for overflows in the algorithm for calculating λ^*. In our simulations, the initial (unquantized) value taken is $\sigma^2(0) =100$.

For the RLS algorithm, the initial value P(0) is also important. Since the bias in the estimates is inversely proportional to P(0), P(0) should be large. However large values can cause the Kalman gain vector K to overflow, and the parameter estimates to grow exponentially in the initial stage [11]. Therefore a compromise value P(0) = 10 I, where I is the identity matrix, was chosen.

Simulation Results. To compare the performance of the OBE algorithm *vis a vis* the RLS and EWLS algorithms, simulations were performed with an AR(4) and an ARX(4,4) process

Example 1. AR(4) process

$$y(t) = -0.6\ y(t-1) -1.58\ y(t-2) - 0.464\ y(t-3) - 0.5576\ y(t-4) + v(t)$$

The noise sequence {v(t)} is generated by a pseudo-random number generator with a uniform probability distribution in [-1.0, 1.0]. The upper bound γ^2 was set equal to 1.0. The parameter estimates were obtained by applying the OBE and the RLS algorithms to 500 point data sequences. Twenty five runs of the algorithm were performed on the same model but with different noise sequences. The number of bits used for the fractional part, *ibit*, was varied from 16 down to 8 bits. The average squared parameter error L is computed for each value of ibit according to the formula

$$L \;=\; \frac{1}{25} \sum_{j=1}^{25} (\theta_j - \theta^*)^T (\theta_j - \theta^*)$$

where θ_j is the final parameter estimate in the j-th run and θ^* is the true parameter. The average tap error for the OBE, RLS and exponentially weighted RLS(EWLS)with forgetting factor $\lambda = 0.99$, is plotted against *ibit* in Fig.1. It can be seen that the performance of the OBE algorithm appears to be constant as the number of bits varies from 16 to 9. In contrast, the performance of the RLS algorithm degrades substantially as the word-length decreases. The performance of the EWLS algorithm is even worse. The RLS and EWLS algorithms overflowed for ibit ≤ 8. The OBE algorithm overflowed for ibit ≤ 7.

Example 2. ARX(4,4) process

$$y(t) = 0.5y(t-1)-0.4y(t-2)+0.6y(t-3)+0.2y(t-4) + u(t)-0.29u(t-1)+0.5u(t-2)-0.7u(t-3) + v(t)$$

The input and noise sequences are generated by a pseudo-random number generator as before. The average tap error L, for the OBE, RLS and EWLS algorithms is plotted against *ibit* in Fig. 2. As before, the average tap error of the OBE algorithm appears constant as *ibit* varies from 16 to 7 bits. The RLS and EWLS algorithms do not work well for ibit ≤ 8.

Simulations were also performed for an ARX(10,10) model. However the large order seems to have caused greater accumulation of round-off errors in both the RLS and OBE algorithms and consequently overflows occurred.

Example 3.

The performance of the EOBE algorithm was evaluated by simulating an ARMA(3,3) process

$$y(t) = -0.4\ y(t-1) + 0.2\ y(t-2) + 0.6\ y(t-3) + w(t) - 0.6\ w(t-1) + 0.2\ w(t-2) + 0.6\ w(t-3)$$

The noise sequence {w(t)} is generated by a pseudo-random number generator with a uniform probability distribution in [-1.0, 1.0]. The upper bound γ^2 was set equal to 25.0. The average parameter error L is plotted for the EOBE and ELS(with forgetting factor $\lambda = 1$ and $\lambda = 0.99$) algorithms in Fig. 3. As in the previous case, it can be seen that while the performance of the EOBE algorithm is fairly constant over a range of word-lengths, the ELS algorithm does not perform properly for *ibit* < 9. The performance of the ELS algorithm with a forgetting factor of 0.99 was worse. The algorithm overflowed for *ibit* < 13.

Discussions. The superior performance of the OBE (EOBE) algorithms, as compared to the RLS (ELS) algorithms, is quite encouraging. One of the reasons could be the selective update strategy of the OBE algorithm. Such an update strategy may be responsible for a slower accumulation of roundoff errors on account of the updates being performed infrequently. Hence, if the RLS(ELS) and OBE(EOBE) algorithms operate on large sets of data, then the OBE(EOBE)algorithm could be less prone to divergence, simply because it does not update as often.

The difference in performance could also result from the differences in the update equation for P(t). The update equation for the RLS (ELS)algorithm with a forgetting factor λ is

$$P(t) = \left[I - \frac{P(t-1)\Phi(t)\Phi^T(t)}{\lambda + \Phi^T(t)P(t-1)\Phi(t)} \right] \frac{P(t-1)}{\lambda} \tag{6}$$

The corresponding equation for the OBE(EOBE) algorithm is (2e), which can be rewritten as

$$P(t) = \left[I - \frac{P(t-1)\Phi(t)\Phi^T(t)}{\dfrac{1-\lambda_t}{\lambda_t} + \Phi^T(t)P(t-1)\Phi(t)} \right] \frac{P(t-1)}{1-\lambda_t} \tag{7}$$

Since 1- λ_t plays the same role in the OBE algorithm as does λ in the RLS algorithm, the only difference between (6) and (7) is that the factor $(1-\lambda_t)/\lambda_t$ appears in the denominator of the term within braces in (7) as opposed to the corresponding term λ in (6). The degradation of performance occurs primarily because the term within braces becomes indefinite(has positive and negative eigenvalues) on account of round-off errors. Since λ_t is usually much smaller than unity, the term which is being subtracted from the identity matrix in (7) is much smaller than the one in (6). Thus P(t) in the RLS(ELS) algorithm has a greater tendency to become indefinite than the P(t) in the OBE(EOBE) algorithm. This observation has been confirmed by examining the eigenvalues of P(t), for runs in which the RLS algorithm performed poorly.

The failure of the RLS algorithm when the order is large(>10) is well known and there exist several methods like the UDU' [2,3] and QR factorization [4] methods, to make the P update numerically stable. For the OBE algorithm, a recenlty proposed systolic array implementation[12] may have better numerical properties. The derivation of other numerically stable recursions for the OBE algorithm is currently under investigation.

Conclusions. The finite-precision performance of the OBE and the EOBE algorithms has been studied through simulations. The performance of these algorithms in small word-length environments is superior to that of the well known RLS and ELS algorithms . The improvement is attributed to differences in the resursion of the matrix P(t) and less number of updates of the OBE algorithm. For large order processes, both the RLS and the OBE algorithm did not work properly when the word-length was small and hence more numerically robust algorithms may be required for such situations.

Acknowledgement. This work has been supported in part by the National Science Foundation under Grant MIP-87-1174 and in part by the office of Naval Research under Contract N00014-87-k-0284. Inquiries may be directed to Dr. Yih-Fang Huang, Department of Electrical & Computer Engineering, University of Notre Dame, Notre Dame, IN 46556.

REFERENCES

[1] J. M. Cioffi and T. Kailath, " Fast RLS transversal filters for adaptive filtering," *IEEE Trans. Acoust., Speech, Signal Process.*, vol. ASSP-32, No. 2, pp. 304-337, April 1984.

[2] G. J. Bierman, *Factorization Methods for Discrete Estimation*, Academic Press, New York 1977.

[3] L. Ljung and T. Soderstrom,*Theory and Practice of Recursive Identification*, Cambridge, Mass: MIT Press, 1983.

[4] J. M. Cioffi, "Limited-precision effects in adaptive filtering," *IEEE Trans. Circuits and Systems*, vol. CAS-34, No. 7, pp. 821-833, July 1987.

[5] S. Dasgupta and Y.F. Huang, " Asymptotically convergent modified recursive least-squares," *IEEE Trans. Information Theory* , vol. IT-33, No.3, pp. 383-392, May 1987.

[6] F. C. Schweppe, " Recursive state estimation : unknown but bounded errors and system inputs" *IEEE Trans. Automatic Control* , vol. AC-13, No.1, pp. 22-28, Feb. 1968.

[7] E. Fogel and Y.F. Huang, " On the value of information in system identification - bounded noise case," *Automatica* , vol. 18, No. 2, pp. 229-238, March 1982.

[8] M. Milanese and G. Belaforte, " Estimation theory and uncertainty intervals," *IEEE Trans. Automatic Control* , vol. AC-27, No. 2, pp. 408-414, April 1982.

[9] A. K. Rao, Y. F. Huang and S. Dasgupta, " An extended OBE algorithm for ARMA parameter estimation, " Proc. 24th Annual Allerton Conf., Urbana-Champaign, Sept. 28-30, 1988.

[10] Y. F. Huang, " A recursive estimation algorithm using selective updating for spectral analysis and adaptive signal processing,"*IEEE Trans. Acoust.Speech and Signal Process.* , vol. ASSP-34, No. 5, pp. 1131-34 , October 1986.

[11] S. H. Ardalan and S. T. Alexander," Fixed-point roundoff error analysis of the exponentially windowed RLS algorithm for time-varying systems", *IEEE Trans. Acoust.Speech and Signal Process.* , vol. ASSP-35, No. 6, pp. 1131-34 , June 1987.

[12] J. R. Deller, " 'Systolic array ' implementations of the optimal bounding ellipsoid algorithm," *IEEE Trans. Acoust.Speech and Signal Process.* , to be published.

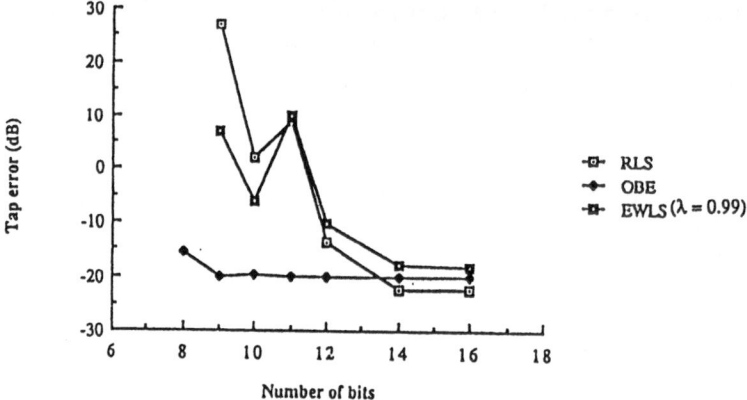

Fig. 1   Average tap error of the OBE and
RLS algorithms for an AR(4) process.

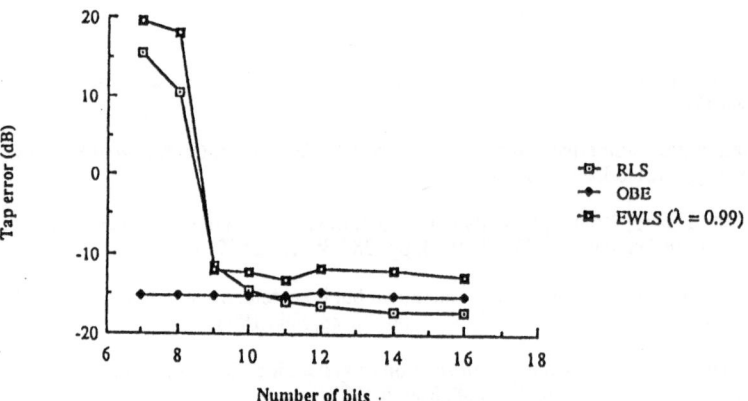

Fig. 2   Average tap error of the OBE and RLS
algorithms for an ARX(4,4) process.

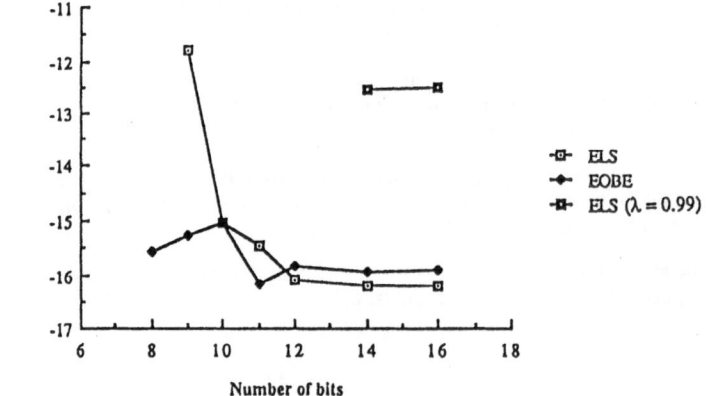

Fig. 3   Average tap error of the EOBE and ELS
algorithms for an ARMA(3,3) process.

ANALYSIS OF A FAST QUASI-NEWTON
ADAPTIVE FILTERING ALGORITHM

Daniel F. Marshall
MIT Lincoln Laboratory

W. Kenneth Jenkins
Department of Elec. and Comp. Engr. and
The Coordinated Science Laboratory

1. Introduction. An important consideration in the design of an adaptive
system is the convergence rate of the system. This is closely related to the
system's ability to track a time-varying optimum. Most basic adaptive filtering
algorithms, such as LMS, give poor convergence performance when the input to
the adaptive system is narrowband or highly colored. There are more sophisti-
cated algorithms which converge very rapidly regardless of the input spectrum,
most notably Recursive Lease Squares (RLS). But these algorithms typically
require $O(N^2)$ computation, which is a significant disadvantage for many real-time
applications, and they often behave poorly in finite-precision implementation. In
this paper, an $O(N)$ adaptive filtering algorithm is described which employs a
quasi-Newton approach to give rapid convergence even with colored inputs. This
Fast Quasi-Newton (FQN) algorithm appears to be quite robust with respect to
finite-precision implementation.

The adaptive filtering structure considered in this paper is the FIR tapped-
delay line shown in Figure 1. The tap vector $w(n)$ and the input vector $x(n)$ are
defined as:

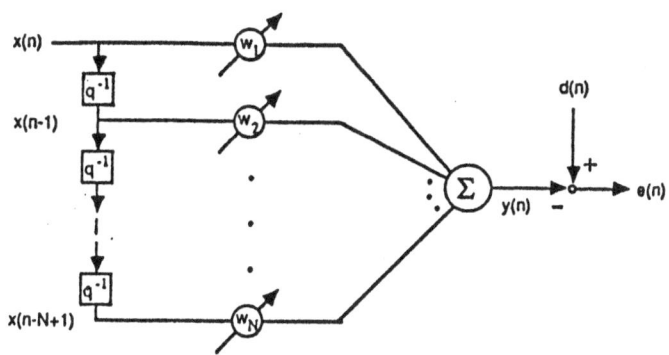

FIGURE 1. Tapped delay line adaptive filter structure.

$$w(n) \triangleq [w_1(n), w_2(n), \ldots, w_N(n)]^T$$
$$x(n) \triangleq [x(n), x(n-1), \ldots, x(n-N+1)]^T \tag{1}$$

The input and desired signal sequences $\{x(n)\}$ and $\{d(n)\}$ are assumed to be jointly stationary or only slowly time varying, and at least partially correlated. An adaptive algorithm is used to adjust the filter taps to minimize the expected value of the squared error $E[e^2(n)]$, so that the filter output $y(n) = x^T(n)w(n)$ will approximate the desired signal.

The convergence rate of basic adaptive algorithms depends on the eigenvalues of the autocorrelation matrix R_x of the input signal:

$$R_x \triangleq E[x(n)x^T(n)] \tag{2}$$

R_x is always positive semidefinite [1], and must in fact be positive definite for a unique solution to the adaptive filtering problem to exist. When the input is real and wide-sense stationary, $[R_x]_{ij}$ depends only on $|i-j|$; then R_x is symmetric and Toeplitz (all elements along a diagonal are equal).

The quasi-Newton approach to avoiding the problems associated with a colored input is to form an estimate $R(n)$ of R_x, and use it to cancel the effect of the input coloring. If the estimate is good then the adaptation process behaves as if the filter input were white (the ideal case) [2]. The general quasi-Newton adaptive filtering algorithm is give by

$$w(n+1) = w(n) + 2\mu e(n)R^{-1}(n)x(n)$$
$$\text{where } R(n) \approx R_x. \tag{3}$$

It can be shown that RLS has this form, with a step size of

$$\mu^{-1} = \epsilon + x^T(n)R^{-1}(n)x(n). \tag{4}$$

The main computational expense in implementing a quasi-Newton algorithm is incurred in inverting the estimated autocorrelation matrix and multiplying it by the input vector at each time step. In the next section, a particular form will be specified for the autocorrelation estimate to allow the use of fast computational techniques in its inversion.

2. The Fast Quasi–Newton (FQN) Algorithm. To specify a quasi-Newton algorithm, it is necessary to provide an expression for the estimate $R(n)$ of R_x in terms of the input samples. Recall that R_x is symmetric and Toeplitz. It is thus completely specified by the elements of its first row, which consists of the first N autocorrelation lags $r_0, r_1, \ldots, r_{N-1}$; $r_i \triangleq E[x(n)x(n-i)]$. To develop a fast quasi-Newton algorithm, this structure is imposed on the autocorrelation estimate to be used in (3): the first N autocorrelation lags are estimated, then used to construct an autocorrelation *matrix* estimate $R(n)$ which is symmetric and Toeplitz. A well-known autocorrelation lag estimator [1] can be put into the following recursive, exponentially weighted form for use in this application:

$$\hat{r}_i(n) = \alpha \hat{r}_i(n-1) + \alpha^{i/2} x(n-1)x(n), \quad i = 0 \text{ to } N-1 \tag{5}$$

This estimator is also positive semidefinite, a necessary property for stability of the adaptive algorithm.

Mathematically, the FQN algorithm is essentially given by equations (3)-(5). The unique feature of this algorithm is that its autocorrelation matrix estimate is Toeplitz and symmetric. This structure permits the fast implementation which we now describe.

Because of the symmetry properties of $R(n)$, the well-known Levinson recursion [3], [4] can be used to construct the gain vector $R^{-1}(n)x(n)$ in $O(N^2)$ operations. To maintain an overall computational requirement of $O(N)$, the Levinson recursion must be performed only once every N time steps. Thus if a particular autocorrelation matrix estimate is inverted using the Levinson recursion at time step n, then this *same* estimate must be used for the next N adaptive tap updates, until the next Levison recursion is performed. This will not seriously affect the adaptive system's performance as long as the lag estimates (5) do not change rapidly. Since the system input statistics are constrained to be only slowly varying, this is a reasonable assumption.

According to the above scheme, The Levinson recursion will generate $R^{-1}(n)x(n)$ at time step n. If another such recursion is not performed until time step $n+N$, then some means of generating the gain vectors $R^{-1}(n)x(n+1), R^{-1}(n)x(n+2), \ldots, R^{-1}(n)x(n+N-1)$ must be provided; i.e., although $R^{-1}(n)$ is held fixed, the input vector x must still be updated at every time step. These gain vectors can be constructed using an $O(N)$ algorithm developed by Manolakis, Kalouptsidis, and Carayannis (for an entirely different application) [4], which will be referred to here as the MKC algorithm. This algorithm again relies on the Toeplitz property of $R(n)$. It uses by-products of the Levinson inversion of $R(n)$ to generate $R^{-1}(n)x(k+1)$ from $R^{-1}(n)x(k)$, where x is a sliding vector as in (1). Thus the MKC algorithm can be used to generate the necessary gain vector for each iteration where the Levinson recursion is not performed. Details of the Levinson recursion and the MKC algorithm are given in Appendices A and B, respectively.

The FQN algorithm, as described so far, achieves an average of $O(N)$ multiplications and divisions per iteration by requiring an $O(N^2)$ Levinson recursion every N time steps. In some applications, this computational bottleneck would make the algorithm no better than an $O(N^2)$ algorithm. However, the Levinson recursion consists of N $O(N)$ stages, which can be spread out over N time steps. This results in the use of an autocorrelation estimate that is already N time steps old when its inverse becomes available. Thus a somewhat greater reliance is placed on the input stationarity assumption.

The FQN algorithm is summarized as a flow diagram in Figure 2. The step size given in (4) is used. The test performed just before the tap update, though

254

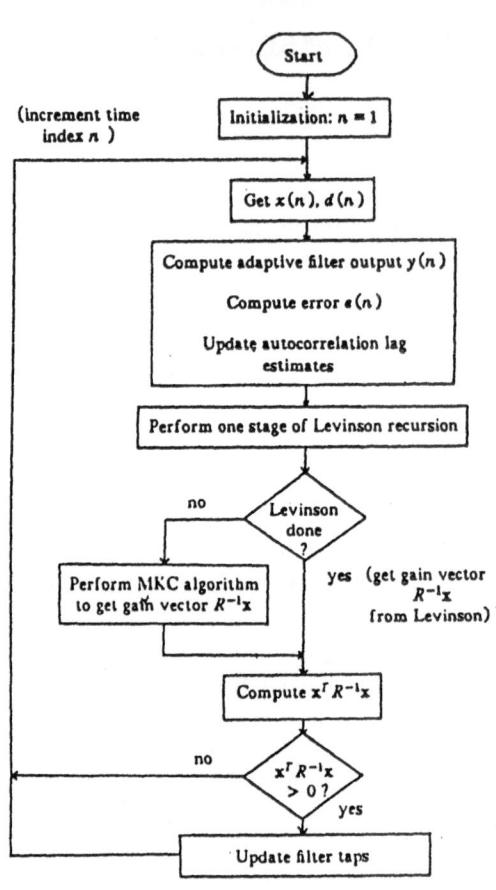

FIGURE 2. FQN flowchart.

rarely necessary, helps to ensure stability of the algorithm in finite-precision implementation. Several of the steps can be performed in parallel if convenient. This algorithm is not simply recursive like most adaptive algorithms, i. e., the same operations are not performed at each iteration. Rather, it is periodic with a period of N time steps. The total number of multiplications and divisions per iteration varies from 8N+1 to 12N−7, and averages 10N−3. The number of additions and subtractions is somewhat less than this. The algorithm might be somewhat complicated to implement in some cases. However, its robust performance in computer simulations suggests that it is well worth the trouble, as shall be shown in the next section.

3. Simulations. Several simulation experiments were conducted to explore the behavior of the new FQN algorithm. The $O(N^2)$ exponentially weighted RLS algorithm [5] was also simulated, for comparison. In the simulations, the adaptive filter performed system identification, with both the filter and the unknown system (plant) having order $N = 8$. White noise, uncorrelated with the system input, was added to the desired signal to give a minimum Mean Squared Error (MSE) of -100 decibels. The system input had a highpass spectrum, with an order-8 autocorrelation matrix eigenvalue spread of 680. The simulations were designed to demonstrate tracking performance: all adaptive system variables were initialized to the values they would have if the system had previously converged to a zero plant vector, using a white input. In the first set of simulations, the dependence of the FQN algorithm's convergence rate on the exponential weighting factor α was studied. The results of these simulations are presented as smoothed learning curves in Figure 3. As α is decreased from 0.99 to 0.95, the convergence

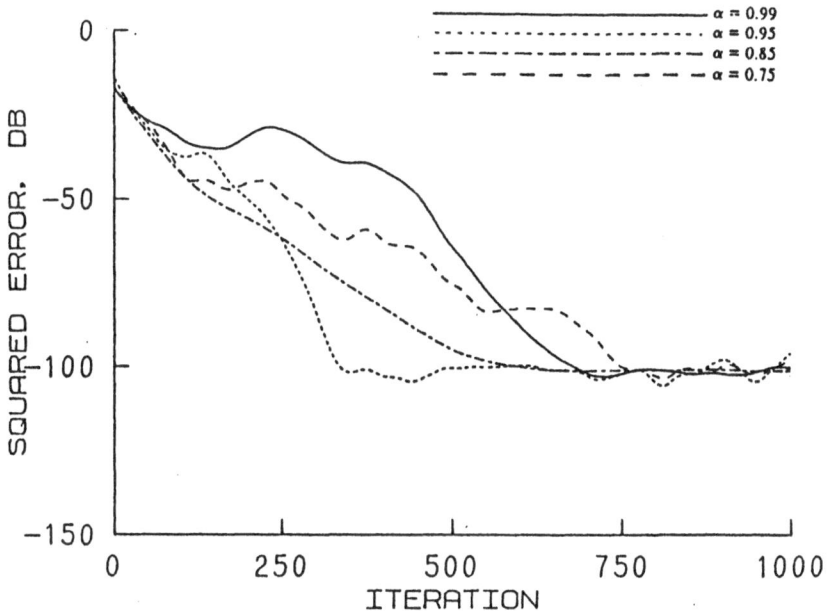

FIGURE 3. Simulated learning curves (FQN).

rate increases as the transient response of the autocorrelation estimator improves — the old input statistics are more quickly forgotten. But as α decreases further to 0.85, the convergence rate slows again as the quality of the autocorrelation estimate is degraded. Thus a tradeoff must be made between these effects; the optimum value of α in this case is about 0.95.

A similar set of simulations was run using the RLS algorithm. The results are shown in Figure 4. Here, it appears that the convergence rate increases indefinitely as α is decreased. This would seem to suggest that a very small value of α be used. But in practice, α is never chosen less than about 0.9. The reason, which can be seen in Figure 4, is that as α is reduced, the noise floor rises. The variation appears to be faily small here, but when the signal-to-noise is -20 or -30 decibels, it can be significant.

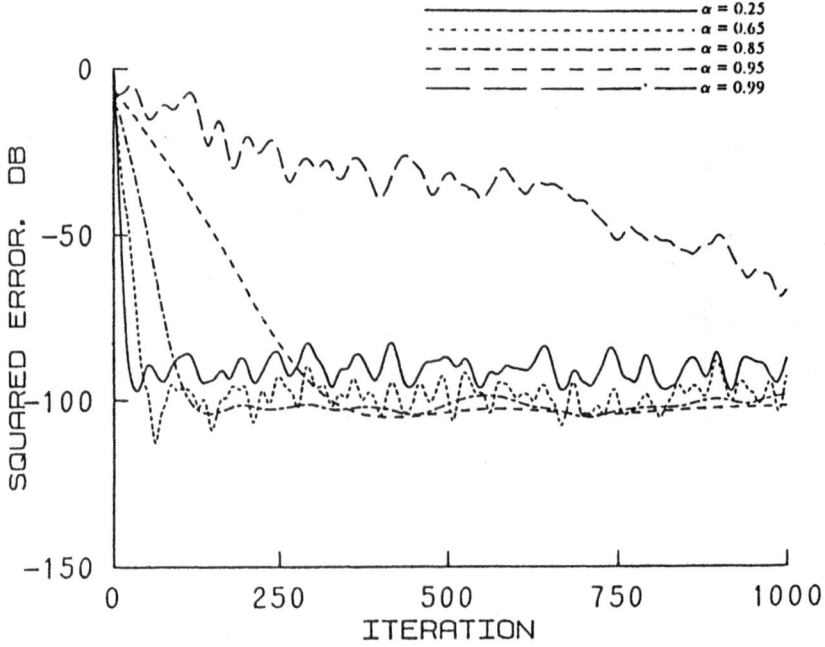

FIGURE 4. Simulated learning curves (RLS).

In these simulations, the FQN and RLS algorithms exhibit roughly similar convergence rates (for typical values of α). By comparison, when the LMS algorithm was used, the error decreased only 20 dB in 8,000 iterations. The FQN and RLS convergence rates were affected very little when different input colorings were used. This is as expected, since both algorithms estimate and cancel the effects of input coloring.

All of the above simulations used the full 45-bit (floating point) wordlength of the computer used for these experiments. In order to study the behavior of the FQN and RLS algorithms under more adverse conditions, simulations were run using 12 and 10 bit arithmetic. The exponential weighting factor α was set at 0.95 for both algorithms.

The result of the first short wordlength FQN simulation, in which 12 bit arithmetic was used, is shown in Figure 5. Since only one curve is plotted, the data has not been smoothed. The noise floor is higher than it was when 45 bits were used, with an average value of -73 decibels. This is a well-known consequence of decreasing the wordlength in an adaptive system. Note that a large number of iterations were run, in order to get a good indication of the stability of the algorithm. The result of an FQN simulation using 10 bits is shown in Figure 6. The noise floor is much higher with an average of -57 decibels.

Simulation results for the RLS algorithm using 12 bit arithmetic are given in Figure 7. The noise floor is -52 dB, much higher than observed with FQN. At iteration 2784, the algorithm quite suddenly failed due to register overflow. With 10 bit arithmetic, the RLS algorithm does not appear to converge at all, as shown

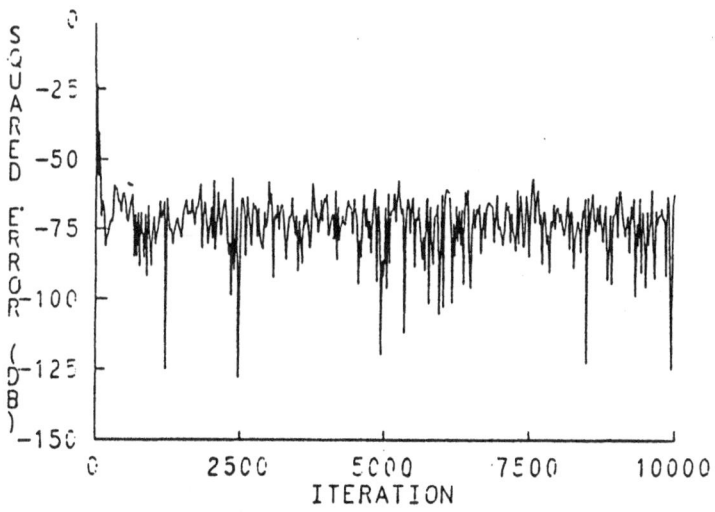

FIGURE 5. Simulated learning curve with quantized arithmetic (FQN, 12 bits).

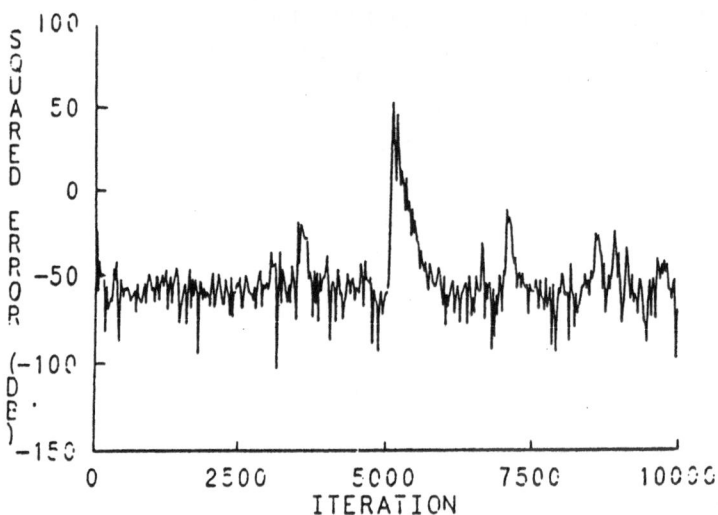

FIGURE 6. Simulated learning curve with quantized arithmetic (FQN, 10 bits).

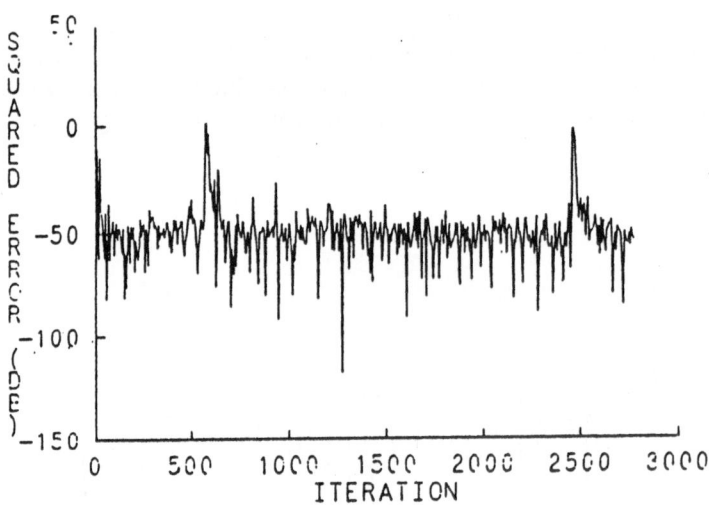

FIGURE 7. Simulated learning curve with quantized arithmetic (RLS, 12 bits).

in Figure 8. Although no divergence was observed, the algorithm again failed quite suddenly, this time at iteration 3014.

It must be acknowledged that major efforts to stabilize the RLS algorithm were avoided here, since the intent was to compare FQN to the standard RLS

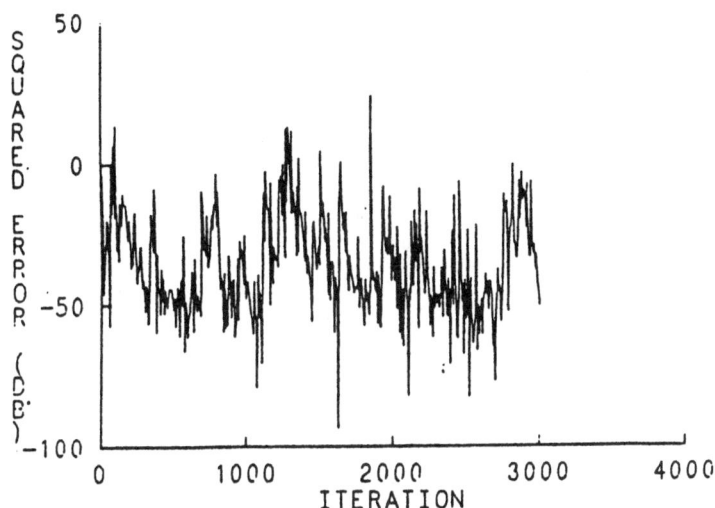

FIGURE 8. Simulated learning curve with quantized arithmetic (RLS, 10 bits).

formulation. Several analyses and stabilization techniques for RLS have recently appeared [6], [7]. But it is probably safe to conclude that FQN has a basic advantage over RLS in terms of robustness, due to its lack of internal recursions. The above simulations demonstrate a great difference in performance between the two algorithms, with FQN being the leader.

4. Conclusion. The Fast Quasi-Newton algorithm described in this paper has been seen to avoid the performance degradation caused in basic adaptive algorithms by colored input signals. The FQN algorithm offers convergence performance far superior to LMS, and is comparable to RLS in tracking ability. At the same time, FQN requires only O(N) computation, and should prove to be robust in finite-precision implementation. The latter expectation results from the fact that the FQN algorithm has no internal variables that are recursively computed for more than N time steps. Thus round-off errors are not allowed to accumulate. The FQN algorithm should prove to be useful in a variety of adaptive system applications.

Acknowledgements. Daniel F. Marshall is with MIT Lincoln Laboratories, Lexington, MA 02174, and W. Kenneth Jenkins is with the Department of Electrical and Computer Engineering and the Coordinated Science Laboratory, 1101 West Springfield Avenue, Urbana, IL 68101. This work was supported in part by the AT&T Industrial Affiliates Program and in part by the Joint Services Electronics Program under contract number N00014-84-C-0149.

REFERENCES

[1] A. Papoulis, *Probability, Random Variables, and Stochastic Processes,* second edition. New York: McGraw-Hill, 1984.

[2] R. D. Gitlin and F. Magee, "Self-Orthogonalizing Adaptive Equalization Algorithms," *IEEE Trans. Comm.,* vol COM-25, no. 7, pp. 666-672, July 1977.

[3] A. Giordano and F. Hsu, *Least Square Estimation with Applications to Digital Signal Processing.* New York: Wiley, 1985.

[4] D. Manolakis, N. Kalouptsidis, and G. Carayannis, "Fast Algorithms for Discrete-Time Wiener Filters with Optimum Lag," *IEEE Trans. Acoust., Speech, Sig. Proc.,* vol. ASSP-31, no. 1, pp. 168-179, Feb. 1983.

[5] G. Goodwin and K. Sin, *Adaptive Filtering, Prediction, and Control.* Englewood Cliffs, NJ: Prentice-Hall, 1984.

[6] J.M. Cioffi, "Limited-Precision Effects in Adaptive Filtering," *IEEE Trans. Circuits and Systems,* vol. CAS-34, no. 7, pp. 821-833, July 1987.

[7] S.H. Ardalan and S.T. Alexander, "Fixed-Point Roundoff Error Analysis of the Exponentially Windowed RLS Algorithm for Time-Varying Systems," *IEEE Trans. Acoust., Speech, Signal Proc.,* vol. ASSP-35, no. 6, pp. 770-783, June 1987.

APPENDIX A
Levinson recursion

<u>INPUT</u>

$x = [x_1, \ldots, x_N]^T$

$R: [R]_{ij} = r_{i-j}, i \& j = 1, \ldots N; \text{ i.e., } r_0, \ldots, r_{N-1}$

<u>OUTPUT</u>

$a = [a_1, \ldots, a_N]^T = R^{-1}x$

Internal variables: $\phi = [\phi_1, \ldots, \phi_{N-1}]^T, \xi, \beta, \gamma$

Initialization: $\xi = r_0 \quad a_1 = \dfrac{x_1}{r_0}$

For k = 2 to N:

$$\gamma = r_{k-1} + [r_1 \cdots r_{k-2}] \begin{vmatrix} \phi_1 \\ \cdot \\ \cdot \\ \cdot \\ \phi_{k-2} \end{vmatrix}$$

$$\begin{vmatrix} \phi_{k-1} \\ \cdot \\ \cdot \\ \cdot \\ \cdot \\ \phi_1 \end{vmatrix} \leftarrow \begin{vmatrix} \phi_{k-2} \\ \cdot \\ \cdot \\ \cdot \\ \phi_1 \\ 0 \end{vmatrix} - \frac{\gamma}{\xi} \begin{vmatrix} \phi_1 \\ \cdot \\ \cdot \\ \cdot \\ \phi_{k-2} \\ 1 \end{vmatrix}$$

$$\beta = -x_k - [x_1 \ldots x_{k-1}] \begin{vmatrix} \phi_1 \\ \cdot \\ \cdot \\ \phi_{k-1} \end{vmatrix}$$

$$\xi \leftarrow \xi + \gamma \phi_1$$

$$\begin{vmatrix} a_1 \\ \cdot \\ \cdot \\ \cdot \\ \cdot \\ a_k \end{vmatrix} \leftarrow \begin{vmatrix} a_1 \\ \cdot \\ \cdot \\ \cdot \\ a_{k-1} \\ 0 \end{vmatrix} - \frac{\beta}{\xi} \begin{vmatrix} \phi_1 \\ \cdot \\ \cdot \\ \cdot \\ \phi_{k-1} \\ 1 \end{vmatrix}$$

APPENDIX B
MKC algorithm

<u>INPUT</u>

$\mathbf{x} = [\, x_1, \ldots, x_N \,]^T$

$R: [R]_{ij} = r_{i-j}, \; i \,\&\, j = 1, \ldots, N; \; \text{i. e.,} \; r_0, \ldots, r_{N-1}$

ϕ and ξ from last Levinson recursion

$\mathbf{a} = [\, a_1, \ldots, a_N \,]^T = R^{-1}\mathbf{x}, \; p = \mathbf{x}^T\mathbf{a}, \; x_0$

<u>OUTPUT</u>

$\hat{\mathbf{a}} = [\hat{a}_1, \ldots, \hat{a}_N]^T = R^{-1}\hat{\mathbf{x}}, \; \hat{p} = \hat{\mathbf{x}}^T\hat{\mathbf{a}}$

where $\quad \hat{\mathbf{x}} = [x_0, \ldots, x_{N-1}]^T$

Internal variables: $\quad \rho_1, \ldots, \rho_{N-1}, \theta$

$$
\begin{bmatrix} \rho_1 \\ \cdot \\ \cdot \\ \cdot \\ \rho_{N-1} \end{bmatrix} = \begin{bmatrix} a_1 \\ \cdot \\ \cdot \\ \cdot \\ a_{N-1} \end{bmatrix} - a_N \begin{bmatrix} \phi_1 \\ \cdot \\ \cdot \\ \cdot \\ \phi_{N-1} \end{bmatrix}
$$

$$
\theta = -x_0 - [x_{N-1} \cdots x_1] \begin{bmatrix} \phi_1 \\ \cdot \\ \cdot \\ \cdot \\ \phi_{N-1} \end{bmatrix}
$$

$$
\begin{bmatrix} \hat{a}_N \\ \cdot \\ \cdot \\ \cdot \\ \hat{a}_1 \end{bmatrix} = \begin{bmatrix} \rho_{N-1} \\ \cdot \\ \cdot \\ \rho_1 \\ 0 \end{bmatrix} - \frac{\theta}{\xi} \begin{bmatrix} \phi_1 \\ \cdot \\ \cdot \\ \phi_{N-1} \\ 1 \end{bmatrix}
$$

$$
\hat{p} = p - \left(a_N + \frac{\theta}{\xi} \right) \left(a_N - \frac{\theta}{\xi} \right) \xi
$$

ADAPTIVE STACK FILTERING UNDER THE
MEAN ABSOLUTE ERROR CRITERION

J.H. Lin
Purdue University

T.M. Sellke
Purdue University

E.J. Coyle
Purdue University

1. Introduction.   Stack filters [1] are a class of sliding window, nonlinear digital filters. Each filter in this class possesses the weak superposition property known as the threshold decomposition [2,3] and an ordering property known as the stacking property [2,3]. These two properties are illustrated in Figure 1 with the particular stack filter known as the window-width three asymmetric median filter.

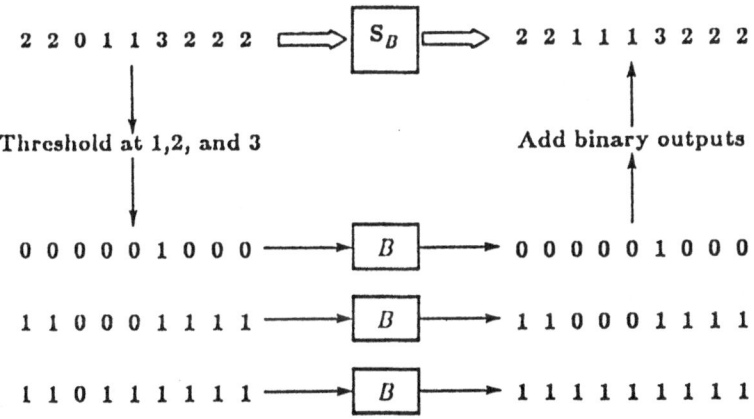

Figure 1: The Asymmetric Median Filter.   The operation performed by the stack filter shown above is

$$S_B(\cdot) = \max(\min(r_1, r_3), r_2)$$

in which the r_i's are the integer-valued samples, in time order, in the filter's window. The positive Boolena function $B(\cdot)$ on each level is

$$B(\cdot) = x_1 x_3 + x_2 = \max(\min(x_1, x_3), x_2)$$

in which the x_i's are the bits, in time order, in the Boolean filter's window, and where mutiplication is the logical AND, and addition is the logical OR.

Many of the nonlinear filters which have been used successfully in a wide variety of signal and image processing applications are included within the class of stack filters. Specifically, the median filter [4], rank-order filters [5], morphological filters [6,7], and all compositions of these filters have representations as stack filters [1,8] and generalized stack filters [9]. Thus, within the very large class of stack filters we are guaranteed that there are many which are of practical importance.

The fact that many previously known filters are stack filters is not, however, what makes stack filters or any other class of nonlinear filters significant. Indeed, there are many filter classes, such as morphological filters, which can claim to include stack filters [8], or even all linear and essentially all nonlinear filters [8,10]. What sets the class of stack filters apart from these other classes of nonlinear filters is that analytical techniques are available for determining the stack filter which minimizes the mean absolute error between its output and some desired signal, given noise corrupted observations of that signal as the input to the filter.

This theory of Minimum Mean Absolute Error (MMAE) stack filtering is analogous to Minimum Mean Square Error (MMSE) linear filtering since both classes of filters are defined by superposition properties, in both filtering theories it is the superposition property which makes the optimization tractable, and both classes of filters are straightforward to implement.

Although these optimal stack filtering results [9,12] do provide a systematic approach for designing an optimal nonlinear filter, knowledge of the joint threshold crossing statistics of the signal and noise processes is required. Such knowledge is rarely available in practice, particularly in image processing applications. Furthermore, the complexity of the optimization procedure increases rapidly with the window width of the filter because of the very extensive statistical characterization that is then required and because the number of variables in the linear program also increases very rapidly.

What is needed is a method of adaptive stack filtering, whereby the optimal filter can be determined via observations of training sequences. The foundation required for this theory of optimal adaptive nonlinear filtering is developed in this paper. Thus, this paper further extends the analogy between the class of linear filters and the class of stack filters to the realm of adaptive filtering.

This approach not only has the same advantages as any other adaptive filtering approach -- such as the ability to deal with unknown statistics -- it is also computationally more efficient than the previous nonadaptive approach. In fact, it is perfectly suited for hardware implementation since the only operations needed are increments, decrements and comparisons. Furthermore, the physical architecture of commercially available machines like the N-cube [13] exactly matches the pattern of local communications required by the adaptation algorithm when the goal is to find an optimal window width N stack filter.

The adaptation algorithm developed in this paper is analogous to the Stochastic Gradient algorithm in adaptive linear filtering [14,15]. The primary difference is that the adjustment of variables in the adaptive stack filtering algorithm requires some coordination. This coordination is necessary for the maintenance of the stacking property; that is, to ensure that the logical operator on each level in the architecture is always a positive Boolean function so that the superposition property is maintained.

The coordination required to maintain the stacking constraint, which at first sight appears to be a complex task, can be achieved via simple arithmetic operations. This is a consequence of the local, structural nature of the stacking constraint. It is, in fact, shown that the arithmetic operations required for the coordinated update of the variables are only simple increment, decrement, and comparison operations. Real-time hardware implementations of the adaptive stack filtering algorithm are thus very feasible.

2. Background Review.

2.1. The Class of Stack Filters.
Let $R(n)$ be the process at the input of a stack filter. Assume $R(n)$ takes on integer values in $Q = \{0,1,2,...,M-1\}$. A window of width b slides, by increments of one sample, across the input process $R(n)$. At each time instant n, the stack filter maps the samples in the window, which are

$$\vec{R_b}(n) = \left[R(n-\frac{b-1}{2}) \quad \cdots \quad R(n) \quad \cdots \quad R(n+\frac{b-1}{2}) \right], \tag{2.1}$$

to some integer $S_f(\vec{R_b}(n))$ in Q. The mapping, $S_f(\cdot):Q^b \rightarrow Q$, defining the stack filter is required to have the threshold decomposition structure

$$S_f(\vec{R_b}(n)) = \sum_{k=1}^{M-1} f(T_k(\vec{R_b}(n))), \tag{2.2}$$

where

$$T_m(\vec{R_b}(n)) = \left[T_m(R(n-\frac{b-1}{2})) \quad \cdots \quad T_m(R(n)) \quad \cdots \quad T_m(R(n+\frac{b-1}{2})) \right], \tag{2.3}$$

$$T_m(R(n)) = \begin{cases} 1 & \text{if } R(n) \geq m \\ 0 & \text{else} \end{cases}, \tag{2.4}$$

and $f(\cdot)$ is the Boolean function used on each level in the architecture.

The stacking property requires that whenever the Boolean function on level l outputs a 1, then the Boolean functions on every level below level l must also output 1's. From this requirement, and the requirement that the same Boolean function be used on every level of the filter, it follows that only *positive* Boolean functions are allowed [1].

Any digital filter that can be realized by (2.2) with $f(\cdot)$ being a positive Boolean function possesses the threshold decomposition property [1]. This is a weak superposition property which is made precise in the second equality in the following equation:

$$S_f(\vec{R_b}(n)) = S_f\left(\sum_{k=1}^{M} T_k(\vec{R_b}(n))\right) = \sum_{k=1}^{M} S_f(T_k(\vec{R_b}(n))) = \sum_{k=1}^{M} f(T_k(\vec{R_b}(n))). \tag{2.5}$$

The class of stack filters and generalized stack filters includes all rank order operators and any Morphological filters which are composition of dilation and erosion operators. Also, the number of stack filters for any fixed window width b grows faster than $2^{2^{b/2}}$ [12].

2.2. Optimal Filtering over The Class of Stack Filters.
The optimal filtering problem over the class of stack filters can be stated as in Figure 2.

The process $R(n)$ at the input of a stack filter is assumed to be a corrupted version of some desired process $S(n)$. The corruption may be caused either by a noise process $N(n)$ or by some intentional operation, such as a modulation scheme.

At each time instant n, the stack filter output is an estimate, called $\hat{S}(n)$, of the desired process $S(n)$. This estimate is based on the observed sequence $\vec{R_b}(n)$ in the window of the stack filter; thus,

$$\hat{S}(n) = S_f\left(\vec{R}_b(n)\right). \tag{2.6}$$

The goal is to pick a stack filter from the class of window width b stack filters such that the average mean absolute error per time unit between the filter's output and the desired signal is minimized. If $S(n)$ and $R(n)$ are jointly stationary, then the cost to be minimized is

$$E\left[\ |S(n)-S_f\left(\vec{R}_b(n)\right)|\ \right]. \tag{2.7}$$

The rationale of using the absolute error criterion is that it nicely reduces the estimation error of the stack filter to the sum of the decision errors incurred by the Boolean filters on each level of the threshold decomposition architecture. Let $s_k(n) = T_k(S(n))$; then

$$E\left[\ |S(n) - S_f\left(\vec{R}_b(n)\right)|\ \right] = \sum_{k=1}^{M} E\left[\ |s_k(n)) - f\left(T_k(\vec{R}_b(n))\right)|\ \right]. \tag{2.8}$$

Because of the above reduction, and since a stack filter is completely defined by the Boolean function on each level of its architecture, the cost function to be minimized can be expressed in terms of a linear function of the output variables of the Boolean function $f(\cdot)$ [9,12].

Let the output of $f(\cdot)$, when the length b binary sequence $\vec{s}_j$ is at its input, be called the decision variable $x_j \in \{0, 1\}$. The Boolean function $f(\cdot)$ can therefore be represented as a length 2^b vector $\vec{X}$, whose k'th entry is x_k. Then the cost to be minimized is

$$Cost = \sum_{j=1}^{2^b} C_j \cdot x_j. \tag{2.9}$$

The coefficients C_j depend on the statistics of the corrupted and the desired processes and can be estimated from sample functions of the corrupted and desired processes under some appropriate statistical assumptions [11]. C_j can be interpreted as the cost incurred by $f(\cdot)$ for deciding a 1 when seeing $\vec{s}_j$. Also the stacking constraint can be expressed as a set of inequality equations in terms of these decision variables,

$$x_i \le x_j \qquad \text{if} \ \ \vec{s}_i \le \vec{s}_j \tag{2.10}$$

where for any two length b real sequences $\vec{s}$ and $\vec{t}$, $\vec{s} \le \vec{t}$ if and only if each entry of $\vec{s}$ is less than or equal to the corresponding entry in $\vec{t}$. The optimal filtering problem over the class of stack filters under the mean absolute error criterion can therefore be formulated as an zero-one integer linear program; the goal being to determine whether each x_j should be 0 or 1.

By exploiting the structure of the constraint matrix, which is TUM (totally unimodular, [16]), this zero-one integer linear program can be formulated as the following linear program [9,12]:

$$minimize\ \sum_{j=1}^{2^b} C_j \cdot x_j \tag{2.11}$$

subject to the constraints:

$$x_i \le x_j \qquad \text{if} \ \ \vec{s}_i \le \vec{s}_j \tag{2.12}$$

$$0 \leq x_i \leq 1 \quad \forall i \tag{2.13}$$

So far we have reviewed a nonadaptive methodology for finding an optimal stack filter. In the rest of this paper, an adaptive methodology for finding an optimal stack filter is developed.

3. Signal and Noise Assumptions. In this section, a statistical assumption is made concerning the desired process and the corrupted process version of this process that appears in the window of the stack filter to be trained. This assumption is significantly weaker than the assumption of joint stationarity discussed previously. It can be thought of as allowing local nonstationarity, which might be quite useful when applying this theory to image processing applications.

From this statistical assumption, it is then straightforward to define an optimal stack filter. There may be multiple optimal stack filters since the optimal filtering problem is solved by the linear program in (2.11)-(2.13). Multiple optimal solutions should be very rare, though, since even the slightest change in the cost coefficients will again yield a unique optimal filter. In this paper, we therefore consider only cases in which the optimal filter is unique.

Let $((Y_n)_{n \geq 0}, Y_n = (\vec{R}_b(n), S(n)))$, be a sequence of random variables on a common probability space $(\Omega, \mathcal{F}, \mathcal{P})$ adapted to an increasing sequence $(\mathcal{F}_n)_{n \geq 0}$ of sub-σ-algebras of $\mathcal{F}$. Let $S_{f_j}(\cdot)$ be the jth stack filter in the set of window width b stack filters, $1 \leq j \leq K$, where K is the total number of positive boolean functions of b variables.

The stack filter $S_{f_*}(\cdot)$ is said to be optimal for recovering $S(n)$ from $\vec{R}_b(n)$ in the following sense. Let $A_j(n)$ be the absolute error between the output of the j'th stack filter and the desired signal $S(n)$ given that the filter's input is $\vec{R}_b(n)$:

$$A_j(n) = |S_{f_j}(\vec{R}_b(n)) - S(n)|, \tag{3.1}$$

and define $A_j(n+1, n+m)$ to be the sum of the absolute errors over the period $[n+1, n+m]$:

$$A_j(n+1, n+m) = \sum_{k=1}^{m} A_j(n+k), \tag{3.2}$$

then $S_{f_*}(\cdot)$ is an optimal stack filter if there exists some positive integer L and a real number $\rho' \geq 0$ such that for any $n \geq 0$,

$$E[A_o(n+1, n+L) \mid \mathcal{F}_n] - E[A_j(n+1, n+L) \mid \mathcal{F}_n] \leq -\rho' \quad \forall \quad j \neq o \tag{3.3}$$

Note that if $\rho' > 0$ then (3.3) implies that the optimal stack filter is unique. As stated previously, we assume throughout this paper that the optimal stack filter is unique; so that ρ' is always greater than zero.

In the following, an expression equivalent to (3.3) is derived for use in the proof of the convergence of the adaptive algorithm.

Let

$$C_k^n(\alpha, j) = \sum_{m=1}^{M-1} \sum_{\substack{(\overline{W}, V) \in Q' \times Q \\ T_m(V) - \alpha, \ T_m(W) - \tau_j}} \pi_k^n(\overline{W}, V) \tag{3.4}$$

where

$$\pi_k^n(\overline{W}, V) = Pr\{\vec{R}_b(n+k) = \overline{W}, S(n+k) = V \mid \mathcal{F}_n\} \tag{3.5}$$

and let

$$\vec{U}(n,\ n+L) = \left[p_1^n,\ \cdots,\ p_j^n,\ \cdots,\ p_{2^i}^n \right] \tag{3.6}$$

where

$$p_j^n = \sum_{k=1}^{L} C_k^n(1,\ j) - C_k^n(0,\ j) \tag{3.7}$$

Let $\vec{f}_j$ be the length 2^b binary vector corresponding to the boolean function in the architecture of the stack filter $S_{f_j}(\cdot)$. The k'th entry of $\vec{f}_j$ is the decision the filter $\vec{f}_j$ makes when it observes the binary sequence $\vec{s}_j$ in its window. Then (3.3) is equivalent to

$$\vec{U}(n,\ n+L)\cdot(\vec{f}_j - \vec{f}_o) \leq -\rho' \qquad \forall\ j \neq o\ ,\ \forall\ n \geq 0 \tag{3.8}$$

Let

$$\vec{W}_j \triangleq \frac{\vec{f}_j - \vec{f}_o}{\|\ \vec{f}_j - \vec{f}_o\ \|}$$

then (3.8) can be rewritten as

$$\vec{U}(n,\ n+L)\cdot\vec{W}_j \leq -\rho \qquad \forall\ j \neq o\ ,\ \forall\ n \geq 0 \tag{3.9}$$

where $\rho = \dfrac{\rho'}{\tau}$, and $\tau = \max_j \|\ \vec{f}_j - \vec{f}_o\ \|$ is some positive integer.

4. Adaptive Approach. In this section, an adaptive filtering approach for finding the optimal stack filter will be presented. Equivalently, the goal is to find the optimal decision vector $\vec{f}_o$, which corresponds to the Boolean function in the optimal stack filter's architecture. In this adaptive approach, it is assumed that no a priori information of the decision costs C_j for each decision variable x_j is known in advance. An initial decision vector is arbitrarily chosen, and then the decision vector is adjusted based on the observations of the corrupted and the desired processes according to an adaptation algorithm. Eventually, the decision vector converges to the optimal decision vector. This approach will be formally described next.

4.1. The Basic Ideas. Let $\vec{X}(n)$ be the decision vector at time n, where its i'th entry, $x_i(n)$, is the decision variable for the binary sequence $\vec{s}_i$ at time n. The proposed adaptation algorithm has two main features. First, each decision variable is allowed to make soft decisions; i.e., each decision variable can take on values in $[0,1]$ with resolution $\dfrac{1}{N}$, where N is some positive integer. Secondly, the decision variables satisfy the stacking constraint for all n. This is necessary to ensure that when the learning process is terminated at any point in time, the resulting Boolean filter's decisions always satisfy the stacking constraint. Equivalently, the decision vector, whose evolution is controlled by the adaptation algorithm, always stays in the polytope described by (2.12) and (2.13) during the adaptation period. In other words,

$$x_i(n) \leq x_j(n) \qquad \text{if} \qquad \vec{s}_i \leq \vec{s}_j \tag{4.1}$$

In addition to enforcing the stacking constraint upon the decision vector, the adaptation algorithm needs to control the drift of the decision vector in such a way that the decision vector is getting closer and closer in the stochastic sense to the optimal decision vector as time passes. When the adaptation process is terminated at some time n, it is possible that $\vec{X}(n)$ is still at some interior point of the polytope. Since the optimal decision vector is always at some vertex of the polytope, a thresholding operator is necessary to convert the final soft decision vector $\vec{X}(n)$ to a hard decision vector $\vec{X}(n)$. Although there are many ways to achieve this goal, the simplest way is the thresholding operation $\vec{X}(n) = T_{\frac{1}{2}}(\vec{X}(n))$. Note that the stacking property is preserved by this thresholding operation, i.e., $\vec{X}(n)$ obeys the stacking constraint if $\vec{X}(n)$ obeys the probabilistic stacking constraint. Furthermore, if $\vec{X}(n)$ is close enough in Euclidean distance to a vertex corresponding to the optimal decision vector, then $\vec{X}(n)$ would be equal to the optimal decision vector. This is true since the thresholding operation converts $\vec{X}(n)$ to the vertex $\vec{X}(n)$ which is nearest to $\vec{X}(n)$ in city-block distance, and city-block distance is equivalent to the Euclidean distance in finite dimensional space.

So far, we have described the basic ideas behind the adaptation algorithm to be proposed. Before presenting the adaptation algorithm, a modification consisting of a scaling by N will be made. Instead of taking values in $[0,1]$ with resolution $\frac{1}{N}$, each decision variable $x_i(k)$ will take integer values in $[0,N]$. Let $\vec{J}_j$ be the length 2^b binary vector corresponding to the boolean function in the architecture of the stack filter $S_{f_j}(\cdot)$. Each entry of $\vec{J}_j$ is now either a 0 or an N, and $\vec{X}(n) = T_{\frac{N}{2}}\vec{X}(n)$.

4.2. The Adaptation Algorithm.

The structure of the proposed adaptation algorithm is as follows. At each time instant n, $Y_n = (\vec{R}_b(n), S(n))$ is observed and used to update $\vec{X}(n)$, the filter decision vector at time n, to $\vec{X}(n+1)$. First, $\vec{R}_b(n)$ and $S(n)$ are thresholded into $M-1$ binary vectors and $M-1$ bits respectively.

$$y_{k,n} \triangleq (\vec{r}_k, s_k) = T_k(Y_n) \tag{4.2}$$

Then $\vec{X}(n)$ is updated sequentially based on $y_{k,n}$, $k=1,2,...,M-1$, to generate $\vec{X}(n+1)$. In other words, each update from $\vec{X}(n)$ to $\vec{X}(n+1)$ based on Y_n takes $M-1$ sub-updates, each based on a thresholded version of Y_n.

To account for the intermediate decision vectors generated in between $\vec{X}(n)$ and $\vec{X}(n+1)$, a new indexing scheme needs to be introduced. An obvious choice is to let the time index be the count of sub-updates based on the thresholded version of observations instead of the multilevel observations. In the new indexing system, y_k represents the thresholded observation $T_m(Y_n)$, where $k = n \cdot (M-1)+m$; and $\vec{X}(k)$ is the decision vector after the update based on y_k. In the sequel, the time index will always refer to the new index unless indicated otherwise.

Let the decision vector at time k be denoted as $\vec{X}(k)$. Then the evolution of $\vec{X}(k)$ can be described by

$$\vec{X}(k+1) = h(\vec{X}(k), y_k) \tag{4.3}$$

where $h(\cdot)$ is the adaptation algorithm.

The following notation is introduced for describing the proposed adaptation algorithm. Let the set of decision variables which are subject to the local stacking constraint with x_i be denoted as I_i. I_i can be partitioned into two subsets I_i^1 and I_i^0 such that if $y \in I_i^1$, then the stacking constraint is $y \geq x_i$, and if $y \in I_i^0$, then the stacking constraint is $y \leq x_i$. The symbols $\wedge$ and $\vee$ denote the minimum and maximum operators, respectively. The proposed algorithm is summarized below.

Input: $\vec{X}(k)$, *Output:* $\vec{X}(k+1)$

1:(Update)

If $\vec{r}_k = \vec{s}_i$ and $s_k = 1$, then $x_i := \{x_i + 1\} \wedge N$.

If $\vec{r}_k = \vec{s}_i$ and $s_k = 0$, then $x_i := \{x_i - 1\} \vee 0$.

Set $m := i$.

2:(Check)

If x_m and x_j obey the stacking constraint $\forall x_j \in I_m^{s_i}$, then return $\vec{X}(k+1)$.

3:(Iteration)

Pick any $x_j \in I_m^{s_i}$, where x_j violates the stacking constraint with x_m, and swap x_m and x_j.

Set $m := j$

Go to 2

The proposed adaptation algorithm has a simple interpretation. Suppose that there is no stacking constraint imposed on the decision variables, then the "best" decision for a binary sequence depends only on the difference of frequencies of occurrence of the "correct" decisions, which can be a "1" or a "0", when that particular binary sequence is observed. The difference of frequencies of occurrence can be interpreted as an estimate of the unnormalized version of the decision cost for that particular binary sequence.

What the adaptation algorithm does first is the bookkeeping, which is counting the difference of the number of occurrences of each binary threshold vector when the correct decision is 1 and when the correct decision is 0. The arithmetic operations required are just increment or decrement.

In addition, the adaptation algorithm must enforce the stacking constraint. If a variable x_i is incremented or decremented, the algorithm only needs to check those variables which are subjected to the local stacking constraint with x_i to see if the stacking constraint is violated. Depending on whether x_i is incremented or decremented, the checking can be further reduced to the decision variables in I_i^1 or I_i^0 respectively. If the decision variable x_i, after the increment based on the observation, does not violate the stacking constraint with any decision variable in I_i^1, then this update has been completed. Otherwise, pick a decision variable x_j in I_i^1 which violates the stacking constraints with x_i and perform a swap between x_i and x_j. The swap between x_i and x_j is equivalent to decrementing x_i by 1 and incrementing x_j by 1. After the swap, x_i has been reset to its previous value, and it no longer violates the stacking constraint with any member in I_i^1. However, x_j might now violate the stacking constraint with the decision variables in I_j^1. Repeated perform similar checking and swap operation on x_j until no stacking constraint is violated, then this update is done.

One obvious question is,"What is the maximum number of iterations before the process of checking the stacking constraint finishes?" The answer can be found from the following trivial fact.

Fact 4.1:

Let the set of all length b binary sequences be partitioned into the subsets $\{\eta_i\}_{i=0}^b$ such that $\forall \vec{s} \in \eta_i$ the number of 1's in $\vec{s}$ is i. If $\vec{x} \in \eta_i$ and there is a $\vec{y}$ with $\vec{x} \leq (\geq) \vec{y}$ and $d_H(\vec{x}, \vec{y}) = 1$, then $\vec{y} \in \eta_{i+1}(\eta_{i-1})$. Also $\forall \vec{x}, \vec{y} \in \eta_i$, $\vec{x}, \vec{y}$ do not stack on each other.

From the above fact, if $\vec{s}_i \in \eta_l$, and x_j is a decision variable in I_i^1, then $\vec{s}_j \in \eta_{l+1}$. Since there are only b+1 subsets $\{\eta_i\}_{i=0}^b$, the swap can be performed at most b times before the binary sequence associated with the decision variable incremented is in η_b. That decision variable is x_{2^b}, and its increment will not be subjected to any constraint. If the update based on the observation is decremented, similar reasoning shows at most b swaps are needed.

It is also clear from the previous discussion that for each thresholded observation, only one decision variable, which is the last one updated, is incremented or decremented by 1. The above discussion is summarized in the following fact.

Fact 4.2:

Let the evolution of $\vec{X}(k)$ be described as (4.3) with $\vec{X}(0)$ an arbitrary decision vector satisfying the stacking constraint, and the learning algorithm $h(\cdot)$ as stated above. Then

1) $\vec{X}(k+1) - \vec{X}(k) = \vec{1}_{j(k)}$ or $-\vec{1}_{j(k)}$ for some $j(k)$, where $\vec{1}_{j(k)}$ is a length 2^b vector whose entries are all 0's except the $j(k)$-th entry is 1.

2) $\vec{X}(k)$ obeys the constraint (4.1) for all k.

3) There are at most b swaps for each update.

5. Convergence Property of the Algorithm.

In the previous section, an adaptation algorithm for solving the optimal filtering problem within the class of stack filters has been proposed. The adaptation algorithm bookkeeps the statistics for each binary sequence while maintaining the stacking constraint among the decision variables associated with the binary sequences. In this section, the convergence property of the algorithm under the signal and noise assumption in section III is investigated. It can be shown that if the step size $\dfrac{1}{N}$ is smaller enough, $\vec{X}(n)$ will converge to $\vec{f}_o$ in probability as $n \to \infty$. The result follows the distance-diminishing property of $\| \vec{X}(nL) - \vec{f}_o \|$ and the fact that if $\vec{X}(nL)$ is close enough to an optimal decision vector, then $\vec{X}(nL)$ must equal to that optimal decision vector. The reason that we only look at the trajectory of $\vec{X}(n)$ for every L period is the result of the assumption made in (3.3).

In the following, the effect of the swap operations for maintaining the stacking constraint on the trajectory of $\vec{X}(k)$ is first examined. The swap operation, from the geometric point of view, is actually a projection operation of the decision vector about the hyperplane describing the stacking constraint between the swapped decision variables. An projection occurs whenever the updated decision vector is drifting outside the polytope (2.11).

For convenience, the time index before Lemma 5.3 will still refer to the sub-update time index. After that, the time index will refer to the actual time index.

Assume that $\vec{X}(k)$ is the decision vector at time k which satisfies the stacking constraint. Let $\vec{X}(k+I)$ be the decision vector at time $k+I$, which is generated from $\vec{X}(k)$ and $(y_m)_{m-k+1}^{k+I}$ according to the adaptation algorithm. Let $\vec{\tilde{X}}(k+I)$ be the unconstrained decision vector, which is generated from $\vec{X}(k)$ and $(y_m)_{m-k+1}^{k+I}$ by step 1 of the adaptation algorithm without regard to the stacking constraint.

$$\vec{\tilde{X}}(k+I) - \vec{X}(k) = \sum_{m-1}^{I} \vec{1}_{j(m)}(-1)^{s_{k+m}+1} \quad \text{if} \quad \vec{r}_{k+m} = \vec{s}_{j(m)} \tag{5.1}$$

Then we have

$$\vec{X}(k+I)-\vec{\tilde{X}}(k+I) = \sum_{m=1}^{I} \vec{1}_{i(m)}-\vec{1}_{j(m)} \qquad (5.2)$$

where $\vec{s}_{i(m)} \geq \vec{s}_{j(m)}$ for some $i(m), j(m)$. If $i(m)=j(m)$, then it implies the update from $\vec{X}(k+m-1)$ to $\vec{X}(k+m)$ does not involve any swap. Otherwise, $x_{j(m)}$ is the first variable updated while $x_{i(m)}$ is the final variable updated at the update instant $k+m$ or vice versa depending on whether s_{k+m} is equal to 1 or 0. Before examining the geometric relationship between $\vec{X}(k+I)$ and $\vec{\tilde{X}}(k+I)$, some notation needs to be introduced. $||\cdot||$ denotes the Euclidean norm, while $\vec{\alpha}\cdot\vec{\beta}$ denotes the inner product of $\vec{\alpha}$ and $\vec{\beta}$.

Lemma 5.1:

$$||\vec{\tilde{X}}(k+I) - \vec{X}(k)|| \leq I \qquad (5.3)$$

proof:

The proof follows from (5.1) and the triangle inequality.

Lemma 5.2:

$$||\vec{X}(k+I)-\vec{\tilde{X}}(k+I)||^2 \leq I\cdot(I+1) \qquad (5.4)$$

proof:

Proven by induction on I. (Details omitted).

Lemma 5.2 gives an upper bound on the Euclidean distance between the constrainted vector $\vec{X}(k+I)$ and the unconstrainted vector $\vec{\tilde{X}}(k+I)$.

Lemma 5.3:

$$(\vec{X}(k+I)-\vec{\tilde{X}}(k+I))\cdot(\vec{\tilde{X}}(k+I)-\vec{f}_o) \leq I^2+\frac{I(I-1)}{2} \qquad (5.5)$$

proof:

In (5.2), suppose $i(m) \neq j(m)$ for some m, then $x_{i(m)}(k)-x_{j(m)}(k) \leq m-1$. For if it were not true, then it takes at least $m+1$ increments or decrements before $x_{j(m)}$ becomes greater than $x_{i(m)}$; i.e., when the swap between $x_{i(m)}$ and $x_{j(m)}$ occurs. But that is impossible within m updates since at each update instant, only one variable is incremented or decremented by 1. As a result, we have $\tilde{x}_{i(m)}(k+I)-\tilde{x}_{j(m)}(k+I) \leq I+m-1$ In addition, for any stack filter $\vec{f}_i$ and any pair of $i(m)$, $j(m)$, $(\vec{1}_{i(m)}-\vec{1}_{j(m)})\cdot\vec{f}_i \geq 0$.

Therefore,

$$(\vec{X}(k+I)-\vec{\tilde{X}}(k+I))\cdot(\vec{\tilde{X}}(k+I)-\vec{f}_o) = \sum_{m=1}^{I} (\vec{1}_{i(m)}-\vec{1}_{j(m)})\cdot(\vec{\tilde{X}}(k+I)-\vec{f}_o) \qquad (5.6)$$

$$\leq \sum_{m=1}^{I} (\vec{1}_{i(m)}-\vec{1}_{j(m)})\cdot\vec{\tilde{X}}(k+I) = \sum_{m=1}^{I} \tilde{x}_{i(m)}(k+I)-\tilde{x}_{j(m)}(k+I) \leq \sum_{m=1}^{I} (I+m-1) = I^2+\frac{I(I-1)}{2}$$

Lemma 5.3 gives a bound on the the inner product between $\vec{X}(k+I)-\vec{\tilde{X}}(k+I)$ and $\vec{\tilde{X}}(k+I)-\vec{f}_o$. The importance of this bound is that it is also independent of N and $\vec{X}(k)$.

In the rest of this section, the time index will refer to the actual time index. Whenever any of the previous Lemmas is quoted, the count of the number of updates needs to be properly converted.

Lemma 5.4:

$$||\vec{X}(nL+L) - \vec{T}_o|| \leq ||\vec{X}(nL+L) - \vec{T}_o|| + \frac{2B^2}{||\vec{X}(nL) - \vec{T}_o|| - B}$$ (5.7)

if $||\vec{X}(nL) - \vec{T}_o|| - B > 0$, where $B = L \cdot (M-1)$ is the number of sub-updates done within the period $[nL+1, \ nL+L]$.

proof:

$$||\vec{X}(nL+L) - \vec{T}_o||^2 = ||\vec{X}(nL+L) - \vec{X}(nL+L)||^2 + ||\vec{X}(nL+L) - \vec{T}_o||^2$$
$$+ 2 \cdot (\vec{X}(nL+L) - \vec{X}(nL+L)) \cdot (\vec{X}(nL+L) - \vec{T}_o)$$

By Lemma 5.2, Lemma 5.3 and the fact that

$$\sqrt{a^2 + x} \leq a + \frac{x}{2a} \quad \text{if} \quad a > 0$$ (*)

we get

$$||\vec{X}(nL+L) - \vec{T}_o|| \leq ||\vec{X}(nL+L) - \vec{T}_o|| + \frac{2B^2}{||\vec{X}(nL+L) - \vec{T}_o||}$$

By Lemma 5.1 and the triangle inequality,

$$||\vec{X}(nL+L) - \vec{T}_o|| \geq ||\vec{X}(nL) - \vec{T}_o|| - ||\vec{X}(nL+L) - \vec{X}(nL)||.$$

Lemma 5.5:

For any $\vec{X}(n) \neq \vec{T}_o$,

$$\frac{\vec{X}(n) - \vec{T}_o}{||\vec{X}(n) - \vec{T}_o||} = \sum_{\substack{j=1 \\ j \neq o}}^{K} \alpha_j \vec{W}_j$$ (5.8)

where $\alpha_j \geq 0$ and $\displaystyle\sum_{\substack{j=1 \\ j \neq o}}^{K} \alpha_j \geq 1$. α_j is implicitly dependent on $\vec{X}(n)$.

proof:

$\vec{X}(n)$ is in the convex polytope described by the constraint equations (2.11), so it can be expressed as the convex combination of the vertices of the polytope. Since each vertex corresponds to a particular stack filter, for any $\vec{X}(n)$, there exist some $\beta_j \geq 0$ and $\displaystyle\sum_{j=1}^{K} \beta_j = 1$ such that

$$\vec{X}(n) = \sum_{j=1}^{K} \beta_j \vec{T}_j$$ (5.9)

Then

$$\vec{X}(n) - \vec{T}_o = \sum_{\substack{j=1 \\ j \neq o}}^{K} \beta_j (\vec{T}_j - \vec{T}_o)$$ (5.10)

If $\vec{X}(n) \neq \vec{J}_o$, let

$$\alpha_j = \beta_j \cdot \frac{||\vec{J}_j - \vec{J}_o||}{||\vec{X}(n) - \vec{J}_o||} \tag{5.11}$$

then (5.8) and $\alpha_j \geq 0$ follows. $\sum\limits_{\substack{j=1 \\ j \neq 0}}^{K} \alpha_j \geq 1$ follows from applying $|| \cdot ||$ to (5.8) and using the triangle inequality.

Lemma 5.6:

$$E\{||\vec{X}(nL+L) - \vec{J}_o|| \mid \mathcal{F}_{nL}\} \leq ||\vec{X}(nL) - \vec{J}_o|| - \rho + \frac{B^2}{2||\vec{X}(nL) - \vec{J}_o||} \tag{5.12}$$

proof:

$$||\vec{X}(nL+L) - \vec{J}_o||^2 = ||\vec{X}(nL+L) - \vec{X}(nL)||^2 + ||\vec{X}(nL) - \vec{J}_o||^2$$
$$+ 2 \cdot (\vec{X}(nL+L) - \vec{X}(nL)) \cdot (\vec{X}(nL) - \vec{J}_o)$$

From (*),

$$||\vec{X}(nL+L) - \vec{J}_o|| \leq ||\vec{X}(nL) - \vec{J}_o|| \tag{5.13}$$

$$+ \frac{||\vec{X}(nL+L) - \vec{X}(nL)||^2}{2 \cdot ||\vec{X}(nL) - \vec{J}_o||} + \frac{(\vec{X}(nL+L) - \vec{X}(nL)) \cdot (\vec{X}(nL) - \vec{J}_o)}{||\vec{X}(nL) - \vec{J}_o||}$$

Take $E[\cdot \mid \mathcal{F}_{nL}]$ on both sides of (5.13) and use Lemma 5.5, the conditional expectation of the third term at the right hand side of (5.13) becomes

$$E[(\vec{X}(nL+L) - \vec{X}(nL)) \cdot \sum_j \alpha_j \vec{W}_j \mid \mathcal{F}_{nL}] = \vec{U}(nL+1, nL+L) \sum_j \alpha_j \vec{W}_j \leq \sum_j \alpha_j(-\rho) \leq -\rho$$

The proof follows by bounding the numerator of the second term at the right hand side of (5.13) using Lemma 5.1.

The following is the main theorem used to prove the convergence of the learning algorithm. The notation $y \nmid x$ in the theorem means the random variable x stochastically dominates the random variable y; namely, $P(x>c) \geq P(y>c)$ for $-\infty < c < +\infty$.

Theorem 5.7 (from [17]):

Let $(z_n)_{n \geq 0}$ be a sequence of random variables on a probability space $(\Omega, \mathcal{F}, \mathcal{P})$ adapted to an increasing sequence $(\mathcal{F}_n)_{n \geq 0}$ of sub-σ-algebras of $\mathcal{F}$, and the following two conditions are satisfied:

1) $||z_{n+1} - z_n|| \mid \mathcal{F}_n \nmid w \; \forall n \geq 0$ and $E[e^{\lambda w}] = D$ is finite for some $\lambda > 0$.
2) $\exists \; \epsilon > 0$, a such that $E[z_{n+1} - z_n, z_n \geq a \mid \mathcal{F}_n] \leq -\epsilon \; \forall \; n \geq 0$.

Then there exist some constants $0 < \eta$ and $0 < \delta < 1$ such that

$$P[z_n \geq b \mid \mathcal{F}_0] \leq \delta^n e^{\eta(z_0 - b)} + \frac{1 - \delta^n}{1 - \delta} D e^{\eta(a - b)}. \tag{5.14}$$

proof:

see [17]

Using the above theorem, we can now prove the convergence of the learning algorithm.

Theorem 5.8:

Let $\vec{X}(n)$ be the decision vector at time n, and the evolution of $\vec{X}(n)$ be described by

$$\vec{X}(n+1) = \tilde{h}(\vec{X}(n), Y_n)$$

given that $(Y_n)_{n \geq 0}$ satisfies the condition in (3.3) and $\tilde{h}(\cdot)$ is the learning algorithm stated in the previous section in terms of the actual time index. Then $\forall \epsilon' > 0$, there exists some N' such that $\forall N \geq N'$,

$$\lim_{n \to \infty} P(\vec{X}(n) \neq \vec{f}_o) \leq \epsilon' \tag{5.15}$$

proof:

Let $z_n = ||\vec{X}(nL) - \vec{f}_o||$. By Lemmas 5.4 and 5.6,

$$E[||\vec{X}(nL+L) - \vec{f}_o|| \mid \mathcal{F}_{nL}] \tag{5.16}$$

$$\leq ||\vec{X}(nL) - \vec{f}_o|| - \rho + \frac{B^2}{2||\vec{X}(nL) - \vec{f}_o||} + \frac{2B^2}{||\vec{X}(nL) - \vec{f}_o|| - B}.$$

Let a be the smallest constant that satisfies

$$\frac{B^2}{a - B} \leq \frac{2\rho}{5},$$

then, for any $z_n \geq a = B + \frac{5B^2}{2\rho}$, the last two terms in (5.16) can be bounded from above by $\frac{\rho}{2}$. Clearly, z_n satisfies the second condition of Theorem 5.7 with $\epsilon = \frac{\rho}{2}$. Also

$$|z_{n+1} - z_n| \leq ||\vec{X}(nL+L) - \vec{X}(nL)|| \leq B.$$

The increment of z_n is uniformly bounded, thus z_n clearly satisfies the first condition of Theorem 5.7 with some finite D. So we have shown that z_n satisfies both conditions of Theorem 5.7 and thus (5.14). From (5.10) and (3.9)

$$z_n \leq \max_j ||\vec{f}_j - \vec{f}_o|| \triangleq r \cdot N$$

Take expectation over z_o on (5.14), $P[z_n \geq b] \leq \delta^n e^{\eta(r \cdot N - b)} + \frac{1 - \delta^n}{1 - \delta} D e^{\eta(a - b)}.$

Let $b = \frac{N}{2} - 3B$, then for any N, $\lim_{n \to \infty} P[z_n \geq \frac{N}{2} - 3B] \leq \frac{1}{1 - \delta} D e^{\eta(a + 3B - \frac{N}{2})}.$

The proof follows from $P(\vec{X}(k) \neq \vec{f}_o) \leq P(z_n \geq \frac{N}{2} - 3B)$ $\forall k \in [nL+1, nL+L]$, since the variation of $||\vec{X}(k) - \vec{f}_o||$ over the interval $[nL+1, nL+L]$ is upper bounded by 3B.

6. Conclusions. An adaptive filtering algorithm is developed for the class of stack filters, which is a class of nonlinear filters obeying a weak superposition property.

The adaptation algorithm can be interpreted as a learning algorithm for a group of decision-making units, the decisions of which are subject to a set of constraints called the stacking constraints. Under a rather weak statistical assumption on the training inputs, the decision strategy adopted by the group, which evolves according to the proposed learning algorithm, can be shown to converge asymptotically to an optimal strategy in the sense that it corresponds to an optimal stack filter under the mean absolute error criterion.

Acknowledgments: This reserach was supported by the National Science Foundation under grant EET87-21333, and by a research grant from the AT&T Foundation.

8. References.

[1] P.D. Wendt, E.J. Coyle, and N.C. Gallagher, "Stack Filters," *IEEE Trans. on Acoustics, Speech and Signal Processing,* vol. ASSP-34, no. 4, pp. 898-911, August 1986.

[2] J.P. Fitch, E.J. Coyle, and N.C. Gallagher, "Median Filtering by Threshold Decomposition," *IEEE Trans. on Acoustics, Speech and Signal Processing,* vol. ASSP-32, no. 6, pp. 1183-89, Dec. 1984.

[3] J.P. Fitch, E.J. Coyle, and N.C. Gallagher, "Threshold decomposition of multi-dimensional rank order operations," *IEEE Trans. on Circuits and Systems,* vol. CAS-32, no. 5, May 1985.

[4] J.W. Tukey, "Nonlinear (nonsuperposable) methods for smoothing data," in *Conf. Rec., 1974 EAS-CON,*

[5] T. Nodes and N.C. Gallagher, "Median Filters: Some Modifications and Their Properties," *IEEE Trans. on Acoustics, Speech and Signal Processing,* vol. ASSP-30, pp. 739-46, Oct. 1982.

[6] J. Serra, *Image Analysis and Mathematical Morphology,* New York: Academic Press, 1982.

[7] G. Matheron, *Random Sets and Integral Geometry,* New York: Wiley, 1975.

[8] P.A. Maragos, R.W. Schafer, "Morphological Filters -- Part II: Their Relations to Median, Order-Statistic, and Stack Filters," *IEEE Trans. on Acoustics, Speech and Signal Processing,* vol. ASSP-35, no. 8, pp. 1170-84, August 1987.

[9] J. H. Lin and E. J. Coyle, "Optimal Nonlinear Filtering under the Mean Absolute Error Criterion," *Proc. IEEE Int. Conf. on Acoustics, Speech, Signal Processing,* New York, NY, April 1988.

[10] P.A. Maragos and R.W. Schafer, "Morphological Filters - Part I: Their Set-Theoretic Analysis and Relations to Linear Shift-Invariant Filters," *IEEE Transactions on Acoustics, Speech and Signal Processing,* vol. ASSP-35, no. 8, pp. 1153-69, August 1987.

[11] E.J. Coyle, "Rank order operators and the mean absolute error criterion," *IEEE Trans. on Acoustics, Speech, and Signal Processing,* vol. ASSP-36, no. 1, pp. 63-76, Jan. 1988.

[12] E.J. Coyle and J.-H. Lin, "Stack Filters and the Mean Absolute Error Criterion," *IEEE Trans. on Acoustics, Speech and Signal Processing,* vol. ASSP-36, no. 8, August 1988.

[13] J.P. Hayes, et al., "Architecture of a Hypercube Supercomputer," *Proceedings 1986 International Conference on Parallel Processing,* pp. 653-660, Aug. 1986.

[14] S. T. Alexander, *Adaptive Signal Processing: Theory and Application,* Springer-Verlag: New York, 1986.

[15] M. L. Honig, D. G. Messerschmitt, *Adaptive Filters: Structures, Algorithms, and Applications,* Kluwer Academic, 1984.

[16] C. Papadimitriou and K. Steiglitz, *Combinatorial Optimization: Algorithms and Complexity,* Prentice Hall: Englewood Cliffs, NY, 1982

[17] B. Hajek, "Hitting-time and Occupation-time Bounds Implied by Drift Analysis with Applications," *Advances in Applied Probability,* 14, pp 502-525, 1982.

THREE-DIMENSIONAL IMAGE RECONSTRUCTION FROM SCATTERING DATA

Andrew E. Yagle

The University of Michigan

1. Introduction. We consider the problem of exact reconstruction of a three-dimensional image from the scattered field produced when the image is illuminated with an impulsive plane wave. "Exact reconstruction" means that all diffraction and multiple scattering effects are accounted for, unlike approaches such as the Born approximation. The image itself is described by a function $V(x), x \in R^3$ which is real-valued, smooth, and has compact support. Two solution procedures are outlined: integral equation-based methods and differential fast algorithms.

The integral equations were first derived in [1]. They require a large amount of computation, and scattering data for all angles of incidence. Furthermore, the large data requirement means the problem is overdetermined and ill-posed. In contrast, the differential fast algorithms, recently developed by the author [2],[3], employ time causality to reduce the amount of computation required. Furthermore, backscattered data over a plane, from a single direction of incidence, is sufficient to reconstruct the image exactly in its entirety.

We also consider a stochastic formulation of the image reconstruction problem. In this formulation the image is modelled as a three-dimensional *random field* $s(x), x \in R^3$, where $s(x)$ again has compact support and also has known covariance. Instead of noise-free scattering data, we have noisy observations of the image over the inside of a sphere containing the image. We wish to compute the linear least-squares estimate of the image given these noisy observations.

This stochastic formulation is selected because the procedure for solving it is almost identical to that for the first problem. The integral equations and differential fast algorithms for the first problem can also be applied to the second problem. This rather astonishing fact is due to a fundamental equivalence between the inverse scattering problem and the linear least-squares estimation problem [4].

This paper is organized as follows. In Section 2 the problem of image reconstruction from scattering data is formulated. In Section 3 integral equation-based methods for solving this problem are specified. These integral equations were first derived in [1]; simpler derivations utilizing the orthogonality of the scattering wave fields were given in [3]. In Section 4 the differential fast algorithms for solving this problem are derived; these algorithms were first presented in [2] and [3].

In Section 5 the stochastic formulation of the image reconstruction problem is formulated; this results in a Wiener-Hopf integral equation identical to one of the inverse scattering problem integral equations. This allows the differential fast algorithms of Section 4 to be applied to this problem as well. Section 6 concludes by summarizing the paper and noting work presently in progress on these problems.

2. Image Reconstruction from Scattering Data.

The image $V(x), x \in R^3$ is assumed to be real-valued, non-negative, smooth and have compact support. It is probed or illuminated with an impulsive plane wave; the response due to another excitation such as a point source can be constructed by superposition of the plane wave response at various frequencies and angles. Given as data is the scattered wave field due to an impulsive plane wave; the goal is to reconstruct the image $V(x)$.

The wave field $u(x, k)$ satisfies the Schrodinger equation

$$(\Delta + k^2 - V(x))u(x, k) = 0 \qquad (2-1)$$

Two different sets of boundary conditions are specified, resulting in two different solutions.

Scattering Solution

The *scattering* solution $\psi(x, k, e_i)$ has the boundary condition

$$\psi(x, k, e_i) = e^{-ike_i \cdot x} + (e^{-ik|x|}/4\pi|x|)A(k, e_s, e_i) + O(|x|^{-2}) \text{ as } |x| \to \infty \quad (2-2)$$

where the *scattering amplitude* $A(k, e_s, e_i)$ is defined as

$$A(k, e_s, e_i) = -\int e^{ike_s \cdot y}V(y)\psi(y, k, e_i)dy \qquad (2-3)$$

and e_i and e_s are unit vectors. In the time domain, $\psi(x, k, e_i)$ is the total wave field (incident plus scattered) resulting from an an incident impulsive plane wave in the direction e_i hitting the image and being scattered in all directions. The scattering amplitude specifies the far-field behavior of the wave field, and constitutes the data from which the image is to be reconstructed.

Regular Solution

The *regular* solution $\phi(x, k, e_i)$ is defined as being the solution to (2-1) that is an entire analytic function of k and is of exponential order $|x|$. $\phi(x, t, e_i) = \mathcal{F}^{-1}\{\phi(x, k, e_i)\}$ has support in t on the interval $[-|x|, |x|]$, and it can be written as

$$\phi(x, k, e_i) = e^{-ike_i \cdot x} - \int_{-|x|}^{|x|} m(x, t, e_i)e^{-ikt}dt \qquad (2-4)$$

where $m(x,t,e_i)$ is the non-impulsive part of $\phi(x,t,e_i)$. Although unlike $\psi(x,k,e_i)$ it has no direct physical interpretation, it is an intermediate result in reconstructing the image $V(x)$.

Jost Operator

The regular and scattering solutions are related by a *Jost operator* $J(k)$. This is an operator on the space $L^2(S^2)$ (S^2 is the unit sphere) with kernel $J(k,e_1,e_2)$. Specifically,

$$\phi(x,k,e_i) = \int_{S^2} \psi(x,k,e_s)J(k,e_s,e_i)de_s \qquad (2-5a)$$

$$\psi(x,k,e_i) = \int_{S^2} \phi(x,k,e_s)J^{-1}(k,e_s,e_i)de_s \qquad (2-5b)$$

where J^{-1} is the inverse Jost operator. Both $J(k)$ and $J^{-1}(k)$ are analytic in the lower half-plane, which corresponds to *causality* in the time domain.

From [1], the Jost operator satisfies

$$J(-k) = QS(k)J(k)Q \qquad (2-6)$$

where $S(k)$ is the scattering operator with kernel

$$S(k,e_1,e_2) = \delta(e_1 - e_2) + (k/2\pi i)A(k,e_1,e_2) \qquad (2-7)$$

and Q is the operator such that $QA(k,e_1,e_2) = A(k,-e_1,e_2)$.

Orthonormality

It is well known [1] that for non-negative real $V(x)$ the scattering solutions $\{\psi(x,k,e_i)\}$ form a complete set. Thus they are orthonormal, in that

$$(2\pi)^{-3} \int_0^\infty \int_{S^2} \psi(x,k,e)\psi^*(y,k,e)k^2 de\, dk = \delta(x-y). \qquad (2-8)$$

It follows that the solutions $\{\phi(x,k,e_i)\}$ are orthonormal with respect to the positive definite *spectral function* $(J^H J)^{-1}$, i.e.,

$$(2\pi)^{-3} \int_0^\infty \int_{S^2} \int_{S^2} \int_{S^2} \phi(x,k,e_1)J^{-1}(k,e_1,e_3)J^{-1}(k,e_3,e_2)^*\phi^*(y,k,e_2)k^2\, de_1\, de_2\, de_3\, dk$$

$$= \delta(x-y) \qquad (2-9)$$

The orthonormality of the solutions $\{\phi(x,k,e_i)\}$ with respect to $(J^H J)^{-1}$ leads directly to the integral equations below [3],[5]. The orthogonality of (2-8) is analogous to the orthonormality of the innovations in linear least-squares estimation; this is why the two problems are so closely related [4].

3. Integral Equation Solutions.

Here we quickly sketch the integral equation based procedures for reconstructing the image $V(x)$ from the scattered field $V(x)$.

Generalized Marchenko Procedure

The image $V(x)$ may be reconstructed as follows. Given far-field scattering data in the form of the scattering amplitude $A(k, e_s, e_i)$, compute

$$G(t, e_s, e_i, x) = -(2\pi)^{-2} \int_{-\infty}^{\infty} e^{ik(t-(e_s-e_i)\cdot x)} ik A(k, e_s, e_i) \qquad (3-1)$$

and then solve the *generalized Marchenko integral equation*

$$v(x, t, e_i) = \int_{S^2} G(t, e_s, e_i, x) de_s + \int_0^{\infty} \int_{S^2} G(t+\tau, e', e_i, x) v(x, t, -e') \, de' \, d\tau \quad (3-2)$$

for the delayed scattered field

$$v(x, t, e_i) = \psi(x, t - e_i \cdot x, e_i) - \delta(t - e_i \cdot x). \qquad (3-3)$$

The image $V(x)$ is then recovered from $v(x, t, e_i)$ using the *miracle* equation [1]

$$V(x) = 2e_i \cdot \nabla v(x, t = 0, e_i). \qquad (3-4)$$

Note that the right side of (3-4) must be independent of the direction of incidence e_i. This is the "miracle" (the term was first used in [1]), and it imposes a constraint on possible $\psi(x, k, e_i)$, and hence on possible scattering amplitudes $A(k, e_s, e_i)$. Since the scattering data $A(k, e_s, e_i)$ has five degrees of freedom and the image $V(x)$ generating it has only three degrees of freedom, this constraint is not surprising. There is no known simple test to determine which functions $A(k, e_s, e_i)$ result in a miraculous solution $\psi(x, k, e_i)$.

The Marchenko integral equation (3-2) can be derived directly from the orthogonality of the $\{\psi(x, k, e_i)\}$ [5]; it can also be derived using Green's function arguments [6]. The miracle equation (3-4) is derived in Section 4 below.

Generalized Gel'fand-Levitan Procedure

As an alternative to the above procedure we may do the following [1],[7]. First, obtain the inverse Jost operator kernel $J^{-1}(k, e_1, e_2)$ from the scattering data $A(k, e_s, e_i)$ using (2-6) and (2-7); this amounts to solving the integral equation

$$L(t, e_s, e_i) = G(t, -e_s, e_i, 0) + \int_0^{\infty} \int_{S^2} L(\tau, -e_s, e') G(t + \tau, e', e_i, 0) \, de' \, d\tau \quad (3-5)$$

where $G(t, e_1, e_2, x)$ is defined in (3-1). $J^{-1}(k)$ is computed from $L(t)$ by taking a Fourier transform:

$$J^{-1}(k, e_1, e_2) = 1 + \int_0^\infty L(t, e_1, e_2) e^{-ikt} dt \qquad (3-6)$$

Although (3-5) looks like (3-2), note that (3-5) need only be solved for $x = 0$, whereas (3-2) must be solved for *all* x.

Next, compute the double inverse spatial Fourier transform of the perturbation of the spectral function away from its free-space representation. This is

$$k(x, y) = \mathcal{F}^{-1}\mathcal{F}^{-1}\{[(J^H J)^{-1} - \delta(e_1 - e_2)](k, e_1 - e_2)\} \qquad (3-7)$$

Then the inverse Fourier transform of the non-impulsive part of the regular solution

$$m(x, y) = \mathcal{F}^{-1}\{\phi(x, k, e_i) - e^{-ike_i \cdot x}\} \qquad (3-8)$$

may be obtained by solving the *generalized Gel'fand-Levitan integral equation* [1],[7]

$$k(x, y) = m(x, y) + \int_{|z| < |x|} m(x, z) k(z, y) dz, \quad |y| < |x| \qquad (3-9)$$

In the (x, t, e_i) domain (3-9) has a more complicated form [7].

The image $V(x)$ is then recovered from the miracle equation for the regular solution

$$V(x) = -2e_x \cdot \nabla |x|^2 m(x, x) \qquad (3-10)$$

Here the "miracle" manifests itself by the vanishing of the right side of (3-10) unless $y = x$.

Although the generalized Gel'fand-Levitan procedure requires that the integral equation (3-5) be solved in addition to (3-9), it should be noted that (3-9) has only a finite range of integration, while the generalized Marchenko equation (3-2) has an infinite range of integration.

The Gel'fand-Levitan integral equation (3-9) can be derived directly from the orthogonality of the $\{\phi(x, k, e_i)\}$ with respect to $(J^H J)^{-1}$ [3]. Even more remarkably, the Gel'fand-Levitan integral equation (3-9) is a Wiener-Hopf integral equation! This allows the fast algorithms we develop next to be applied to either recovering the image from scattering data, or to estimating it from noisy observations.

4. Differential Fast Algorithm Solutions. As an alternative to solving either of the above integral equations, we may use either of the following fast algorithms. The generalized Levinson algorithm recursively reconstructs the regular

solution, while the generalized Schur algorithm recursively reconstructs the scattering solution.

Generalized Levinson Algorithm

The major computation in the generalized Gel'fand-Levitan procedure (3-5)-(3-10) is the solution of the Gel'fand-Levitan integral equation (3-9). The Levinson algorithm is a fast algorithm for solving this integral equation. Nomenclature for the algorithms follows that of their one-dimensional counterparts in [8].

The generalized Levinson algorithm is derived as follows. Recall that the regular solution $\phi(x, k, e_i)$ solves the Schrodinger equation (2-1). The inverse spatial Fourier transform of (2-1) can easily be rewritten as the coupled set of partial differential equations [3],[4]

$$\left(\frac{\partial}{\partial|x|} + \frac{\partial}{\partial|y|}\right) |x||y|m(x, y) = |x||y|Q(x, y) \qquad (4-1a)$$

$$\left(\frac{\partial}{\partial|x|} - \frac{\partial}{\partial|y|}\right) |x||y|Q(x, y) = |x||y|H(x, y) \qquad (4-1b)$$

$$H(x, y) = (\Delta_y^T - \Delta_x^T)m(x, y) + V(x)m(x, y) \qquad (4-1c)$$

$$V(x) = -2Q(x, x) \qquad (4-1d)$$

Here $m(x, y)$ is defined in (3-8), $Q(x, y)$ and $H(x, y)$ are auxiliary quantities defined in (4-1a) and (4-1c), Δ^T is the transverse Laplacian in spherical coordinates

$$\Delta_x^T = \frac{1}{x^2 \sin \theta} \frac{\partial}{\partial \theta} \left(\sin \theta \frac{\partial}{\partial \theta}\right) + \frac{1}{x^2 \sin^2 \phi} \frac{\partial^2}{\partial \phi^2}, \qquad (4-2)$$

and (4-1d) simply restates (3-10).

Equations (4-1) can be propagated in increasing $|x|$ and $|y| < |x|$, yielding the image $V(x)$ in increasing $|x|$. However, the computation (4-1d) requires the boundary value on the characteristic $|x| = |y|$ that is not automatically updated in $|x|$. Hence it must be computed from the other values which have been updated in $|x|$, using the integral equation (3-9):

$$V(x) = -\frac{2}{|x|^2} \frac{d}{d|x|} |x|^2 m(x, x) = -\frac{2}{|x|^2} \frac{d}{d|x|} |x|^2 \left(k(x, x) - \int_{|z|<|x|} m(x, z)k(z, x)dz\right).$$
$$(4-3)$$

The *generalized Levinson algorithm* consists of propagating (4-1)-(4-3) in increasing $|x|$ and $|y| < |x|$. The algorithm is initialized with $m(0, 0) = Q(0, 0) = 0$;

non-zero values are introduced into the algorithm via (4-1c). The image $V(x)$ is computed in increasing $|x|$ from $m(x,y)$ and $Q(x,y)$ using (4-3).

Comments

This algorithm is a generalization of the one-dimensional Krein-Levinson algorithm of [8]. The major difference is that the one-dimensional Levinson algorithm involves *waves* propagating in opposite directions in a lattice structure; this wave decomposition is not possible for the scattered field produced by a three-dimensional image. This algorithm is in fact a three-dimensional *split* Levinson algorithm. This follows since the split Levinson algorithm recursively computes the regular solution to the discrete Schrodinger equation in one dimension, just as the present algorithm recursively computes the regular solution in three dimensions.

This algorithm is much faster computationally than solving a discretized version of (3-9) by Gaussian elimination: If each spatial coordinate is discretized to N values, the reduction in computation is from $O(N^{12})$ to $O(N^8)$. Compare the reduction of computation $O(N^{12})$ to $O(N^8)$ to the reduction in computation $O(N^3)$ to $O(N^2)$ attained by the Levinson algorithm.

The problem with this algorithm lies in the computation of the boundary value (4-3). This is a large, non-parallelizable computation that accounts for a significant amount of the total computation required by the algorithm. The corresponding computation in the one-dimensional Levinson algorithm is the "inner product" computation of the reflection coefficient that accounts for roughly one-third of all the computation required by the algorithm. It would be extremely desirable to avoid this computation.

Generalized Schur Algorithm

The Schur algorithm avoids the computation (4-3) by propagating the scattering solution, rather than the regular solution. It also uses near-field scattering data, instead of far-field data; this is often more realistic. Most importantly, it requires backscattered data for only *one* direction of incidence e_i; the other methods discussed require scattering data for a hemisphere of both incident and scattered directions.

The generalized Schur algorithm is derived as follows. The scattering solution $\psi(x,k,e_i)$, like the regular solution $\phi(x,k,e_i)$, solves (2-1). Let $z = e_i \cdot x$ be the axis normal to the incident impulsive plane wave used to illuminate the image, and let y be the two directions normal to z. Then the inverse *temporal* Fourier transform of (2-1) can be rewritten as the coupled set of partial differential equations [2],[3]

$$\left(\frac{\partial}{\partial z} + \frac{\partial}{\partial t}\right) v(z, y, t) = q(z, y, t) \tag{4 − 4a}$$

$$\left(\frac{\partial}{\partial z} - \frac{\partial}{\partial t}\right) q(z, y, t) = (V(z, y) - \Delta_y) v(z, y, t) \tag{4 − 4b}$$

$$V(z, y) = -2q(z, y, t = z) \tag{4 − 4c}$$

Here $v(z, y, t)$ is defined in (3-3), $q(z, y, t)$ is an auxiliary quantity defined in (4-4a), and Δ_y is the transverse Laplacian (i.e., with respect to y), and (4-4c) simply restates (3-4).

Equations (4-4) can be propagated in increasing z and $t \geq z$, yielding the image $V(z, y)$ in increasing z. Note that (4-4c) requires the boundary value on the characteristic $t = z = e_i \cdot x$, and that this is automatically updated in z. This is a significant difference from the Levinson algorithm, in which this boundary value must be computed from other updated values using (4-3).

The *generalized Schur algorithm* consists of propagating (4-4) in increasing $z = e_i \cdot x$ and $t \geq z$. The algorithm is initialized on the plane $z = 0$ using the backscattered field $v(z = 0, y, t)$ and its derivative $q(z = 0, y, t)$.

Comments

This algorithm is a generalization of the one-dimensional Schur algorithm of [8]. The major difference again is that the scattering solution, rather than waves, is propagated, making it properly a generalized split Schur algorithm.

The computational savings of the Schur algorithm over solving the Marchenko integral equation (3-2) is even greater than that of the Levinson algorithm over solving the Gel'fand-Levitan integral equation. Solving (3-2) by discretization and Gaussian elimination requires $O(N^{12})$ computations; the Schur algorithm requires only $O(N^6)$ computations, since only one direction of incidence e_i of an illuminating plane wave is needed. Also, the computations in (4-4) can be performed in parallel over y, since (4-3) is no longer needed.

Furthermore, the "miracle" equation (3-4) is easily understood in this context. The image $V(x)$ is reconstructed in slices normal to the direction of incidence from the gradient of the scattered field at the wave front $t = e_i \cdot x$. This eliminates all diffraction and multiple scattering–at the wave front there hasn't been *time* for any of these to occur. The "miracle" is simply a statement that this is true for any direction of illumination. Multiple scattering is accounted for in the coupling between (4-4a) and (4-4b); neglect of this coupling is equivalent to the Born or single-scattering approximation [2].

5. Stochastic Formulation of Image Reconstruction.

We now reformulate the image reconstruction problem in stochastic terms. We are given noisy observations $w(y)$ of the image $s(y)$ inside a sphere of radius $|x|$. These observations could be the $V(y)$ obtained from scattering data using the above procedure, but this is not necessary; to emphasize this we use $s(y)$ rather than $V(y)$. The goal is to compute the *prediction* filter for the linear least-squares estimate of $s(x)$ given $\{w(y), |y| < |x|\}$. The *smoothing* filter for estimating $s(x)$ from $\{w(y), |y| < T\}$, where $T > |x|$, can then be obtained from the prediction filter [9].

Specifically, we observe

$$w(y) = z(y) + v(y), \quad |y| < |x| \tag{5-1}$$

where $v(x)$ is a zero-mean real-valued white noise field with covariance

$$E[v(x)v(y)] = \delta(x - y) \tag{5-2}$$

We assume $s(x)$ has zero mean (any a priori notion of the image has been subtracted out) and known covariance

$$E[s(x)s(y)] = \dot{k}(x, y) \tag{5-3}$$

where the function $k(x, y)$ is positive definite and has the *generalized displacement property*

$$(\Delta_x - \Delta_y)k(x, y) = 0. \tag{5-4}$$

The structure of $k(x, y)$ implied by (5-4) reduces the number of degrees of freedom in the function $k(x, y)$ from six to five. Note that both homogeneous and isotropic random fields are included as special cases of the property (5-4).

The estimate $\hat{s}(x)$ of $s(x)$ has the form

$$\hat{s}(x) = \int_{|y| \le |x|} h(x, y)w(y)dy. \tag{5-5}$$

By the orthogonality principle the filter $h(x, y)$ is determined by the three-dimensional Wiener-Hopf integral equation

$$k(x, y) = h(x, y) + \int_{|z| \le |x|} h(x, z)k(z, y)dz, \quad |y| \le |x|. \tag{5-6}$$

Note that this integral equation has the same form as the generalized Gel'fand-Levitan integral equation (3-9). Without loss of generality, we define $h(x, y) = 0$ for $|y| > |x|$.

Applying the operator $(\Delta_x - \Delta_y)$ to the integral equation (5-6) and using the generalized displacement property (5-4), Green's theorem, and the unicity of solution to (5-6) when $k(x, y)$ is positive definite yields, after some algebra (see [10]),

$$(\Delta_x - \Delta_y)h(x, y) = \int_{S^2} V(x, e)h(|x|e, y)\,de \qquad (5-7)$$

where the *non-local potential* $V(x, e)$ is defined as

$$V(x, e) = -\frac{2}{|x|^2}\frac{d}{d|x|}|x|^2 h(x, |x|e). \qquad (5-8)$$

A spatial Fourier transform of (5-7) results in the Schrodinger equation with a non-local potential

$$(\Delta + k^2)h(x, k, e_i) = \int_{S^2} V(x, e)h(|x|e, k, e_i)\,de \qquad (5-9)$$

The non-local potential $V(x, e)$ arises since the covariance $k(x, y)$ has five degrees of freedom; since the $V(x, e)$ characterize the prediction filter $h(x, y)$ associated with $k(x, y)$ via (5-6), they too must have five degrees of freedom. The $V(x, e)$ characterize the prediction filter $h(x, y)$ in three dimensions much as the reflection coefficients characterize the prediction filter in one dimension.

Although differential fast algorithms similar to the generalized Levinson and Schur algorithms can be derived for non-local potentials (see [5]), we here make the simplifying assumption that the covariance $k(x, y)$, in addition to satisfying (5-4), has the additional structure to make $V(x, e)$ a *local* potential $V(x, e) = V(x)\delta(x/|x| - e)$. This reduces $k(x, y)$ to three degrees of freedom; an exact characterization of this additional structure is unknown. Indeed, this is a major unsolved problem in inverse scattering theory: the scattering amplitude $A(k, e_s, e_i)$ (five degrees of freedom) overdetermines the image $V(x)$ (three degrees of freedom). What additional constraints does $A(k, e_s, e_i)$ have?

With this provision, (5-9) becomes identical to the Schrodinger equation (2-1). Hence the integral equations of Section 3 and the differential fast algorithms of Section 4 can also be applied to the stochastic formulation of the image reconstruction problem. The scattering data is replaced by the covariance $k(x, y)$; the connection between them is specified by (3-7) (note the suggestive notation used in Section 3).

6. Conclusion. The image reconstruction problem has been formulated in two different ways. The first problem is to reconstruct an image from the scattered wave field produced by illuminating impulsive plane waves at various directions of incidence. The second problem is to reconstruct an image from noisy observations of it inside a sphere; these observations could be data from the first problem. Amazingly, both problems may be solved using the same procedures. Two integral-equation based procedures, and two differential fast algorithms, were specified; all are generalizations of similar equations for the one-dimensional inverse scattering problem.

Work in progress on the subjects of this paper include: (1) investigation of the numerical performance of the algorithms of Section 4; (2) investigating the additional constraints on the scattering amplitude $A(k, e_s, e_i)$ produced by an image $V(x)$; and (3) investigating the additional constraints on the covariance $k(x, y)$ that result in a local potential $V(x)$, rather than a non-local potential $V(x, e)$ (see (5-8)). The latter two problems are closely related, as are the problems of three-dimensional inverse scattering and linear least-squares estimation of random fields.

Acknowledgments. This work was supported in part by the National Science Foundation under grant number MIP-8858082. The author's current address is: Dept. of Electrical Engineering and Computer Science, the University of Michigan, Ann Arbor, MI 48109-2122.

References

1. R.G. Newton, "Inverse Scattering. II. Three Dimensions," *J. Math. Phys.* 21(7), 1698-1715 (1980).

2. A.E. Yagle and B.C. Levy, "Layer Stripping Solutions of Multi-dimensional Inverse Scattering Problems," *J. Math. Phys.* 27(6), 1701-1710 (1986).

3. A.E. Yagle, "Multi-Dimensional Inverse Scattering: An Orthogonalization Formulation," *J. Math. Phys.* 28(7), 1481-1491 (1987).

4. A.E. Yagle, "Connections between Three-Dimensional Inverse Scattering and the Linear Least-Squares Estimation of Random Fields," to appear in *Acta Appl. Mathe.*

5. A.E. Yagle, "Differential and Integral Methods for Three-Dimensional Inverse Scattering Problems with a Non-Local Potential," *Inverse Problems* 4(2), 549-566 (1988).

6. J.H. Rose, M. Cheney, and B. DeFacio, "The Connection between Time and

Frequency Domain Three-Dimensional Inverse Scattering Methods," *J. Math. Phys.* 25(10), 2995-3000 (1984).

7. R.G. Newton, "Inverse Scattering. IV. Three Dimensions: Generalized Marchenko Construction with Bound States, and Generalized Gel'fand-Levitan Equations," *J. Math. Phys.* 23(4), 594-604 (1982).

8. A.M. Bruckstein, B.C. Levy, and T.Kailath, "Differential Methods in Inverse Scattering," *SIAM J. Appl. Math.* 45(2), 312-335 (1985).

9. A.E. Yagle, "A Fast Algorithm for Linear Estimation of Three-Dimensional Homogeneous Anisotropic Random Fields," *Proc. IEEE Int'l Conf. on Acoustics, Speech, Sig. Proc.*, Dallas, TX, April 6-9, 1987.

10. A.E. Yagle, "Generalized Split Levinson, Schur, and Lattice Algorithms for Three-Dimensional Random Field Estimation Problems," submitted to *SIAM J. Appl. Math.*

11. A.E. Yagle, "Generalized Levinson and Fast Cholesky Algorithms for Three-Dimensional Random Field Estimation Problems," *Proc. IEEE Int'l Conf. on Acoustics, Speech, Sig. Proc.*, New York, NY, April 11-14, 1988.

A TOMOGRAPHIC FORMULATION OF BISTATIC SYNTHETIC APERTURE RADAR

Orhan Arikan and David C. Munson, Jr.
University of Illinois at Urbana-Champaign

I. INTRODUCTION

A bistatic radar is a radar with separate transmitting and receiving antennas that are separated by a distance comparable to the target distance. The concept of bistatic radar is not a new one. Early experiments with bistatic radar led to the development of monostatic radar in the late 1930's. Further development was put aside after the demonstration of a monostatic radar using a single antenna for both transmitting and receiving. Monostatic radar is used almost exclusively because of its simplicity, its ability to scan a hemispherical volume in space, and the relative ease with which useable target information can be extracted from the received signal. Bistatic radar lay dormant for about fifteen years until it was "reinvented" in the early 1950's and received new interest [1].

Synthetic aperture radar (SAR) is used to provide high resolution microwave images, often of terrain, using a small antenna mounted on an aircraft or spacecraft [2]. High resolution in range is achieved through traditional pulse compression methods, while high azimuth resolution is obtained by illuminating the target area of interest from many different vantage points. The returned signals are then coherently processed to form an image having much higher resolution than that ordinarily dictated by the antenna size.

In one form of SAR, known as spotlight-mode SAR, the radar antenna is continuously steered to illuminate the entire scene of interest with each transmitted microwave pulse. This type of SAR, which we shall refer to as monostatic spotlight-mode SAR (MSSAR), collects different angular views of a scene and works on the same principle as computer-aided tomography (CAT) [3]. In recent years there has been considerable military interest in bistatic spotlight-mode SAR (BSSAR) where a high powered transmitter can be operated at a safe distance from a target area while a covert receiver is flown close to the target area, collecting returned signals and forming high resolution imagery.

Although much work has been performed on BSSAR, little of it is documented in the open literature, and most of it has taken place from a traditional Doppler radar viewpoint. In this paper we give a basic derivation of BSSAR, starting from first principles. Similar to the MSSAR case [3], we show that BSSAR can be explained using the projection-slice theorem [4,5] from CAT and that BSSAR does not rely on the Doppler effect. It is shown that at each look angle the demodulated BSSAR data are approximate band-limited samples of a slice of the two-dimensional (2-D) Fourier transform of the target reflectivity. The locations of the Fourier domain samples are found and the shape of the Fourier grid is examined for several special cases of transmitter and receiver motion. We also analyze wavefront curvature within the bistatic setting and compare it with that occuring in the monostatic case.

II. TOMOGRAPHIC FORMULATION OF BISTATIC SPOTLIGHT-MODE SAR

CAT is an imaging technique for providing a 2-D cross-sectional view of a 3-D object by processing many 1-D projectional views collected from different angles. The principle underlying CAT is the projection-slice theorem [4,5]. Using this theorem from tomography, Munson, O'Brien, and Jenkins have shown that MSSAR is a tomographic imaging system and that it can be easily understood from this viewpoint [3]. In this section we develop a tomographic formulation of BSSAR in a similar fashion. We first review the projection-slice theorem from CAT, and then use it in the derivation of BSSAR. We then consider several cases of transmitter and receiver motion, which result in different Fourier domain data grids.

A. Projection-Slice Theorem

The projection of a 2-dimensional (2-D) function, $g(x,y)$, onto a line passing through the origin (Fig. 1) is defined as

$$p_\theta(u) = \int_{-\infty}^{\infty} g(u\cos\theta - v\sin\theta, \ u\sin\theta + v\cos\theta)dv \qquad (1)$$

where $p_\theta(u)$ is a series of line integrals taken at angle θ with respect to the x-axis.

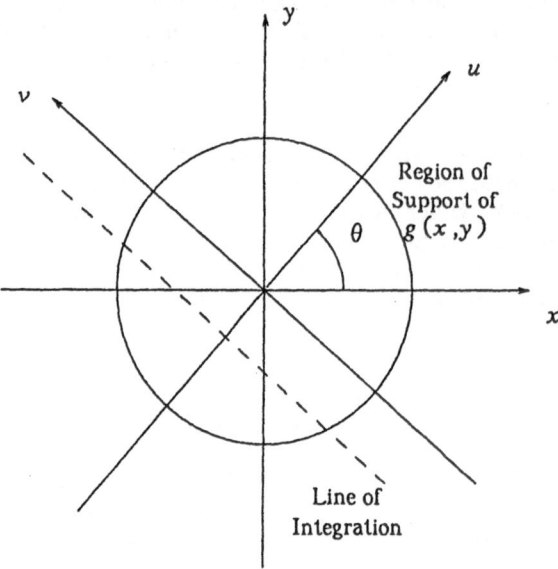

Fig. 1. Projection of $g(x,y)$ at an angle θ.

Now, define the 2-D Fourier transform of $g(x,y)$ as

$$G(X,Y) = \int_{-\infty}^{\infty} \int_{-\infty}^{\infty} g(x,y) \exp\{-j(xX+yY)\}dxdy \ .$$

Then the projection-slice theorem states that [4]

$$P_\theta(U) = G(U\cos\theta, U\sin\theta) \qquad (2)$$

where $P_\theta(\cdot)$ is the 1-D Fourier transform of the projection at angle θ. That is, the Fourier transform of the projection at angle θ is a slice through the origin of the 2-D transform $G(X,Y)$ taken at angle θ with respect to the X-axis.

B. Bistatic Spotlight-Mode SAR Derivation

The geometry for data collection in BSSAR is shown in Fig. 2 where (x_t,y_t) is the transmitter location, (x_r,y_r) is the receiver location, and the ground patch to be imaged is circular with radius L. Here, we have taken the altitude of the radar to be zero for simplicity; the effect of a true nonzero altitude can be accounted for in a manner described later. For ground points lying on the ellipse with foci (x_t,y_t) and (x_r,y_r) in Fig. 2, the total distance from transmitter to a ground point to the receiver is constant. Thus, a signal emanating from the transmitter will be seen at the receiver as simultaneously reflecting off all points on the ellipse. The u-axis and v-axis are the normal and tangent of this ellipse at the origin of the ground patch coordinates. Since the normal of a point on an ellipse bisects the angle between focal lines drawn to the same point, β is one-half of the angle between lines joining the transmitter and receiver to the origin.

We are interested in imaging the reflectivity, $g(x,y)$, of the ground patch, where the reflectivity is assumed to be independent of both the viewing angle and the frequency of incident radiation. We assume a linear FM transmitted signal of the form $Re(s(t))$ with

$$s(t) = \begin{cases} \exp\{j(\omega_0 t + \alpha t^2)\} & |t| \leq T/2 \\ \\ 0 & else \end{cases}$$

where ω_0 is the carrier frequency and 2α is the FM rate.

The returned signal is a sum of returns from all ground points and is given by the real part of

$$r_\theta(t) = A \int_{-L}^{L} \int_{-L}^{L} g(u\cos\theta - v\sin\theta, u\sin\theta + v\cos\theta) s\left(t - \tau(u,v)\right) du\, dv \qquad (3)$$

where A is the free space attenuation factor (assumed constant for all points in the ground patch) and $\tau(u,v)$ is the total propagation delay from the transmitter to the ground point (u,v) to the receiver. The delay $\tau(u,v)$ is constant for all ground points lying on any ellipse having foci (x_t,y_t) and (x_r,y_r). Assuming that $R_t,R_r \gg L$, such ellipses are well approximated over the ground patch by lines orthogonal to the u-axis in Fig. 2. Thus, $\tau(u,v)$ is nearly independent of v, and as shown in the Appendix, is well approximated by

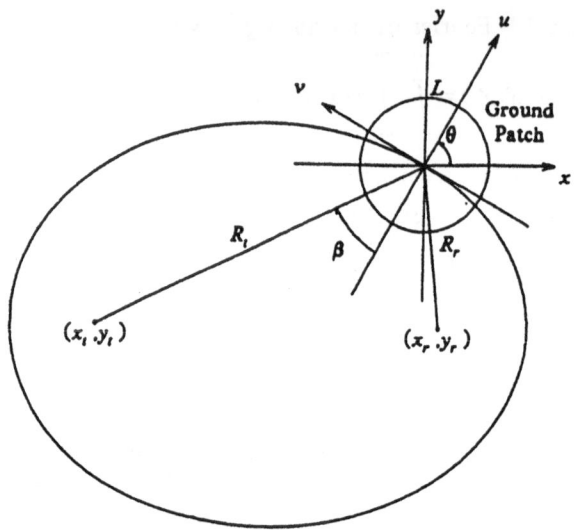

Fig. 2. Ground-plane geometry for data collection in BSSAR.

$$\tau(u) = \frac{R_t + R_r + u \cdot w_{tr}}{c}$$

where

$$w_{tr} = 2 \mid cos(\beta) \mid \qquad (4)$$

Therefore, (3) can be rewritten as

$$r_\theta(t) = A \int_{-L}^{L} \left| \int_{-L}^{L} g \; (ucos \; \theta - vsin \; \theta, \; usin \; \theta + vcos \; \theta)dv \right| s \left(t - \frac{R_t + R_r + u \cdot w_{tr}}{c} \right) du$$

$$= A \int_{-L}^{L} p_\theta(u) \; s \left(t - \frac{R_t + R_r + u \cdot w_{tr}}{c} \right) du$$

which is recognized as a convolution of the projections of the reflectivity at angle θ with the signal $s(t)$.

Multiplying the received signal, $Re(r_\theta(t))$, by $Re(s(t))$, and $Im(s(t))$, each delayed by $(R_t + R_r)/c$, (quadrature demodulation) and low-pass filtering to remove terms centered at $2\omega_0$ gives the real and imaginary components of the baseband signal

$$C_\theta(t) = \frac{A}{2}\int_{-L}^{L} p_\theta(u)\exp\left\{ j\left[-\omega_0\frac{uw_{tr}}{c}+\alpha\frac{(uw_{tr})^2}{c^2}-\alpha\frac{2uw_{tr}}{c}\left(t-\frac{R_t+R_r}{c}\right)\right]\right\} du \; .$$

Neglecting the quadratic phase term, and regrouping the remaining terms gives

$$C_\theta(t) \approx \frac{A}{2}\int_{-L}^{L} p_\theta(u)\exp\left\{-j\frac{w_{tr}}{c}\left[\omega_0+2\alpha\left(t-\frac{R_t+R_r}{c}\right)\right]u\right\} du \; .$$

This expression is the Fourier transform $P_\theta(X)$ of the projection $p_\theta(u)$ evaluated at

$$X = \frac{w_{tr}}{c}\left[\omega_0+2\alpha\left(t-\frac{R_t+R_r}{c}\right)\right] , \tag{5}$$

so that

$$C_\theta(t) \approx \frac{A}{2}P_\theta\left(\frac{w_{tr}}{c}\left[\omega_0+2\alpha\left(t-\frac{R_t+R_r}{c}\right)\right]\right) . \tag{6}$$

Now, according to the projection-slice theorem, $C_\theta(t)$ is a radial slice at angle θ of the 2-D transform G of the unknown reflectivity density, i.e., (6) is equivalent to

$$C_\theta(t) \approx \frac{A}{2}G\left(\frac{w_{tr}}{c}\left[\omega_0+2\alpha\left(t-\frac{R_t+R_r}{c}\right)\right]\cos\theta, \frac{w_{tr}}{c}\left[\omega_0+2\alpha\left(t-\frac{R_t+R_r}{c}\right)\right]\sin\theta\right) . \tag{7}$$

This expression holds only for a restricted range of t. Tracing the development back to (3), we find that (3) is valid only for the time interval over which the transmitted pulse completely overlaps the ground patch, namely for

$$-\frac{T}{2}+\frac{(R_t+R_r+Lw_{tr})}{c} \leqslant t \leqslant \frac{T}{2}+\frac{(R_t+R_r-Lw_{tr})}{c}$$

(For other values of t, $r_\theta(t)$ is not correctly described by (3).) Using (5), this restriction on the range of t provides a corresponding restriction on the range of frequencies over which the transform of the projection $P_\theta(X)$ is acquired. $P_\theta(X)$ is determined for $X1 \leqslant X \leqslant X2$, where

$$X1 = \frac{w_{tr}}{c}\left[\omega_0+2\alpha\left(-\frac{T}{2}+\frac{Lw_{tr}}{c}\right)\right] \tag{8}$$

$$X2 = \frac{w_{tr}}{c}\left[\omega_0+2\alpha\left(\frac{T}{2}-\frac{Lw_{tr}}{c}\right)\right] \tag{9}$$

Thus, only a segment of each Fourier slice is available because of the limited bandwidth of the transmitted pulse. Assuming that L is not too large, the boundaries, $X1$ and $X2$, of each radial slice are nearly proportional to the lowest and the highest frequencies in the transmitted chirp pulse, and they are given approximately by

$$X1 \approx \frac{w_{tr}}{c}[\omega_0-\alpha T]$$

$$X2 \approx \frac{w_{tr}}{c} [\omega_0 + \alpha T].$$

We will use these simplified expressions for $X1$ and $X2$ throughout the remainder of the paper. If the assumption used in stating these approximations is not valid (L is too large), then (8), and (9) should be used instead and some of the following results (notably Fig. 3, below) will change.

For MSSAR, with $\beta=0$ and $w_{tr}=2$, $X1$ and $X2$ are constant for all look angles θ, in agreement with the analysis in [3]. For BSSAR where $\beta>0$, we have $w_{tr}<2$ so that $X2-X1$ is smaller than in the monostatic case, i.e., less Fourier domain data is acquired on each radial trace. Furthermore, β, and hence w_{tr}, will generally vary with θ as the transmitter and receiver move so that $X1$ and $X2$ will not be constant. Figure 3 gives polar plots of $X1$ and $X2$ as a function of β. It can be shown that $X1$ and $X2$ are circles with centers on the $\beta=0$ axis. This figure can be used to help determine the shape of the Fourier domain region where data are collected, since the acquired Fourier samples lie on the interval $[X1,X2]$ corresponding to the angle 2β between the focal lines drawn to the origin of the ground patch in Fig. 2.

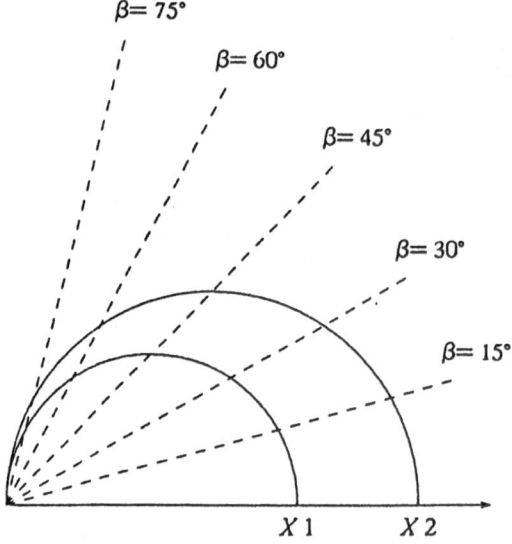

Fig. 3. Polar plot of the boundaries of the Fourier domain
data interval as a function of β.

C. Some Different Possibilities for Data Acquisition

Besides providing for a passive receiver near the target area, which is important in some military applications, BSSAR offers the possibility of obtaining Fourier data on grids that might simplify image reconstruction when compared with the polar format algorithm used in MSSAR. In this section we examine some of the different data grids that can be obtained using BSSAR. In each case the shape of the Fourier grid can be determined using the projection-slice theorem along with (6) and Fig. 3. The Fourier data acquired from each returned pulse is (samples of) a band-limited slice of the Fourier transform of the reflectivity function, and the location and extent of this slice is fully characterized by the angle θ between the u-axis and x-axis and the angle 2β between the focal lines joining the transmitter and receiver locations to the origin.

With a stationary transmitter, moving receiver, and uniform sampling, the obtained Fourier data lies on a grid like that shown in Fig. 4. The arcs of the grid are circular, and their centers are on the X-axis. To reconstruct high resolution images, 2-D high-order interpolation is necessary to provide approximate Cartesian Fourier data so that the FFT can be used for Fourier inversion. Thus, this case of data acquisition suffers from the same (and possibly worse) computational difficulties arising in MSSAR, where high-order polar-to-Cartesian interpolation is required.

If the transmitter and receiver both rotate around the origin with the same angular velocity, then the obtained Fourier data lie on a uniform polar grid, such as in Fig. 5. (Here we assume that $s(t)$ is transmitted at equally spaced increments of θ.) In this case the angle 2β between the lines joining the origin to the transmitter and receiver is constant so that $X1$ and $X2$ are also constant. The traditional MSSAR reconstruction algorithm employing polar-to-Cartesian interpolation followed by a 2-D IFFT could be applied in this case.

If we can obtain Fourier domain data on the trapezoidal grid of Fig. 6, application of 1-D interpolation along vertical lines, prior to the IFFT, is sufficient to produce high resolution images [6]. This results in a computational savings that is important because interpolation is a bottleneck in the reconstruction algorithm. Also, there is a newly proposed

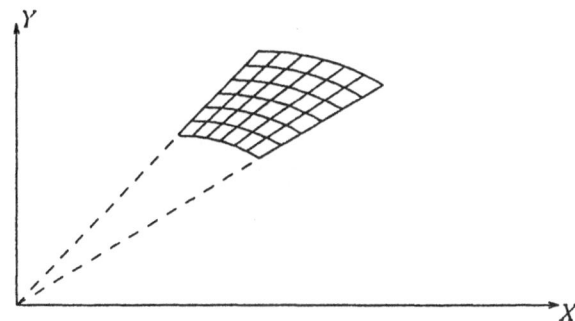

Fig. 4. Fourier domain data grid for BSSAR with stationary transmitter
and uniform sampling.

reconstruction algorithm that begins with trapezoidal samples and purportedly reconstructs images using much less computation than polar format algorithms [7].

There appear to be only two reasonable ways of obtaining Fourier data on a trapezoidal grid. The first involves the same scenario as in the polar format case (constant
β), except that the sampling period in the A/D converter is adjusted to sample each
demodulated return $C_\theta(t)$ at a uniform rate determined by the desired horizontal spac-
ing on the corresponding radial trace in Fig. 6. Thus, the sampling rate in the A/D converter must be varied from one radial trace to the next. This is feasible, but the same
idea can be employed in MSSAR. A second possibility for obtaining trapezoidal data is to
use a constant sampling rate in the A/D converter, but adjust the sample spacings on
each radial trace by properly varying β, which has the effect of linearly stretching or

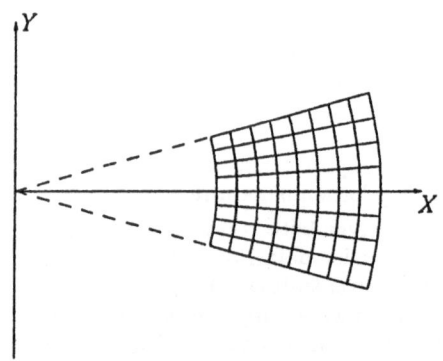

Fig. 5. Uniform polar grid in Fourier domain.

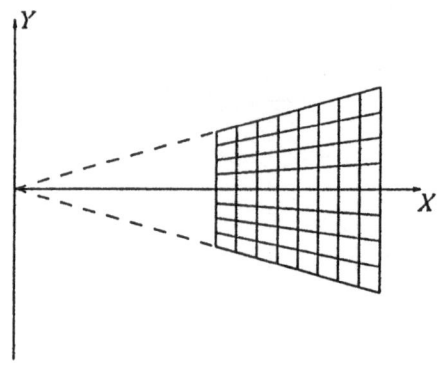

Fig. 6. Trapezoidal grid in Fourier domain.

compressing the acquired Fourier data. The necessary variation in β and the collection of radial traces at uniformly spaced increments of θ can be simultaneously accomplished in either of two ways of practical interest: (a) Independently controlling the angular velocities of both the transmitter and receiver, or (b) Using a transmitter with fixed angular velocity (possibly zero) with a variable pulse repetition interval, and a receiver with a variable angular velocity.

It is interesting to consider the approximation of a trapezoidal grid using a constant sampling rate in the A/D converter and constant, but unequal, transmitter and receiver angular velocities. For example, if the transmitter follows the receiver in a counterclockwise direction in Fig. 2 with a higher angular velocity than that of the receiver, then the angle β will decrease with time as θ increases. Hence, the acquired Fourier data will lie on larger intervals as θ increases. If the difference between angular velocities is small, the curvature of the data grid is considerablly reduced, where the curvature at a point of a second-order differentiable function is defined to be $1/\rho$, where ρ is the radius of the circle passing through that point and having the same first and second-order derivatives as the function at that point. We have studied this problem and found that $v_\theta^{trans} \approx 1.14 v_\theta^{rec}$ minimizes the curvature of the Fourier grid under the conditions described above. Using this ratio of velocities and beginning at $\theta=0$, the top half of the trapezoidal grid in Fig. 6 can be approximated. The bottom half of the trapezoidal grid can be approximated in a similar fashion. Although this scheme does not produce an exact trapezoidal grid, it is possible that it would simplify the required interpolation operation.

Finally, we might consider the possibility of directly acquiring Cartesian Fourier data, so that only a 2-D IFFT would be required for image formation. However, any practical scheme for accomplishing this is immediately ruled out by the observation that BSSAR, like MSSAR, can collect Fourier data only on lines passing through the origin.

D. Additional Comments

In practice it is helpful to have an expression for w_{tr} directly in terms of the transmitter and receiver locations. Such an expression is easily derived by noting that

$$
\begin{aligned}
w_{tr} &= 2 \, |\cos (\beta)| \\
&= \left| 2 + 2\cos(2\beta) \right|^{1/2} \\
&= \left| 2 + 2 \, \hat{R}_t \cdot \hat{R}_r \right|^{1/2} \\
&= \left| 2 \left(1 + \frac{x_t}{R_t} \frac{x_r}{R_r} + \frac{y_t}{R_t} \frac{y_r}{R_r} \right) \right|^{1/2}
\end{aligned}
$$

where in the third line, $\hat{R}_t$ and $\hat{R}_r$ are the unit vectors pointing to the transmitter and receiver, respectively, with an angle of 2β between them, and "$\cdot$" represents a scalar dot product.

As a second comment we note that our BSSAR derivation assumed that the transmitter, receiver, and ground patch all lie in the same plane. For the true 3-D situation this derivation can be modified by considering intersections of the ground patch with ellipsoids whose foci are the transmitter and receiver locations. These intersections can again be replaced by line segments so that a similar derivation holds, but with the modifications

$$R_t = \left(x_t^2 + y_t^2 + z_t^2\right)^{1/2},$$

$$R_r = \left(x_r^2 + y_r^2 + z_r^2\right)^{1/2},$$

and

$$w_{tr} = \left[2\left(1 + \frac{x_t}{R_t}\frac{x_r}{R_r} + \frac{y_t}{R_t}\frac{y_r}{R_r} + \frac{z_t}{R_t}\frac{z_r}{R_r}\right)\right]^{1/2}$$

III. WAVEFRONT CURVATURE IN BSSAR

The basic assumption made in the tomographic formulation of BSSAR is that the family of ellipses with foci at the transmitter and receiver locations, has line segment intersections with the terrain patch. However, this is an approximation that must be used with care. In practice, if the phase error due to this approximation is small, e.g. less than $\pi/10$, then the approximation is considered valid. If the resulting phase error is more than tolerable, then the terrain patch must be subdivided into smaller subsections for processing such that there is a valid approximation in each subsection.

It was shown earlier that

$$C_\theta(t) \approx \frac{A}{2}\int_{-L}^{L} p_\theta(u)\exp\left\{-j\frac{w_{tr}}{c}\left[\omega_0 + 2\alpha\left(t - \frac{R_t + R_r}{c}\right)\right]u\right\}du$$

Therefore, an error Δu in the u-coordinate of a reflector on the terrain patch introduces the phase error

$$\Delta\phi = 2\alpha\left(t - \frac{R_t + R_r}{c}\right)\Delta u\frac{w_{tr}}{c}.$$

Thus, over the time window of data collection

$$|\Delta\phi| \leqslant 2\alpha\frac{T}{2}|\Delta u|\frac{w_{tr}}{c}. \tag{10}$$

The phase error is maximum when the deviation Δu attains its maximum. To find the maximum of Δu, we approximate the ellipse over the ground patch by a circle having the same radius of curvature ρ, as depicted in Fig. 7. For an ellipse having the geometry shown in Fig. 8 we have

$$\rho = \frac{\left(a^4 y_0^2 + b^4 x_0^2\right)^{3/2}}{a^4 b^4}$$

where a and b are the lengths of the major and minor axii. From Fig. 7, with $\rho \gg L$, it follows that

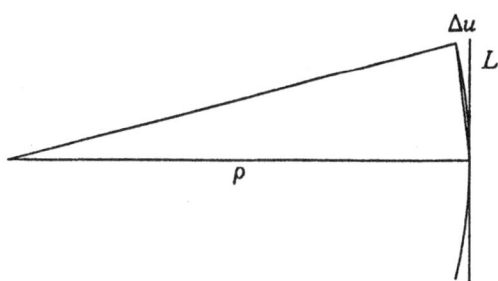

Fig. 7. Deviation of a circular arc from a line due to curvature.

$$\Delta u \leqslant \frac{L^2}{2\rho} = L^2 \cdot \frac{a^4 b^4}{2\left(a^4 y_0^2 + b^4 x_0^2\right)^{3/2}} \; .$$

After considerable algebra, w_{tr} in (10) can also be expressed in terms of the parameters in Fig. 8 as

$$w_{tr} = 2 \cdot \frac{ab}{\left(a^4 - (a^2 - b^2)x_0^2\right)^{1/2}}$$

so that from (10)

$$|\Delta\phi| \leqslant \alpha \frac{T}{c} L^2 \frac{a^5 b^5}{\left(a^4 y_0^2 + b^4 x_0^2\right)^{3/2}\left(a^4 - (a^2 - b^2)x_0^2\right)^{1/2}} \; . \tag{11}$$

Thus the phase error will be smaller than $\pi/10$ (which is conservative) if the size of the terrain patch satisfies

$$L \leqslant \left|\frac{\pi}{10} \cdot c \cdot \frac{\left(a^4 y_0^2 + b^4 x_0^2\right)^{3/2}\left(a^4 - (a^2 - b^2)x_0^2\right)^{1/2}}{\alpha T a^5 b^5}\right|^{1/2} \; .$$

For a MSSAR at distance $a = (R_t + R_r)/2$ from the target patch, the phase error is bounded by

$$|\Delta\phi| \leqslant \frac{\alpha T}{c} \frac{L^2}{a} \; .$$

Thus, comparing this expression with (11) it can be determined whether MSSAR or BSSAR will have more serious curvature effects as a function of the BSSAR imaging geometry. Making this comparison, it can be shown with algebra that the curvature effects are less for BSSAR whenever

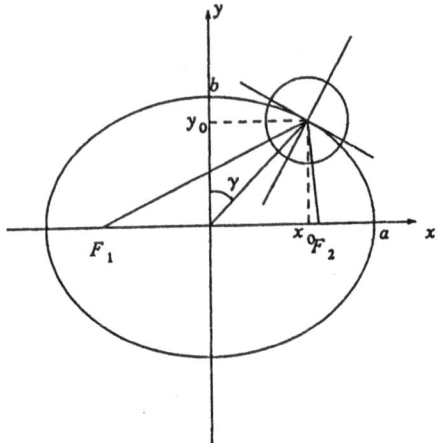

Fig. 8. Ellipse centered coordinate system
with transmitter at location F_1 and receiver at F_2.

$$|\gamma| \le \tan\left(\left[\frac{b}{a}\right]^{-3/2}\right)$$

where the angle γ is defined in Fig. 8. Figure 9 shows the boundary of this region as a function of b/a, the ratio of the minor axis to the major axis.

III. CONCLUSIONS

A formulation of BSSAR was developed using the projection-slice theorem from computer-aided tomography. Just as in the monostatic case, the imaging principle employed is tomographic in nature and relies on geometric considerations, rather than the Doppler effect. For the general case, with both a moving transmitter and moving receiver, the locations were found for the acquired Fourier samples of the terrain reflectivity. These locations were derived assuming a sufficiently small terrain patch; for larger scenes Eqs. (8) and (9) must be used and the results will vary somewhat from those presented. The possibility of designing the Fourier grid to simplify image reconstruction was examined, but no significant advatages over MSSAR were found in this regard.

The phase error due to wavefront curvature was studied for the bistatic case. The maximum phase error in the bistatic case can be either greater or less than the phase error in the monostatic case depending on whether the terrain patch is near the major or minor axis of the ellipse in Fig. 8. A boundary of the region where BSSAR exhibits less curvature than MSSAR was found.

Although, BSSAR has strategic advantages over MSSAR, the image reconstruction problem appears to be just as difficult as in the monostatic case.

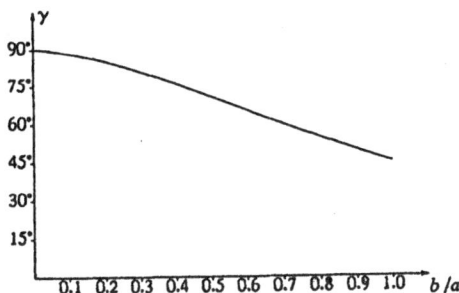

Fig. 9. The boundary of γ, under which the curvature in BSSAR is less than in MSSAR, as a function of b/a .

APPENDIX

In this appendix we derive an approximate expression for $\tau(u,v)$, the total time required for a signal to propagate from the transmitter to a ground point at (u,v) and then to the receiver. We assume that $R_t, R_r \gg L$ so that $\tau(u,v)$ is nearly independent of v. Thus, it suffices to find $\tau(u)$, the time required to traverse the outer focal lines terminating at point C with coordinates $(u,0)$ in Fig. 10. By drawing perpendiculars to these two lines the propagation distance $d(u)$ is divided into 4 parts with

$$d(u) = |AB| + |BC| + |CD| + |DE| .$$

Since $|AB| \approx R_t$, $|DE| \approx R_r$, $|BC| \approx |CD|$, and the two focal lines form an angle of approximately 2β, we have

$$d(u) \approx R_t + R_r + 2u \cos(\beta)$$

so that

$$\tau(u) \approx \frac{R_t + R_r + u \cdot w_{tr}}{c} .$$

This same expression can be found in a more systematic and rigorous way by approximating $d(u)$ with the first two terms in its Taylor series approximation around $u=0$ where

$$d(u) = \left[(u \cos\theta - x_t)^2 + (u \sin\theta - y_t)^2 \right]^{1/2} + \left[(u \cos\theta - x_r)^2 + (u \sin\theta - y_r)^2 \right]^{1/2} .$$

Here, the magnitude of the derivative of $d(u)$ at the origin equals w_{tr}. Although this approach is less intuitive, it is straightforward and easily extendible to nonplanar BSSAR imaging.

ACNOWLEDGMENT

This work was supported by the SDIO/IST and managed by the U.S. Army Research Office under Contract DAAL03-86-K-0111. The authors are with the Coordinated Science Laboratory and Department of Electrical and Computer Engineering, University of Illinois at Urbana-Champaign, 1101 W. Springfield Avenue, Urbana, IL 61801.

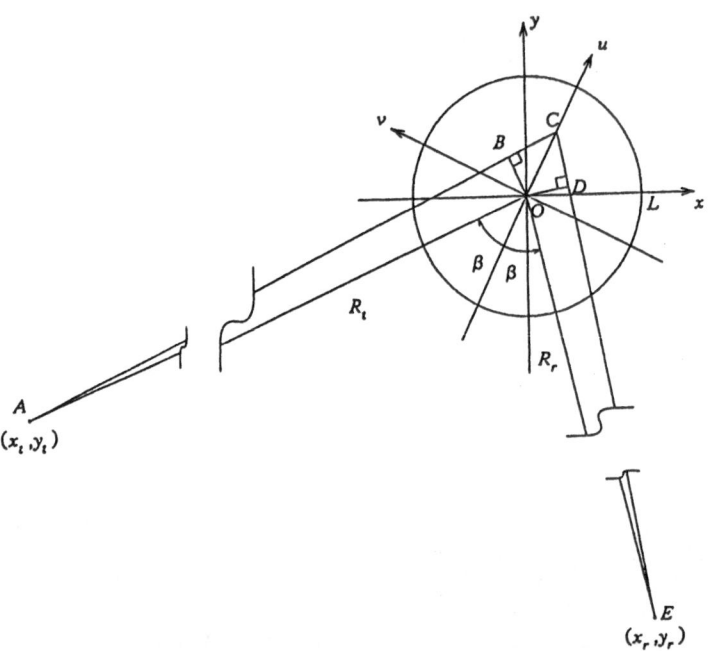

Fig. 10. Detail of ground-plane geometry.

REFERENCES

[1] J. C. Kirk Jr., "Bistatic SAR motion compensation," *IEEE 1985 International Radar Conference Record*, pp. 360-365.

[2] D. A. Ausherman, A. Kozma, J. L. Walker, H. M. Jones, and E. C. Poggio, "Developments in radar imaging," *IEEE Trans. Aero. Electron. Syst.*, vol. AES-20, pp. 363-400, July 1984.

[3] D. C. Munson Jr., J. D. O'Brien, W. K. Jenkins, "A tomographic formulation of spotlight mode synthetic aperture radar," *Proc. IEEE*, vol. 71, pp. 917-925, Aug. 1983

[4] G. T. Herman, Ed., *Image Reconstruction from Projections*. New York: Springer-Verlag, 1979.

[5] A. V. Kak and M. Slancy, *Computerized Tomographic Imaging*. New York: IEEE Press, 1988.

[6] B. C. Mather, "Fourier Domain Interpolation Techniques for Synthetic Aperture Radar," Ph.D. dissertation, University of Illinois at Urbana-Champaign, 1986.

[7] W. Lawton, "A new polar Fourier transform for computer-aided tomography and spotlight synthetic aperture radar," *IEEE Trans. Acoust., Speech, Signal Processing*, vol. 36, pp. 931-933, June 1988.

PHASE RETRIEVAL BY OPTIMAL WEIGHTED NORM EXTRAPOLATION

Sergio D. Cabrera and Chienshu Chen
The Pennsylvania State University

Introduction. The problem of recovery of a finite support signal from a given set of samples and the Fourier transform (F. T.) magnitude is considered. The problem is dealt with from an extrapolation point of view. That is, in a manner similar to the bandlimited extrapolation problem and its extensions using a weighted frequency domain norm. Should one decide to set the weight function exactly equal to the given magnitude, it would give a resulting estimate with magnitude that resembles, but does not equal, the general shape of the weight function. Therefore, this would not be a phase retrieval technique. In order to get an exact match to the given magnitude, one must devise a scheme to accomplish this. An algorithm is presented here which is designed to satisfy the constraints of the given time samples at each iteration (extrapolation) and, in the limit, match the Fourier transform magnitude.

Recent work has shown that for one and two dimensional finite support sequences, the Fourier transform magnitude and one boundary time or object domain sample is enough to determine certain signals uniquely [1]. The proposed scheme in [1] depends on a very crucial constraint which could very easily be broken in the presence of noise in the known edge-sample or in the F. T. magnitude.

In this paper the motivation to find a tradeoff between the amount of available data and the necessary signal-to-noise ratio for successful recovery leads us to consider the case where a variable set of time samples is available. It is expected that an increase in the number of samples available will result in more accurate recovery of the true signal. The general philosophy of this paper is that we always seek signal estimates using all a priori information whether or not this information uniquely determines the signal of interest.

Bandlimited Signal Extrapolation. Defining the continuous-time Fourier transform of a complex valued signal x(t) as

$$X(f) = \int_{-\infty}^{\infty} x(t) \ e^{-j2\pi ft} \ df$$

we say that this signal is bandlimited to the bandwidth $[f_1, f_2]$ if $X(f)=0$ for f not in the interval $[f_1, f_2]$. The bandlimited signal extrapolation algorithm of Papoulis/ Gerchberg [2],[3] provides for recovering a bandlimited signal x(t) given only the signal in the interval $[t_1, t_2]$ and the bandwidth end points f_1 and f_2. It is shown in [2] that the signal can be recovered exactly, for all time t, after an infinite number of iterations of an algorithm. Mathematically, this is so because the bandlimited signal x(t) is analytic.

A more practical situation in a digital signal processing system is if we only have access to a finite number of equally spaced samples within $[t_1,t_2]$. The problem can then be posed directly as a best estimate of a bounded energy bandlimited (to less than the sampling frequency) discrete-time signal given these samples and the bandwidth, see [4]. In this case, the estimate can be obtained in one step and has the additional property of being the minimum energy bandlimited signal that goes through the data samples. The use of other norms has been investigated by several researchers where the minimum norm extrapolation is the desired estimate in any case. One useful norm selected in the time domain is the energy outside of a time interval that is larger than that consisting of the given samples, see [4].

<u>Use of Frequency Weighted Norm in Extrapolation</u>. Another norm that has proven useful involves use of a frequency domain non-negative weight Q(f). The norm for a discrete-time signal x[n] is defined in this case as:

$$\|x[n]\|^2 = \int_{f \in S} \frac{|X(f)|^2}{Q(f)} \, df \qquad (1)$$

where Q(f)>0 for f in the bandwidth region $S \subset [-.5,.5]$. This approach has been studied extensively by Byrne and Fitzgerald in [6] and other works and they denote it the PDFT. In this case, the minimum norm extrapolation that goes through the data has a Fourier transform magnitude that is similar in shape to Q(f) in the region S and is equal to zero outside S. Alternatively one can think of defining Q(f)=0 outside of S and thus the bandlimitation of the estimate is imposed by Q(f). Should one have access to spectral shape information, it can be incorporated into the definition of Q(f) to obtain some resemblance in shape.

A benefit of this approach, over the minimum energy bandlimited solution, is that the problem of high sensitivity to additive noise on the samples can be greatly reduced by allowing Q(f) to be non-zero for all frequencies including the bandwidth region. This large sensitivity encountered in bandlimited extrapolation is well documented in [5]. The specification of Q(f) by the user is a powerful idea used in [6]. In [7] and [8] an approach to defining a useful frequency weight Q(f) from the available data samples is developed and exploited to obtain a useful estimate that is characterized as a frequency stationary extrapolation of the data samples. In this paper, the specification of Q(f) is used to better match F. T. magnitude information when it is available.

If detailed knowledge of the signal's Fourier transform magnitude (Modulus) is actually available *a priori* , the weighted norm extrapolation method can be used to provide an estimate with Modulus resembling but not equal to the given one. Better results can be obtained by parametrizing the choice of Q(f) and optimizing over the parameter. Next, a scheme to do this is developed and illustrated in this paper. Another attempt that promises exact match is also provided here and consists of an iterative algorithm that gives a sequence of extrapolations which in the limit gives an estimate which match the given Modulus. No rigorous proof of this is provided, however, simulations and a careful look at the algorithm demonstrates its potential.

Multiple Optimalities of Weighted Norm Extrapolation. The optimality of the weighted frequency domain norm extrapolation is two-fold. In order to see this, it is necessary to consider the nature of the given data which is the set of time samples $\{x[m_i],$ $i=1,2,\ldots,L\}$. Any set of bounded energy band limited signals is a Reproducing Kernel Hilbert Space, therefore the sampling operation is a linear functional which is bounded and continuous [9]. This operation can therefore be expressed as an inner product with a representer that is bandlimited to the same region as $x[n]$

$$x[m_i] = \big\langle x[n], \varphi_i[n]\big\rangle_Q$$

The form of $\varphi_i[n]$ depends on the form of the inner product which must agree with the norm of equation (1).

From [10], the minimum norm solution that satisfies this set of constraints must lie in the space spanned by the representers. The time sample $x[m_i]$ is expressed using the inverse Fourier transform as:

$$x[m_i] = \int_{f\in S} X(f) \; e^{\,j2\pi f m_i} \; df$$

Expressing this as an inner product proceeds as follows:

$$x[m_i] = \int_{f\in S} \frac{1}{Q(f)} X(f) \Big(Q(f) \; e^{\,-j2\pi f m_i}\Big)^{*} \; df$$

$$x[m_i] = \big\langle x[n], q[n-m_i]\big\rangle_Q$$

where $q[n]$ is the inverse Fourier transform of $Q(f)$ when it is extended to be zero outside the region S. From this point on, we make this assumption in the definition of $Q(f)$.

The set of signals which go through the given time samples forms a Linear Variety in the vector space of bandlimited signals. The vector with smallest norm is known to be the unique signal in the span of the representers. This vector is therefore:

$$\hat{x}[n] = \sum_{i=1}^{L} w_i \; q[n-m_i] \tag{2}$$

with the L constraints:. $\hat{x}[m_i]=x \ [m_i]$, i=1,2,....,L providing the values of the L weights w_i.

The second optimality property of the weighted norm extrapolation is given by the equation used to derive this technique by Byrne and Fitzgerald in [6]. Starting with an estimate model of the form:

$$\hat{X}(f) = Q(f) \sum_{i=1}^{L} w_i \ e^{-j2\pi f m_i} = Q(f) \ W(f) \tag{3}$$

we set out to minimize the approximation error of the difference between the true Fourier transform X(f) and the model of equation (3) above. This is done using the norm of (1) and by choosing the weights w_i to accomplish this. This criterion is therefore:

$$\int_{f \in S} \frac{1}{Q(f)} \left| X(f) - Q(f) \sum_{i=1}^{L} w_i \ e^{-j2\pi f m_i} \right|^2 df \tag{4}$$

and the required information about X(f) turns out to be $x[m_i]$, i=1,2,...,L. It is not difficult to see that the inverse F. T. of (3) gives (2).

The Phase Retrieval Problem. The general problem of recovering a signal from the Fourier transform magnitude and some additional constraints/data is known as Phase Retrieval [11], [12,chap. 6]. Defining the Fourier transform for a discrete-time signal x[n] as

$$X(f) = \sum_{n=-\infty}^{\infty} x[n] \ e^{-j2\pi f n}$$

its polar representation is

$$X(f) = M(f) \ e^{j\phi_x(f)} \tag{5}$$

where $M(f)=|X(f)|$ is the F. T. magnitude or Modulus. It is of interest to note that the inverse Fourier transform of $M(f)^2$ is the autocorrelation of the signal x[n]. Phase retrieval is therefore equivalent to reconstruction from autocorrelation.

It is well known [12, chap. 6] that the solution is not unique and, therefore, one must have additional information (or constraints) about a signal in order to recover it from its Modulus. The approach usually taken is to accept some ambiguities as "trivial" and assume that in practice one would consider the recovery to within these ambiguities as successful. The time (or object) domain equivalent of these ambiguities are: time shifts, time reversals and sign changes. In this paper, these ambiguities are not ignored.

A usual practical extra assumption that is made is that the desired signal has finite time support. This reduces the ambiguity almost completely in the two or higher dimensional case due to the irreducibility of multidimensional polynomials. On the other hand, in the one-dimensional case, the ambiguity is still 2^N possible signals of length N [12, chap. 6]

Exact use of Fourier Transform Magnitude in Extrapolation. Incorporating known time (object domain) sample values in Phase Retrieval has also been considered recently. Methods of recovery when samples are available at the edges of discrete finite-support images have been studied by Hayes and Quatieri in [13]. Direct linear reconstruction algorithms which use these samples as well as the autocorrelation data are described and tested. Further analysis of this algorithm has been done recently by Fienup [14] and some improvements suggested.

Recently [1] a one-dimensional version of this problem was also considered when an edge value of a finite duration sequence is given and the conclusion was that it is "Rich in Phase Information". The claim is that for all but a set of signals of measure zero, knowledge of this one sample and the Modulus is enough information for unique characterization of finite length sequences. The availability of time samples as well as frequency domain information readily suggests that the weighted norm extrapolation method be considered. It is useful to note the direct effect of Q(f) on the resulting Fourier transform estimate form in (3), and that the term $|W(f)|$ is a multiplicative error if one's goal is to make $|\widehat{X}(f)|=Q(f)$.

Use of Optimal Powers of the Fourier Transform Magnitude. The feasibility of deriving algorithms using the weighted norm extrapolation framework is considered. The given magnitude M(f) is to be used to define Q(f) such that the estimate x[n] and its Fourier transform X(f) satisfy as close as possible the constraints:

$$\widehat{x}[m_i] = x[m_i] \; ; \quad i=1,2,3,....,L \qquad (6)$$

$$|\widehat{X}(f)| = M(f) \; ; \quad f \in [-.5,.5] \qquad (7)$$

We investigate here the choice

$$Q(f) = Q_\alpha(f) = M(f)^\alpha \; ; \quad \alpha>0 \qquad (8)$$

and choosing α to make the resulting Magnitude of the F. T. of the resulting estimate as close as possible to the true Modulus. For a given α, the estimation is done and $\widehat{X}_\alpha(f)$ is obtained. In this case, an error occurs and is defined as:

$$\Delta_\alpha = \int_{f \in S} \left| M(f) - \left| \widehat{X}_\alpha(f) \right| \right|^2 df \qquad (9)$$

We have parametrized the problem and can therefore minimize with respect to α. The simplest approach is to vary α near the value α=1 and look for a minimum and call it α*. The resulting best solution is therefore:

$$\widehat{X}(f) = \widehat{X}_{\alpha^*}(f)$$

Example 1: To illustrate this scheme, we choose a test signal x[n] consisting of a two sinusoids at frequencies .1 and .25 Hz. and amplitudes 2 and 3 respectively which is then multiplied by a Gaussian window centered at index 256. The resulting signal has duration 40 as shown in Figure 1. Since the sequence is of duration 40, its autocorrelation is of duration 79. We thus require to use a DFT length larger than this to represent the given Modulus and derived quantities accurately. Using N=512, we have given data

$$M(k) = \left| X(\tfrac{k}{N}) \right| \quad ; \quad k=0,1,2,...,N-1$$

as shown in Figure 2. We also assume we are given the 10 time samples: {x[252], x[253], x[254], x[255], x[256], x[257], x[258], x[259], x[260], and x[261]}.

A series of estimations are performed using values of α varying from 0.1 to 2.0. The error Δ_α is calculated for each α and plotted in Figure 3. As suspected, a minimum error results for α*=0.8 and the optimal estimate for this value is shown in Figure 4 a comparison to the original shows very little difference. The resulting F. T. magnitude of the estimate (not shown) is also indistinguishable from that of the original signal. A second example is done using noisy samples and noisy F. T. magnitude values both at approximately 20 dB SNR. The noisy magnitude is clipped to avoid negative values and is shown in Figure 5. The optimal parameter is again α*=0.8 and the F. T. magnitude of the resulting estimate is shown in Figure 6.

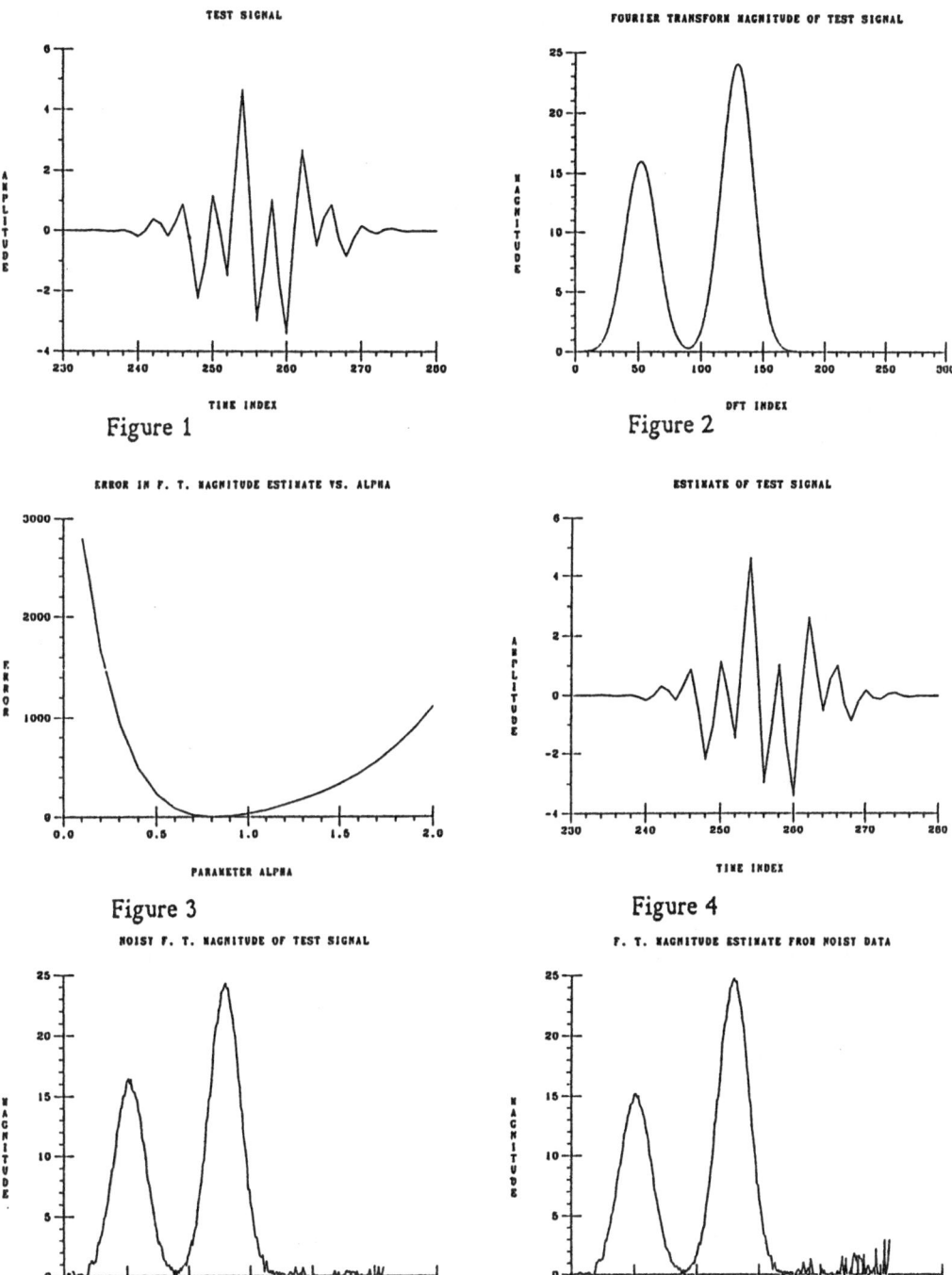

Figure 1

Figure 2

Figure 3

Figure 4

Figure 5

Figure 6

In general, one cannot hope to do this well. For a more general signal (without the Gaussian window effect), the optimal choice of α will give the best estimate in this form of equation (8) but not necessarily something as close to the true signal as in this example. The computational burden of this method can be reduced by finding a more efficient method of obtaining $\alpha*$ while at the same time obtaining $\widehat{X}_{\alpha*}(f)$ without any trial and error. However, the burden will always be that of a single Weighted Norm Extrapolation Problem times a small integer (≈ 10). This burden is comparable to that of an iterative algorithm that converges fast. With this in mind, we next consider an iterative scheme to obtain better results while not increasing the computational burden significantly over this method.

Proposed Iterative Weight Modification Scheme. An iterative modification of $Q(f)$ can be designed to accomplish the desired solution to the problem of interest. One possible way would have the estimate satisfy (6) at each iteration and (7) in the limit. In particular at iteration p, we would have

$$\hat{x}_p[m_i] = x[m_i] \; ; \; i=1,2,3,...,L \tag{11}$$

and the estimate is of the form of (3), thus

$$\widehat{X}_p(f) = Q_p(f) \sum_{i=1}^{L} w_i^p \, e^{-j2\pi f m_i} \equiv Q_p(f) \, W_p(f) \tag{12}$$

with $Q_p(f)$ defined in a way that will lead to an exact F. T. magnitude match

$$\lim_{p \to \infty} |\widehat{X}_p(f)| = |X(f)| = M(f) \tag{13}$$

From our previous analysis, $Q_p(f)=M(f)$ does not work because $|W_p(f)|$ will not have unit magnitude. If we are to achieve the correct magnitude, we should divide by this factor between iterations. Now assuming that we design this algorithm correctly, we would expect $W_p(f) \approx W_{p-1}(f)$ for large enough p, thus:

$$\left| \frac{W_p(f)}{W_{p-1}(f)} \right| \approx 1 \tag{14}$$

One way to accomplish this is to make the modified weight take the form

$$Q_p(f) = \frac{M(f)}{|W_{p-1}(f)|} \tag{15}$$

In practice, some care must be taken to avoid having the denominator vanish, such as adding a small constant $\delta > 0$. This form for $Q_p(f)$ is used in this proposed iterative phase retrieval algorithm so the estimate takes the form

$$\widehat{X}_p(f) = \frac{M(f)}{|W_{p-1}(f)|} W_p(f) \tag{16}$$

Assuming a successful algorithm, $W_p(f) \to \overline{W}(f)$; $\widehat{X}_p(f) \to \overline{X}(f)$ which is the resulting estimate with the magnitude constraint satisfied

$$|\overline{X}(f)| = M(f) = |X(f)| \tag{17}$$

and the phase would be fully contained in $\overline{W}(f)$. A final detail of the algorithm is the starting value for $Q_1(f)$ and clearly it is important that $\widehat{X}_1(f)$ have magnitude close to $M(f)$ itself and therefore we choose $Q_1(f)=M(f)$ (or $W_0(f)=1$). Before testing the algorithm, we do some brief analysis to understand its operation better.

Algorithm Analysis. At the $p^{\underline{th}}$ iteration, the PDFT minimizes the criterion of equation(4) which for the special form of $Q(f)$ becomes:

$$E_p \equiv \int_{f \in S} \frac{|W_{p-1}(f)|}{M(f)} \left| M(f)\, e^{j\phi_x(f)} - \frac{M(f)}{|W_{p-1}(f)|} W_p(f) \right|^2 df \tag{18}$$

where $X(f)$ has been represented in polar form as in (5). This expression can easily be manipulated to another optimal approximation error:

$$E_p \equiv \int_{f \in S} M(f)\, |W_{p-1}(f)| \left| e^{j\phi_x(f)} - \frac{|W_p(f)|}{|W_{p-1}(f)|} e^{j\phi_p(f)} \right|^2 df \tag{19}$$

where we express $W_p(f)$ in polar form:

$$W_p(f) \equiv |W_p(f)|\, e^{j\phi_p(f)} \tag{20}$$

Equation (19) appears to have the right form in order to accomplishment exactly what we want:

$$|W_p(f)| \approx |W_{p-1}(f)| \;;\; \phi_p(f) \approx \phi_x(f) \tag{21}$$

The magnitude tends towards a constant and the phase towards the true phase of x[n]. In order for this to be true, we must have $E_p \to 0$ as $p \to \infty$ and thus we must eventually consider the conditions for this to happen. Results on this matter are a subject of future research.

Example 2. Using the same test signal as in Example 1 and shown in Figure 1, we first use the iterative approach using the same 10 samples with no noise. The result is practically an exact recovery of the sequence as shown in Figure 7 where the estimate is plotted using the dashed line for comparison with the true signal shown in solid line. Next, we choose only 3 samples with no noise {x[252], x[253], x[254]}. The resulting estimate is not exact but remains very close to the true signal. Figure 8 shows the estimate in dashed line and the true signal in solid line. Figure 9 shows the estimated F. T. magnitude in dashed line and the true F. T. magnitude in solid line. It is of interest to note that the estimations for 10 and 3 samples are very good but not of equal quality. The first may be considered practically an exact recovery while the second one a very good estimate. The difference could very easily be attributed to computational limitations and not to the existence of an ambiguous solution that close to the true solution. In particular the division carried out in equation (15) cannot be done perfectly but instead some slight adjustment must be made.

Another simulation is done using the same 10 noisy samples with noisy F. T. magnitude as in Example 1. Figure 10 shows the estimated signal in dashed lines while the true signal is shown in solid line. Finally, the estimated F. T. magnitude is shown in Figure 11. For this case, the shape of the resulting F. T. magnitude is correct while the actual magnitude is roughly 2.5 time larger than the actual magnitude. The time domain estimate does go through the given noisy samples as required but is much longer than the known 40 sample support.

The cause of this behavior seen here is currently being investigated. The problem seen in this last case reminds us of the difficulty of the phase retrieval problem. It should be mentioned that by proper monitoring of the algorithm this difficulty can be lessened by stopping the iteration when the energy of the difference signal between iterations starts to increase after coming close to converging early in the iteration. By stopping the iteration after 3 steps, the same estimation problem which gave rise to Figures 10 and 11 will produce a closer estimate shown in Figure 12. Finally, the known support of the signal has not been used so far and in fact is seen to be violated in Figure 10 A method of incorporating it would also help avoid problems such as the one seen here using noisy data.

313

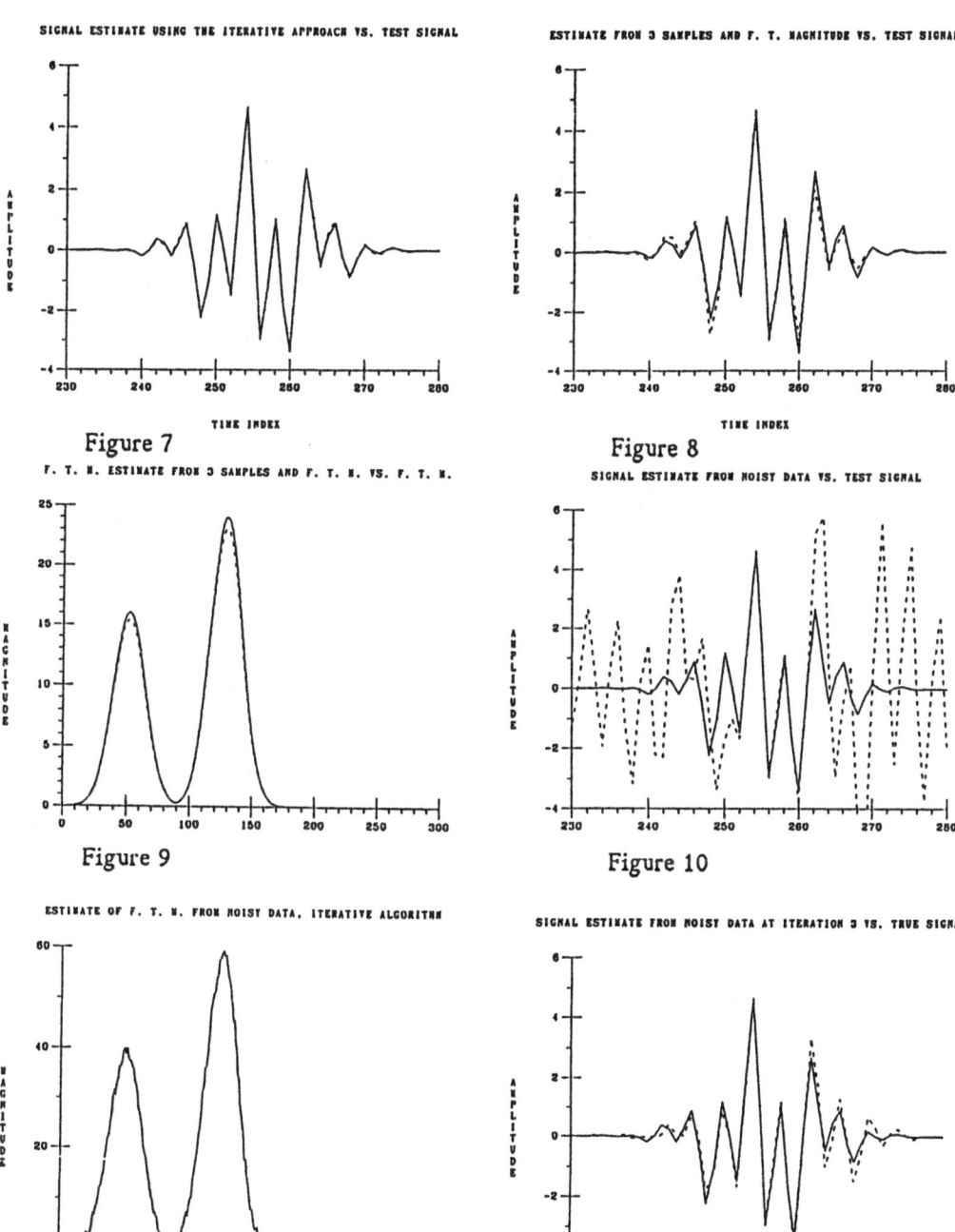

SIGNAL ESTIMATE USING THE ITERATIVE APPROACH VS. TEST SIGNAL

Figure 7

ESTIMATE FROM 3 SAMPLES AND F. T. MAGNITUDE VS. TEST SIGNAL

Figure 8

F. T. M. ESTIMATE FROM 3 SAMPLES AND F. T. M. VS. F. T. M.

Figure 9

SIGNAL ESTIMATE FROM NOISY DATA VS. TEST SIGNAL

Figure 10

ESTIMATE OF F. T. M. FROM NOISY DATA, ITERATIVE ALGORITHM

Figure 11

SIGNAL ESTIMATE FROM NOISY DATA AT ITERATION 3 VS. TRUE SIGNAL

Figure 12

Summary and Conclusions. The problem of recovery of a finite duration sequence from a given set of samples and the Fourier Transform (F. T.) magnitude was considered. The problem is dealt using a weighted frequency domain norm designed to obtain estimates with close match to the given F. T. magnitude. First the feasibility was demonstrated with an approximate method which selects the weight as the power of the given F. T. magnitude which minimizes an error resulting from an inexact match in the magnitude. An example showed promising performance with an without additive noise. Next, an iterative algorithm was presented which was designed to match the Fourier transform magnitude in the limit. An example showed practically exact recovery without additive noise. When noise was added, the algorithm without any careful monitoring converged to an incorrect solution. Modifications to the scheme for noisy data are the topic of current research.

We would like to thank N. K. Bose for calling to our attention the existence of reference [1]. The authors can be reached at: 121 Electrical Engineering East, Dept. of Electrical Engineering, The Pennsylvania State University, University Park, PA 16802, tel. (814)865-2228.

REFERENCES

[1] L. Xu, P. Yan and T. Chang, "Phase Information is Richly Contained in the End Point of a Finite Length Digital Signal," Proc. of the IEEE Intl. Symp. on Circuits and Systems, Philadelphia PA, Vol. 3, pp. 851-854, 1987.

[2] A. Papoulis, "A New Algorithm in Spectral Analysis and Bandlimited Extrapolation," IEEE Trans. Circuits Syst., Vol. CAS-22, pp. 735-742, Sept. 1975.

[3] R. W. Gerchberg, "Super-Resolution Trough Error Energy Reduction," Optica Acta, Vol. 21, No. 9, pp. 709-720, 1974.

[4] D. P. Kolba and T. W. Parks, "Optimal Estimation for Bandlimited Signals Including Time Domain Considerations," IEEE Trans. ASSP, Vol. ASSP-31, pp. 113-122, Feb. 1983.

[5] K. Stewart, T. S. Durrani, J. B. Abbiss, "The Effects of Bandwidth Misestimation in Bandlimited Signal Extrapolation," Proceedings of the IEEE Intl. Conference on Acoustics Speech and Signal Processing, pp. 1497-1500, Tampa FL, March 1985.

[6] C. L. Byrne and R. M. Fitzgerald, "Spectral Estimators that Extend the Maximum Entropy and Maximum Likelihood Methods," SIAM J. Appl. Math., Vol. 44, No. 2, pp. 425-442, April 1984.

[7] S. D. Cabrera and T. W. Parks, "Estimation of Sequences in a Signal Class Determined from the Data," Proceedings of the IEEE Intl. Conference on Acoustics Speech and Signal Processing, pp. 1348-1351, Tampa FL, March 1985.

[8] S. D. Cabrera, "Optimal Recovery of Signals from Linear Measurements and Prior Knowledge," Ph. D. Dissertation, Dept. of Electrical and Computer Engineering, Rice University, Houston Texas, May 1985.

[9] A. Aronszajn, "The Theory of Reproducing Kernels," Trans. Amer. Math. Soc., Vol. 68 (1950), pp. 337-404.

[10] D. Luenbeger, Optimization by Vector Space Methods, New York: John Wiley & Sons, 1969.

[11] M. Hayes, "The Reconstruction of a Multidimensional Sequence from the Phase or Magnitude of its Fourier Transform," IEEE Trans. ASSP, Vol. ASSP-30, No. 2, pp. 140-154, April 1982.

[12] Henry Stark ed., Image Recovery: Theory and Application, Academic Press Inc., Orlando FL, 1987.

[13] M. Hayes and T. Quatieri, "Recursive Phase Retrieval Using Boundary Conditions," Journal of the Optical Society of America, Vol. 73, No. 11, pp. 1427-1433, Nov. 1983.

[14] J. R. Fienup, "Phase Retrieval Using Boundary Conditions," J. Opt. Soc. Am A: Optics and Image Science, Vol. 3, No. 2, pp. 284-288, Feb. 1986.

RECONSTRUCTION OF HIGH RESOLUTION IMAGE FROM NOISE UNDERSAMPLED FRAMES

S. P. KIM   N. K. BOSE   H. M. VALENZUELA
                    The Pennsylvania State University

Motivation and Introduction. The results of the research reported here, though motivated from the need to satisfactorily process LANDSAT satellite pictures, are, however, applicable to imagery from any source which produces multiple images of the same scene in different wavelength bands or at different times and in different wavelength bands. The problem is to restore a high resolution image from a sequence of low-resolution, undersampled, discrete frames.

The pictures taken usually do not coincide exactly. First, the displacement of each frame relative to an arbitrarily chosen reference frame has to be measured. This is the problem of image registration. The images are sometimes degraded and noise is usually present. The modeling of the degrading phenomena is done from knowledge of the nature of imagery and the nature as well as type of noise present is usually inferred from the understanding of the various physical processes which combine to form the digital image. The formation of a digital image requires several components: an illumination source, a set of reflecting surfaces, a sensor and a quantizing sampler (A/D converter). Each of these can be a source of noise. The noise could be additive and in more complicated cases it could also be multiplicative signal-dependent or of the replacement type. Therefore, following the task of image registration, it is important to implement filtering on the registered image frames. Finally, samples on a set of grids with high sampling rate are to be estimated in order to reconstruct the high resolution image.

Tsai and Huang [1] recently considered the problem described above subject to the assumption that the low resolution frames have neither been corrupted by noise nor degraded by a blurring phenomenon. Let $f(x,y)$ be a continuous object and $f_k(x,y)$, $k=1,2,\ldots,p$ be a set of p spatially shifted versions of $f(x,y)$. Then,

$$f_k(x,y) = f(x + \delta_{xk}, y + \delta_{yk}) \tag{1}$$

where δ_{xk} and δ_{yk} are arbitrary but known shifts of $f(x,y)$ along the x and y coordinates, respectively.

The sequence of p frames, denoted by $\{f_k(k,j)\}$ $i=0,1,\ldots,M-1$, $j=0,1,\ldots,N-1$, $k=1,2,\ldots,p$ corresponds to discrete versions of the shifted images $f_k(x,y)$, after uniform sampling, with sampling periods T_x and T_y along the x and y coordinate axes, respectively. Therefore, the discrete frames $\{f_k(i,j)\}$ are defined as in equation (2).

$$f_k(i,j) = f(iT_x + \delta_{xk}, jT_y + \delta_{yk}), \tag{2}$$

$i=0,1,\ldots,M-1$, $j=0,1,\ldots,N-1$, and $k=1,2,\ldots,p$.

By taking the continuous 2-D Fourier transform of $f_k(x,y)$ and making use of the shifting property of this transformation, the following relation is obtained

$$F_k^C(u,v) = \exp[j2\pi(\delta_{xk}u+\delta_{yk}v)]F^C(u,v) \tag{3}$$

where $F_k^C(u,v)$ and $F^C(u,v)$ are continuous 2-D Fourier transforms of $f_k(x,y)$ and $f(x,y)$, respectively.

Let $F_k(m,n)$ be the discrete Fourier transform (DFT) of the kth frame $f_k(i,j)$. Then it can be expressed as

$$F_k(m,n) = \sum_{i=0}^{M-1} \sum_{\ell=0}^{N-1} f_k(i,\ell) \exp\left(-2\pi j\left(\frac{mi}{M} + \frac{n\ell}{N}\right)\right), \tag{4}$$

for $m=0,1,\ldots,M-1$, $n=0,1,\ldots,N-1$ and $k=1,2,\ldots,p$.

From the aliasing relationship, the continuous and discrete Fourier transforms are related by

$$F_k(m,n) = \frac{1}{T_x T_y} \sum_{i=-\infty}^{\infty} \sum_{\ell=-\infty}^{\infty} F_k^C\left(\frac{2\pi m}{MT_x} + i\omega_x, \frac{2\pi n}{NT_y} + \ell\omega_y\right), \tag{5a}$$

where

$$\omega_x = 2\pi/T_x \text{ and } \omega_y = 2\pi/T_y. \tag{5b}$$

If the original continuous object is bandlimited then, for some finite integers L_x and L_y, its Fourier transform $F^C(u,v)$ satisfies

$$|F^C(u,v)| = 0, \ |u| > L_x \omega_x \text{ and } |v| > L_y \omega_y \tag{6}$$

After substituting the condition (6) in the aliasing relation given in (5), the discrete Fourier transform $F_k(m,n)$ of the kth undersampled frame can be expressed as a finite aliased version of the continuous Fourier transform $F_k^C(u,v)$ of the kth shifted object $f_k(x,y)$ by equation (7).

$$F_k(m,n) = \frac{1}{T_x T_y} \sum_{i=-L_x}^{L_x-1} \sum_{\ell=-L_y}^{L_y-1} F_k^C\left(\frac{2\pi m}{MT_x} + i\omega_x, \frac{2\pi n}{NT_y} + \ell\omega_y\right), \tag{7}$$

for $k=1,2,\ldots,p$.

To write (7) in matrix form, choose the lexicographical ordering associated with row-wise scanning:

$$(-L_x,-L_y), (-L_x+1,-L_y), (-L_x+2,-L_y),\ldots(L_x-2,L_y-1), (L_x-1,L_y-1)$$

The matrix to vector mapping is defined by the 2-D to 1-D index map,

$$\left(\frac{2\pi m}{MT_x} + i\omega_x, \frac{2\pi n}{NT_y} + \ell\omega_y\right) \rightarrow r \tag{8}$$

$$i = (r-1) \bmod (2L_x) - L_x \tag{9a}$$

$$\ell = \lfloor (r-1)/(2L_x) \rfloor - L_y \tag{9b}$$

for $r=1,2,\ldots,4L_x L_y$.

Then the matrix form of (7) for k=1,2,...,p is given below.

$$
\begin{bmatrix} F_1(m,n) \\ F_2(m,n) \\ \vdots \\ F_p(m,n) \end{bmatrix} = \frac{1}{T_x T_y} \begin{bmatrix} \phi_{1,1} & \phi_{1,2} & \cdots & \phi_{1,4L_xL_y} \\ \phi_{2,1} & \phi_{2,2} & \cdots & \phi_{2,4L_xL_y} \\ \vdots & & & \vdots \\ \phi_{p,1} & \phi_{p,2} & \cdots & \phi_{p,4L_xL_y} \end{bmatrix} \begin{bmatrix} F^C(1) \\ F^C(2) \\ \vdots \\ F^C(4L_xL_y) \end{bmatrix}
\tag{10a}
$$

where

$$
F^C(r) = F^C(\frac{2\pi m}{MT_x} + i\omega_x, \frac{2\pi n}{NT_y} + \ell\omega_y)
\tag{10b}
$$

for r=1,2,..., $4L_xL_y$, and

$$
\phi_{k,r} = \exp[j2\pi(\delta_{xk}(m/MT_x + i/T_x) + \delta_{yk}(n/NT_y + \ell/T_y))]
\tag{10c}
$$

Via use of (9), equation (10c) can be rewritten as,

$$
\phi_{k,r} = \exp[j2\pi(\delta_{xk}(\frac{m}{MT_x} - \frac{L_x}{T_x}) + \delta_{yk}(\frac{n}{NT_y} - \frac{L_y}{T_y}))]
$$
$$
\exp[j2\pi\{(\delta_{xk}/T_x)(r-1)\bmod(2L_x) + (\delta_{yk}/T_y)(\lfloor \frac{r-1}{2L_x} \rfloor)\}]
\tag{11}
$$

For the sake of notational brevity, we rewrite (10a) in the form

$$
\underline{F}_p = \frac{1}{T_x T_y} \phi \, \underline{F}^C
\tag{12}
$$

where the vectors $\underline{F}_p$ and $\underline{F}^C$ and the matrix ϕ are defined through (10a). There-fore, we can write the matrix $\phi = [\phi_{kr}]$ as a product of two matrices.

$$
\phi = [\phi_{kr}] = DH
\tag{13a}
$$

where

$$
D = \text{diag } [d_1 \ d_2 ... d_p]
\tag{13b}
$$

$$
d_k = \exp[j2\pi(\delta_{xk}(m/MT_x - L_x/T_x) + \delta_{yk}(n/NT_y - L_y/T_y))].
\tag{13c}
$$

Furthermore, if h_{kr} denotes the element in the kth row and the rth column of H, then

$$
h_{kr} = \exp[j2\pi\{(\delta_{xk}/T_x)(r-1)\bmod(2L_x) + (\delta_{yk}/T_y)(\lfloor (r-1)/(2L_x) \rfloor)\}]
$$
$$
= W_{xk}^\alpha W_{yk}^\beta
\tag{13d}
$$

where

$$
W_{xk} = \exp(j2\pi\delta_{xk}/T_x), \ W_{yk} = \exp(j2\pi\delta_{yk}/T_y)
\tag{13e}
$$

and

$$
\alpha = (r-1) \bmod(2L_x), \ \beta = \lfloor (r-1)/(2L_x) \rfloor.
\tag{13f}
$$

The problem of solving (10a) for the column vector on the right-hand side has been considered when p = $4L_xL_y$ by exploiting the structure of the matrix decomposition in (13). A procedure is advanced for recursive updating of the solution with improved signal-to-noise ratio because of the availability of an increasing number of noisy frames. This procedure is generalized to a weighted least-squares based algorithm that leads to further improvement of the signal-to-noise ratio under appropriate assumptions about the noise, especially with the availability of a knowledge base concerning one or more low resolution frames. The effectiveness of the algorithm described in the previous section is illustrated through its implementation on several classes of image. Finally, conclusions are drawn especially with regard to the feasibility of tackling certain types of blur in addition to the noise.

Recursive Reconstruction of High Resolution Image From Noisy Frames. At this point, we assume that we have $k>4L_xL_y$ low resolution frames which have been corrupted by noise. The noise is modeled as an additive error term in the frequency domain. At frequency point (m,n) we have

$$Z_i(m,n) = F_i(m,n) + N_i(m,n) \ , \ i=1,2,\ldots,k \tag{14}$$

where $Z_i(m,n)$ is the discrete Fourier transform of the i-th noisy undersampled frame and $N_i(m,n)$ is an error term at the frequency point (m,n). In the interpolation algorithm, the frequency indices (m,n) do not change. Therefore, for notational convenience, those are dropped. For the i-th frame, the equation extracted from (10) is written below.

$$F_i = \underline{Y}_i^t \ \underline{F}^c \ , \ i=1,2,\ldots,k \tag{15a}$$

where

$$\underline{Y}_i = \frac{1}{T_xT_y} \ \underline{X}_i \tag{15b}$$

with $\underline{X}_i^t$ being the i-th row of the matrix ϕ in (12) and $\underline{F}^c$ is defined in (12).

Therefore, equation (14) can be rewritten as,

$$Z_i = \underline{Y}_i^t \ \underline{F}^c + N_i \ , \ i=1,2,\ldots,k \tag{16}$$

and the system of equations in (10), in the noisy case, is replaceable by

$$\underline{Z}_k = \phi_k \ \underline{F}^c + \underline{N}_k \tag{17a}$$

where ϕ_k is a $(k \times 4L_xL_y)$ matrix defined as

$$\phi_k = [\underline{Y}_1 \ \underline{Y}_2 \ \cdots \ \underline{Y}_k]^t \tag{17b}$$

and

$$\underline{Z}_k = [Z_1 \ Z_2 \ \cdots \ Z_k]^t \ , \tag{17c}$$

$$\underline{N}_k = [N_1 \ N_2 \ \cdots \ N_k]^t \ . \tag{17d}$$

We want to find the vector $\hat{F}(k)$ that minimizes the square of the error norm

$$\| \underline{E}_k \|^2 = (\underline{Z}_k - \phi_k \, \underline{\hat{F}}^{(k)})^* \, (\underline{Z}_k - \phi_k \, \underline{\hat{F}}^{(k)}) \, , \tag{18}$$

for $k \geq p$, where $p = 4L_xL_y$. The star superscript in (18) denotes "complex conjugate transpose".

Using the orthogonality principle, we have

$$(\phi_k \, \underline{\hat{F}}^{(k)})^* \, (\underline{Z}_k - \phi_k \, \underline{\hat{F}}^{(k)}) = 0 \, . \tag{19}$$

Whence,

$$\underline{\hat{F}}^{(k)} = (\phi_k^* \, \phi_k)^{-1} \, \phi_k^* \underline{Z}_k \tag{20}$$

For $k = p = 4L_xL_y$, the initial estimate, $\hat{F}^{(p)}$ is obtained as

$$\underline{\hat{F}}^{(p)} = (\phi_p^* \phi_p)^{-1} \, \phi_p^* \underline{Z}_p \tag{21}$$

where it is assumed that the delays are such that the square matrix ϕ_p is invertible. The objective now is to derive a recursive scheme for updating $\hat{F}^{(k)}$ for $k > p$, given $\hat{F}^{(p)}$. The procedure to do this is based on standard sequential least square estimation theory [2]. Define,

$$R(k) = \phi_k^* \, \phi_k \, , \tag{22a}$$

$$\underline{r}(k) = \phi_k^* \, \underline{Z}_k \, . \tag{22b}$$

After the substitution of (22a) and (22b) in (20), $\underline{\hat{F}}^{(k)}$ can be rewritten as

$$\underline{\hat{F}}^{(k)} = R^{-1}(k) \, \underline{r}(k) \, . \tag{23}$$

After substituting (17b) and (17c) in equations (22a) and (22b), $R(k+1)$, and $r(k+1)$ become

$$R(k+1) = R(k) + \underline{\overline{Y}}_{k+1} \, \underline{Y}_{k+1}^t \tag{24a}$$

and

$$\underline{r}(k+1) = \underline{r}(k) + \underline{\overline{Y}}_{k+1} \, Z_{k+1} \, . \tag{24b}$$

The bar superscript in (24) denotes complex conjugation. After substituting (24a) in (23) the $(k+1)$-th estimate, $\underline{\hat{F}}^{(k+1)}$, is written as

$$\underline{\hat{F}}^{(k+1)} = (R(k) + \underline{\overline{Y}}_{k+1} \, \underline{Y}_{k+1}^t)^{-1} \, \underline{r}(k+1) \tag{25}$$

By applying to (25) the identity,

$$(A+BCD)^{-1} = A^{-1} - A^{-1}B(DA^{-1}B+C^{-1})^{-1} DA^{-1} \tag{26}$$

after setting

$$A = R(k), \; B = \underline{\overline{Y}}_{k+1} \, , \; C = 1 \text{ and } D = \underline{Y}_{k+1}^t,$$

the inverse of $R(k+1)$ in (24) can be written as

$$R^{-1}(k+1) = R^{-1}(k) - \frac{R^{-1}(k) \ \underline{Y}_{k+1} \ \underline{y}_{k+1}^t \ R^{-1}(k)}{1 + \underline{y}_{k+1}^t \ R^{-1}(k) \ \underline{Y}_{k+1}} . \tag{27}$$

Define

$$P(k) = R^{-1}(k) , \tag{28a}$$

$$Q(k+1) = P(k) \ \underline{Y}_{k+1} \tag{28b}$$

and

$$\alpha(k+1) = 1 + \underline{y}_{k+1}^t \ P(k) \ \underline{Y}_{k+1} . \tag{28c}$$

From (23), (27) and (28), the (k+1)-th least square theory based estimate, $\underline{\hat{F}}^{(k+1)}$, can be written as

$$\underline{\hat{F}}^{(k+1)} = P(k+1) \ \underline{r}(k+1) , \tag{29a}$$

where

$$P(k+1) = P(k) - Q(k+1) \ Q^*(k+1)/\alpha(k+1) \tag{29b}$$

and

$$\underline{r}(k+1) = \underline{r}(k) + \underline{Y}_{k+1} \ Z_{k+1} . \tag{29c}$$

The matrix $P(k)$ and the scalar $\alpha(k+1)$, defined in (28a) and (28c), respectively, are real-valued, while the matrix $Q(k+1)$ in (28b) is complex-valued. Following some routine matrix manipulations, an equivalent expression for the (k+1)-th least square theory based estimate is obtainable in the form,

$$\underline{\hat{F}}^{(k+1)} = \underline{\hat{F}}^{(k)} + P(k+1) \ \underline{Y}_{k+1} \ (Z_{k+1} - \underline{y}_{k+1}^t \ \underline{\hat{F}}^{(k)}) \tag{30}$$

where the matrix $P(k+1)$ can be computed from $P(k)$ by equations (29b) and (29c). The steps involved in arriving at (30) are briefly indicated. From (23) it is clear that,

$$R(k) \ \underline{\hat{F}}^{(k)} = \underline{r}(k) \tag{31}$$

$$R(k+1) \ \underline{\hat{F}}^{(k+1)} = \underline{r}(k+1). \tag{32}$$

Let

$$\underline{\hat{F}}^{(k+1)} = \underline{\hat{F}}^{(k)} + \underline{E}^{(k+1)} \tag{33}$$

Substituting the expressions for $R(k+1)$, $\underline{r}(k+1)$, and $\underline{\hat{F}}^{(k+1)}$ from (24a), (24b) and (33), respectively, in (32), then using (31) and simplifying, we get

$$\underline{E}^{(k+1)} = P(k+1) \ \underline{Y}_{k+1}[Z_{k+1} - \underline{y}_{k+1}^t \underline{\hat{F}}^{(k)}] \tag{34}$$

From (33) and (34), equation (30) follows. From equation (30), it is clear that the (k+1)-th least squares estimate, $\underline{\hat{F}}^{(k+1)}$, can be recursively computed from its previous estimate, $\underline{\hat{F}}^{(k)}$, after incorporating an additional (k+1)-th frame through the observation vector $\underline{Z}_{k+1}$. The matrix $P(k+1)$ can also be updated from $P(k)$, without matrix inversions, after the new frame has been considered as is evident from (29b) and associated equations.

__Weighted Least Square Algorithm.__ If a priori knowledge about the measurements is available, it can be incorporated in the recursive algorithm by assigning weights to the observations. As a result, the problem, again, reduces to a sequential least square estimation problem. The weight matrix W defined below will depend on the a priori information available. The square error norm, in this case, becomes

$$|| \underline{E}_k ||^2 = (\underline{Z}_k - \phi_k \hat{\underline{F}}^{(k)})^* W (\underline{Z}_k - \phi_k \hat{\underline{F}}^{(k)}) \tag{35a}$$

where

$$W = \text{diag} (w_1 \ w_2 \ \cdots \ w_k). \tag{35b}$$

The weights w_i i=1,2,...,k will depend on the a priori knowledge about the i-th observation Z_i.

By repeating the procedure outlined in the previous section on the error norm in (35a), expressions shown in (36) below for the (k+1)-th estimate are obtained, in the present case, as counterparts of (29).

$$\hat{\underline{F}}^{(k+1)} = P(k+1) \ \underline{r}(k+1) \tag{36a}$$

where

$$P(k+1) = P(k) - Q(k+1) \ Q^*(k+1)/\alpha(k+1) \tag{36b}$$

and

$$\underline{r}(k+1) = \underline{r}(k) + \overline{Y}_{k+1} \ Z_{k+1} \ . \tag{36c}$$

The matrix Q(k+1) and the scalar α(k+1) in (36) are defined below.

$$Q(k+1) = P(k) \ \overline{Y}_{k+1} \tag{37a}$$

and

$$\alpha(k+1) = w_{k+1}^{-1} + \underline{Y}_{k+1}^t \ P(k) \ \overline{Y}_{k+1} \tag{37b}$$

An equivalent expression for $\hat{F}$(k+1), which is a counterpart of (30), is given below

$$\hat{\underline{F}}^{(k+1)} = \hat{\underline{F}}^{(k)} + P(k+1) \ \overline{Y}_{k+1} \ (Z_{k+1} - w_{k+1} \ \underline{Y}_{k+1}^t \ \hat{\underline{F}}^{(k)}) \tag{38}$$

where the matrix P(k+1) can be computed from P(k) using equations (36b), (37a) and (37b).

__Computer Simulations.__ In this simulation example, a higher resolution (256x256)-pixel image is recursively reconstructed from a set of low resolution (128x128)-pixel noisy frames. In order to reduce the size of the arrays to be processed, the input image was partitioned into 16 non-overlapping sections of (32x32) pixels each and the recursive algorithm was independently applied to each one of them.

For each one of these 16 sections, the interpolation problem corresponds to the restoration of a (64x64)-pixel image from a sequence of shifted low resolution (32x32)-pixel noisy input frames. This set of noisy low resolution frames with nontrivial shifts (not an integer multiple of the sampling period) was simulated and the simulation procedure is described next.

First, a (64x64)-pixel image that corresponds to one of the 16 non-overlapping sections of an original (256x256)-pixel image is considered. This (64x64) array is interpolated to obtain a (256x256) array by appending zeros in the DFT domain and then taking the inverse DFT. By resampling the interpolated image, several low resolution (32x32)-pixel frames with nontrivial shifts are obtained. Subsequently, a uniformly distributed zero mean noise was added to each input frame. The noise sequences were statistically independent and different signal-to-noise ratios (SNRs) were chosen.

From the resulting (32x32)-pixel low resolution frames a (64x64)-image was reconstructed using the recursive weighted least squares algorithm described. This same procedure was repeated for all 16 non-overlapping sections of the original image and after putting together the 16 resulting (64x64)-images, the high resolution (256x256)-pixel image was obtained.

The first simulation example uses the image of an F16 AIR FORCE airplane. The reconstruction steps are illustrated in Figure 1 through Figure 6. Figure 1 shows 4 low resolution (128x128) noisy input frames with different SNRs. The SNRs of the frames in Figure 1 are, respectively, 10 db for the upper left, 3 db for the lower left, 8 db for the upper right and 5 db for the lower right frame. From these four frames an initial estimate of the reconstructed image was obtained in the DFT domain by solving equation (12) with $p=4$. The corresponding spatial image is shown in Figure 2. From this Figure, it becomes clear that the first estimate gives a high resolution image with poor output SNR (SNR=10.3 db).

In order to improve the quality of the reconstructed image, by the recursive weighted least square algorithm in equations (36)-(38), additional low resolution frames were used as new measurements. Figure 3 has four additional shifted low resolution (128x128)-pixel noisy frames with respective SNRs known to be 8 db for the upper left, 6 db for the lower left, 12 db for the upper right and 3 db for the lower right frame. With each one of these input frames a new estimate of the reconstructed image was recursively computed and the resulting SNRs of these estimates were 13.4 db, 15.2 db, 18.8 db and 18.8 db, respectively. The last estimate (the fifth) in this sequence, having an SNR of 18.8, is shown in Figure 4.

This procedure was continued until all 16 input frames were used in the reconstruction. A graph of the SNR of the input frames and of the output high resolution estimates is given in Figure 5. From this plot it is clear that, even though the SNR is an increasing function, after the first few estimates are obtained, the improvement in the SNR for the subsequence estimates becomes very small. The final reconstructed high resolution frame is given in Figure 6. It is important to note that from the low resolution frames in Figure 1, the plane is not identifiable. After the initial reconstruction, as shown in Figure 2, the resolution has doubled along each coordinate axis but the details concerning the plane are still buried in the noise. These details become clear after a few iterations. From the fifth estimate shown in Figure 4, the aircraft is clearly identifiable. Also from Figures 4 and 5, it becomes clear that after the first 8 frames are processed the additional improvement obtained by increasing the number of iterations of the algorithm is not considerable. Therefore, the algorithm could be stopped, resulting in a very good estimate with a lesser number of frames.

Finally, the effect of the different shift patterns in the reconstruction procedure was investigated by carrying out the recursive reconstruction from sets of input frames with the same SNR but different shift patterns. It was concluded that the intermediate estimates of the high resolution image may be dependent on the delay pattern used but the final reconstructed image seems to be independent of the delay patterns.

<u>Conclusions</u>. In this paper a recursive procedure, based on the weighted least square theory, is introduced for reconstructing a high resolution image from available low resolution noisy undersampled frames. The recursive updating proceeds from the solution of a structured set of linear equations. The complete procedure is carried out in the transform domain and the recursive updating is done at each frequency point independently. This suggests the feasibility of massive parallelism in the recursive implementation scheme.

The convergence issue of the recursive procedure may be addressed through the use of standard assumptions employed in linear mean-square sequential estimation theory [2]. This requires the additive noise term in (14) to be an independent and identically distributed sequence having zero mean and finite variance. It is easy to establish that when $\{N_i(m,n)\}$ is characterized by the property just alluded to, the associated sequence in the spatial domain also has like property. Therefore, for the estimation to be consistent, it is required that the additive noise sequence superposed on the true image forms an independent and identically distributed sequence having zero mean and finite variance.

It should be noted that the procedure developed applies directly only to noisy undersampled images. The simulation studies clearly support the fidelity of reconstruction in cases when no blur is present. Certain types of blur, however, can be tackled by adapting the method proposed in this paper. Consider, for instance, the case when the blur is to be modeled by a linear shift-invariant system. Let the 2-D Fourier transform of the unit impulse response of this system be $H^c(u,v)$. Then, in the present situation equation (3) is replaced by

$$F_k^c(u,v) = \exp[j2\pi(\delta_{xk}u + \delta_{yk}v)]H^c(u,v)F^c(u,v)$$

and, correspondingly, after defining

$$H^c(r) = H^c(\frac{2\pi m}{MT_x} + i\omega_x, \frac{2\pi n}{NT_y} + \ell\omega_y)$$

equation (12) is replaced by

$$\underline{F}_p = \frac{1}{T_x T_y} \phi \text{ diag}[H^C(1)....H^C(4L_x L_y)] \underline{F}^C .$$

Subject to the assumption that

$$H^C(r) \neq 0, \; r=1,2,...,4L_x L_y ,$$

the preceding equation can be solved for $\underline{F}^C$ in a manner similar to that done for (12). In the case of the type of blur under consideration, reconstruction is, thus, achieved through recursive interpolation (as in the pure noise case) and inverse filtering [3].

For more general blurs, previous research conducted to handle such situations need be combined with the procedure proposed here. For example, when in addition to noise, blurs that can be modeled by linear shift-variant systems are present, deblurring of the low resolution frames can be effected through the use of a state-space model proposed in [4]. When the blur is modeled by a shift-variant bilinear system, deblurring is possible through the schemes proposed in [5]. Feasibility of doing the same by adapting the method advanced in [4] is currently under investigation.

Acknowledgement

This research was partially supported by a grant from the Singer Foundation given to N. K. Bose, Singer Professor.

The authors are with the Department of Electrical Engineering, Spatial and Temporal Signal Processing Center, at The Pennsylvania State University, University Park, PA 16802.

References

[1] R. Y. Tsai and T. S. Huang, "Multiframe image restoration and registration", in "Advances in Computer Vision and Image Processing", vol. 1, edited by T. S. Huang, Jai Press, Inc., Greenwich, CT, 1984, pp. 317-319.

[2] G. N. Saridis, Self-Organizing Control of Stochastic Systems, Marcel Dekker, Inc., New York, 1977, pp. 96-99.

[3] R. C. Gonzalez and P. Wintz, "Digital Image Processing", Addison-Wesley Publishing Co., Reading, MA, 1977, pp. 199-207.

[4] H. M. Valenzuela and N. K. Bose, "Linear shift-variant multidimensional systems", in "Multidimensional Systems Theory: Progress, Directions, and Open Problems", edited by N. K. Bose, D. Reidel Publishing Co., 1985, pp. 147-183.

[5] H. M. Kim and N. K. Bose, "Approaches towards restoration of bilinearly degraded images", IEEE Trans. Acoustics, Speech, and Signal Processing, vol. ASSP-35, no. 2, Feb. 1987, pp. 181-197.

Figure 1: First four low resolution noisy frames of F16 airplane with different SNRs.

Figure 2: First estimate of the reconstructed image of the F16 airplane with SNR=10.3 db.

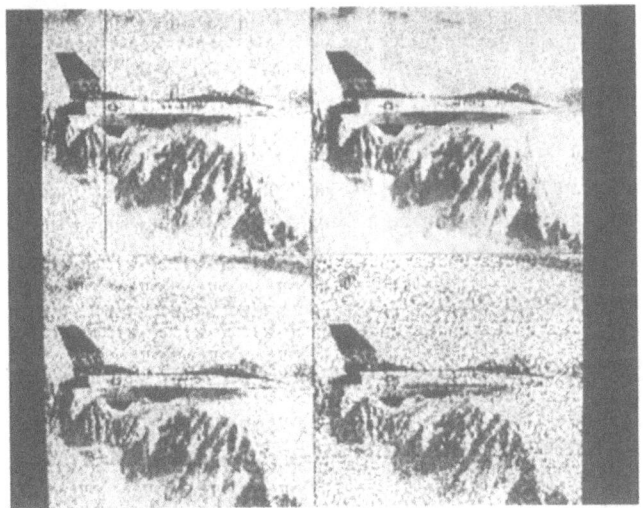

Figure 3: Additional four low resolution noisy frames of F16 airplane.

Figure 4: Fifth estimate of the reconstructed image of the F16 airplane with SNR=18.8 db.

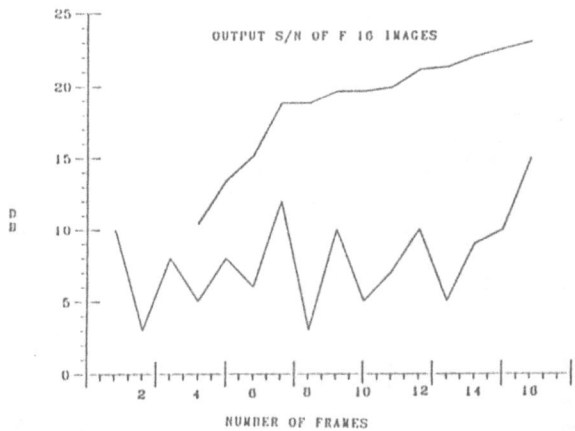

Figure 5: SNRs of the input frames and output images vs. the number of frames used in the reconstruction.

Figure 6: Final reconstructed image of the F16 airplane with SNR=23.0 db.

RECONSTRUCTION OF CONTINUOUS-TONE FROM HALFTONE
BY PROJECTIONS ONTO CONVEX SETS

Jan P. Allebach

Purdue University

Introduction. Halftone images are binary images that are perceived as continuous-tone because of the inability of the human viewer to resolve the binary texture. The actual gray value that is perceived is determined by a local spatial average over the binary image. Image detail is rendered by modulating the texture pattern. Figure 1 contains an example of a halftone image at several different magnifications. At normal viewing distances, the largest magnification does not really give the impression of continuous-tone because the dot pattern is so coarse as to be easily resolved by the viewer. As the magnification decreases, the spatial frequency components of the periodic dot pattern move out along the viewer's contrast sensitivity curve and become less perceptible [1]. Eventually, they will fall below threshold; and the halftone image will be indistinguishable from the continuous-tone original at normal viewing distances.

Halftone images are widely used because they are well suited for many printing and display mechanisms. For example, it is relatively difficult to vary the density of ink on paper in a controlled manner that would produce a range of shades from white (absorptance 0) to black (absorptance 1). At each point on the paper, it is much easier to either have no ink, or to have ink with some maximum density. Similar considerations apply to the toner used with electrographic printing processes, such as xerography. With displays, there are many mechanisms that are well suited for bistable operation. Examples include liquid crystal and plasma panel technologies. In addition, halftone images are compatible with the inherently binary source material that may comprise the bulk of the images reproduced by a given imaging system, such as a copier or a facsimile machine. This type of material includes text, graphs, charts, and line drawings.

The question of reconstructing the continuous-tone original from its halftone representation is one that is of both theoretical and practical interest. Loosely speaking, one may regard the halftone image as a sampled version of the original. Reconstruction may thus be viewed as a problem in sampling theory. From a more practical point of view, reproduction of an existing halftone image by a digital imaging system often leads to severe moire artifacts due to interference between the frequency components associated with the periodic dot patterns in the original image, and the frequency components associated with the scanning and printing rasters of the digital imaging system. One way to avoid this problem is to scan the halftone image at a sufficiently high resolution to prevent aliasing, and then to reconstruct the continuous-tone original. This then becomes the input to the digital printing process. Based on the largest magnification in Fig. 1,
one might conclude that there is little hope for reconstructing the original image. However, inspection of the smaller versions of this image reveals a surprising amount of image detail.

As with any signal reconstruction problem, it is important to understand how the observed data is related to the original signal. There are actually many different techniques for halftoning images [2]-[4]. Figure 2 contains examples of images halftoned with the three most commonly used approaches:
pattern printing, screening, and error diffusion. With pattern printing, the output surface is

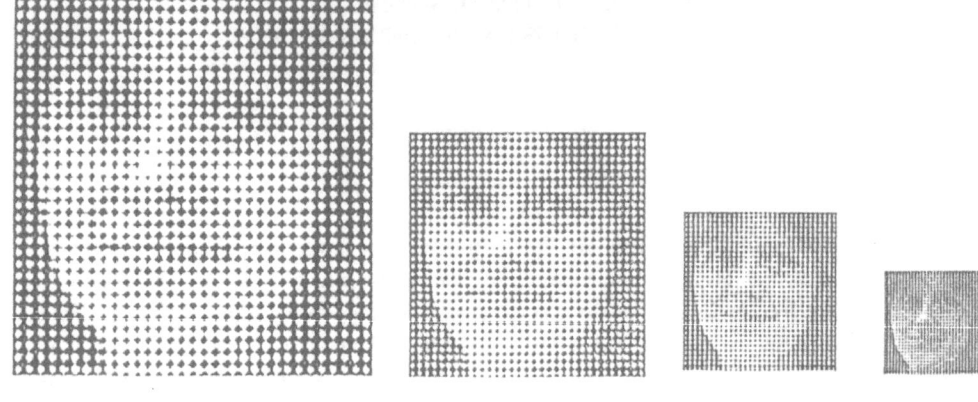

Fig. 1. Typical halftone image at four different magnifications.

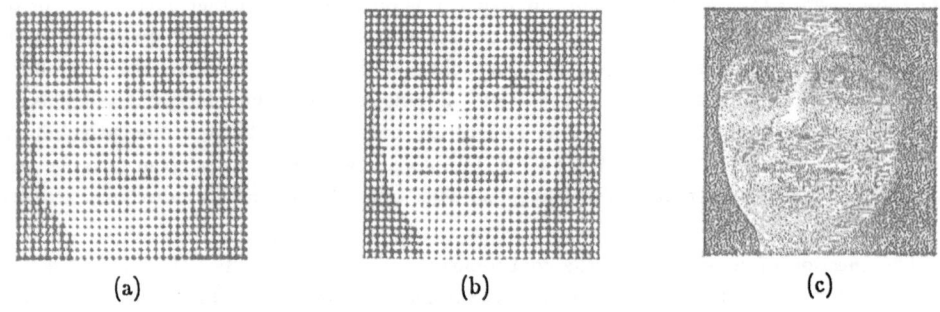

(a)   (b)   (c)

Fig. 2. Images generated by three principal halftone processes:
(a) pattern printing, (b) screening, (c) error diffusion.

partitioned into an array of cells referred to as *halftone cells*. Within each halftone cell, a dot or other binary pattern is printed so that the average absorptance within the cell is proportional to the gray value of the image at the center of the cell. This is analogous to pulse width modulation with uniform sampling [5]. Since the image is only sampled once per halftone cell, no detail smaller than a halftone cell can be reproduced. Figure 2 illustrates the relatively poor detail rendition of pattern printing.

With the screening process, the continuous-tone image is compared point-by-point with a spatially periodic threshold function referred to as the *halftone screen*. Where the screen is exceeded, the halftone image has absorptance 1, and where it is not exceeded, the absorptance of the halftone image is 0. The screening process represents gray values with the same periodic textures produced by pattern printing. However, the point-to-point screening process allows the halftone image to represent detail smaller than a halftone cell by using partial dots. This is also illustrated in Fig. 2, where we see that the quality of the screened image is much better than that obtained by pattern printing. The screening process corresponds to pulse width modulation with natural sampling [5]. The discrete-parameter version of this process is also known as ordered dither, in reference to the fact that the screen function may be viewed as a constant threshold with a dither signal added to it.

With error diffusion [6],[7], the absorptance of the halftone image is also determined by a point-by-point comparison with a continuous-tone image. However, in this case, the threshold is fixed at 0.5; and the comparison is with a modified version of the original image. Initially, the modified image is identical to the original image. However, at each point, after the threshold step, the error is computed between the halftone and modified image values. This error is then used to further modify the values of the continuous-tone image at neighboring points which have not yet been thresholded. The correction is made in such a way as to tend to reduce the error in the locally averaged gray value. Although error diffusion yields excellent detail rendition and generally low texture visibility, the binary textures that it produces are often difficult to print accurately. The result can be an image of relatively poor quality.

Because it produces well controlled easily printable binary textures, provides good detail rendition, and is extremely simple to implement, screening is the most widely used of the halftone processes. In the remainder of this paper, we shall only be concerned with this process which is described in more detail in the following section. The reconstruction method that we propose is specifically tailored to the characteristics of this process, and is not generally applicable to other halftone processes. For simplicity, we work with continuous-parameter functions, even though implementation of the reconstruction method would require discretization of the independent variables.

In practice, low-pass filtering is frequently used as a means of reconstructing the continuous-tone original image from its halftone version. One objective of this paper is to point out the limitations of such an approach. We then show how alternating projections onto constraint sets may be applied to this problem, and why it may be expected to yield results that are superior to those obtained by low-pass filtering. Alternating projections onto constraint sets is a powerful method for signal reconstruction and synthesis that has been used in a wide range of applications. In our own work, we have employed it for image reconstruction from nonuniformly spaced samples [8], reconstruction of signal magnitude from Fourier offset data [9], and synthesis of computer-generated holograms [10]. Additional discussion of the method may be found in [11].

Halftone Screening Process. Figure 3 illustrates the halftone screening process for one spatial dimension.
At each point, the continuous-tone original image f is compared with the screen s. The halftone image g has value 1 at that point if the threshold is exceeded, and 0 otherwise. If f varies smoothly over each screen period X, pulses occur at interval X with width proportional to f. Where f changes abruptly, the dot structure is modified to represent this change, as illustrated in Fig. 3 by the additional pulse centered at 1.5X. In two dimensions, the screening process may be expressed as

$$g(x,y) = \begin{cases} 1, & f(x,y) \geq s(x,y), \\ 0, & \text{else.} \end{cases} \tag{1}$$

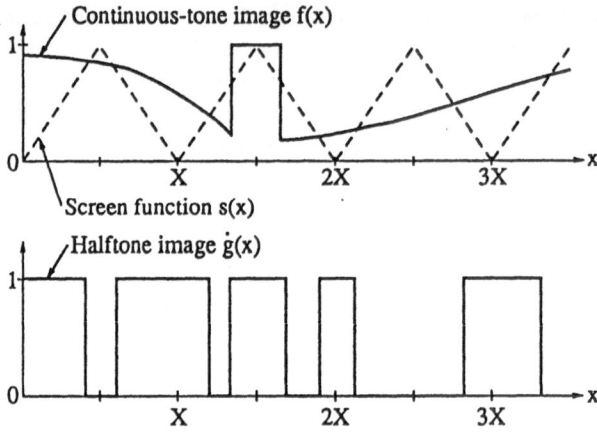

Fig. 3. Screening process for 1-D signals.

Fourier analysis has traditionally been an important tool for understanding the effect of sampling of continuous-parameter signals. Because conventional sampling may be modeled as multiplication of the continuous-parameter signal by a comb function, it is straightforward to determine the spectrum of the sampled signal in terms of that of the continuous-parameter signal. Despite the fact that (1) is a highly nonlinear relation, it is still possible to determine the spectrum of the halftone image in terms of that of the continuous-tone original image because of the spatial periodicity of the screen [12],[13]. The spectrum [5] of a signal pulse-width modulated with natural sampling is a special case of these results.

To carry out the analysis, it is helpful to express the screening process in an alternate manner. Let $p[x,y;b]$ denote the periodic binary texture used to represent an absorptance of b. This function is referred to as the *dot profile* [13]. Since it is the halftone image that would result if the continuous-tone image had a constant absorptance of b, the dot profile function may be expressed in terms of the screen function according to

$$p[x,y;b] = \begin{cases} 1, & b \geqq s(x,y), \\ 0, & \text{else.} \end{cases} \tag{2}$$

Figure 4 shows, for several values of b, the dot profile function corresponding to the screen used to generate Fig. 1.

The halftone image may be expressed directly in terms of the dot profile according to

$$g(x,y) = p[x,y;f(x,y)]. \tag{3}$$

Let

$$F(u,v) = \int\int f(x,y)\, e^{-i2\pi(ux + vy)}dxdy, \tag{4}$$

be the Fourier spectrum of the continuous-tone original image. The Fourier spectrum of the halftone image is given by

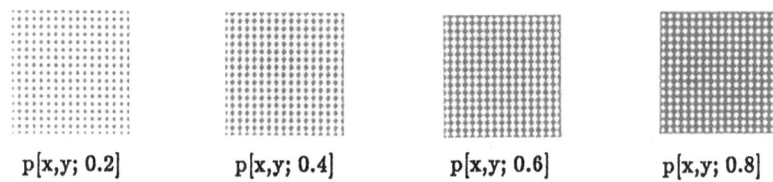

$$p[x,y; 0.2] \qquad p[x,y; 0.4] \qquad p[x,y; 0.6] \qquad p[x,y; 0.8]$$

Fig. 4. Dot profile function.

$$G(u,v) = \sum_m \sum_n F_{mn}(u - m/X,\ v-n/X). \qquad (5)$$

As with the spectrum of an image sampled at interval X, the spectrum of the halftone image consists of image spectra centered at interval $1/X$. However, in contrast to the spectrum of a sampled image, these spectra are not that of the original image. Instead, the spectrum $F_{mn}(u,v)$ centered at $(u,v) = (m/X, n/X)$ is the Fourier transform of the pointwise nonlinearly transformed image

$$f_{mn}(x,y) = P[m,n; f(x,y)]. \qquad (6)$$

The (m,n)-th nonlinearity is given by the (m,n)-th Fourier coefficient

$$P[m,n;b] = \frac{1}{X^2} \int\int_{-X/2}^{X/2} p[x,y;b]\ e^{-i2\pi(mx + ny)/X} dxdy, \qquad (7)$$

of the periodic dot profile function as b ranges from 0 to 1.

For correct tone reproduction, it is necessary that the absorptance of the dot profile function $p[x,y;b]$ averaged over each $X \times X$ halftone cell be b. It follows that $P[0,0;b] = b$, and hence that $F_{00}[u,v] = F(u,v)$. Since the baseband spectrum is that of the original image, we might expect, as is the case with sampled images, that the original image could be recovered by low-pass filtering. However, at screen frequencies such as that shown in Fig. 1, where the continuous-tone original image may vary significantly within $X \times X$ halftone cells, and partial dotting plays an important role in detail rendition, there will be considerable overlap of the higher order spectra $F_{mn}(u,v)$, $(m,n) \neq (0,0)$ in (5) with the baseband spectrum. Even if the continuous-tone image were bandlimited to highest spatial frequencies $|u|, |v| \leq 1/(2X)$, the nonlinear transformations (6) would introduce dispersion causing the higher order spectra to spread beyond the $1/(2X) \times 1/(2X)$ support region of $F(u,v)$, and hence overlap the baseband.

Figure 5a shows the spectrum of the continuous-tone original image screened to produce Fig. 1.

Figure 5b shows the spectrum of the resulting halftone image. Extensive overlap of spectral terms is evident. To attempt recovery of the original image by low-pass filtering would require a passband with support small enough to exclude all higher order spectra in Fig. 5b. From Fig. 5a, it may be concluded that a large amount of image detail would also be lost, since the spectrum of the original image contains significant energy well beyond this region.

Based on the results of this section, we conclude that low-pass filtering will satisfactorily recover the continuous-tone original image only if the screen frequency is somewhat higher than twice the highest frequency in the continuous-tone image. In the following section, we consider

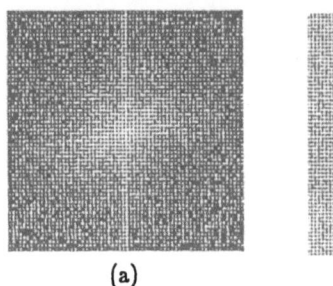 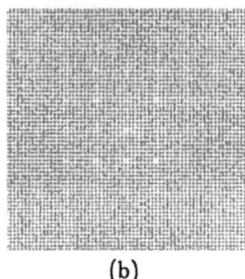

(a)   (b)

Fig. 5. Log-magnitude spectra of (a) the continuous-tone original image, and (b) the halftone image.

an alternate approach to image recovery that is not subject to this limitation.

Image Recovery by Alternating Projections onto Constraint Sets. Before embarking on a search for alternatives to low-pass filtering, it would be prudent to consider whether it is even reasonable to expect to recover image components at spatial frequencies (u,v) that lie outside the baseband region $0 \leq |u|, |v| \leq 1/(2X)$ corresponding to sampling the continuous-tone original image at interval X in both x and y. Returning to Fig. 3, suppose that the halftone image $g(x)$ makes the transition between 0 and 1, or vice versa at some coordinate x_0, $i.e.$ $\lim_{x \to x_0^-} g(x) = 0$ or 1 and $\lim_{x \to x_0^+} g(x) = 1$ or 0. If the screen function $s(x)$ and the original image $f(x)$ are continuous at x_0, then we have that $f(x_0) = s(x_0)$. Thus, in the 1-D case, we can typically determine the value of the continuous-tone original image $f(x)$ at two points within each halftone cell of duration X, simply by examining the screen function $s(x)$ at points where the halftone image $g(x)$ is discontinuous. Of course, the location of these points within each halftone cell varies from cell to cell in an image dependent fashion; and knowledge of the screen function used to generate the halftone image is assumed.

In the 2-D case described by (1), we can similarly determine the value of $f(x,y)$ within each $X \times X$ halftone cell at all points (x,y) lying on the closed contour(s) corresponding to the boundary of the dot(s) within that cell. However, there is still more information about the continuous-tone image available from the halftone image. At all points (x,y) where $g(x,y) = 0$, we know that $f(x,y)$ is bounded from above by the screen function. At all points where $g(x,y) = 1$, $f(x,y)$ is bounded from below by the screen function. Thus, we know that the continuous-tone image lies within a bounding envelope illustrated in Fig. 6 for the 1-D example of Fig. 3.

We conclude that a typical halftone image generated with a screen having period $X \times X$ contains a great deal more information about the continuous-tone image than that provided by one sample per $X \times X$ interval. The problem is to find a method of image recovery that can fully utilize this information.

The method of alternating projections onto constraint sets is precisely suited to this task. With this method, all information known about the signal to be recovered is expressed in the form of constraint sets. The constraint set corresponding to the bounding envelope may be expressed as

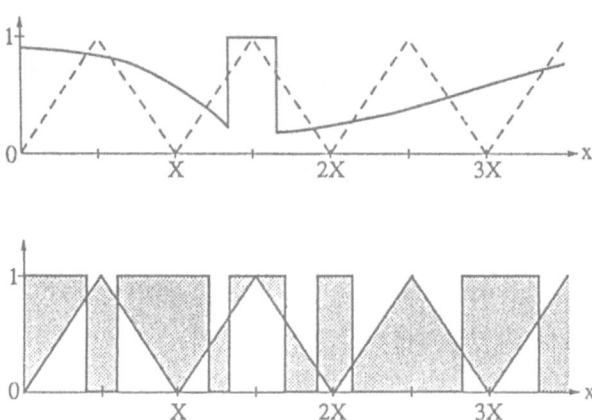

Fig. 6. Bounding envelope that contains every continuous-tone image which
when screened would result in the halftone image shown in Fig. 3.

$$C_1 = \{h(x,y): \quad 0 \leq h(x,y) < s(x,y) \text{ if } g(x,y) = 0, \text{ and}$$
$$s(x,y) \leq h(x,y) \leq 1 \text{ if } g(x,y) = 1\}. \tag{8}$$

From Fig. 6, it is apparent that this is not sufficient information to, by itself, permit recovery of $f(x,y)$. Many different images lie within the bounding set, including the halftone image $g(x,y)$! What distinguishes the continuous-tone original $f(x,y)$ from most of these potential candidates is its smoothness. There are a number of ways to express the fact that we are looking for a generally smooth image, which when screened will yield the known halftone image. In many image recovery problems, this smoothness constraint is accounted for by requiring that the recovered image be bandlimited. However, the spectra of typical continuous-tone images such as that shown in Fig. 5, do not exhibit a sharp cutoff. Instead, the spectrum rolls off gradually as a function of increasing spatial frequency. This type of behavior is better expressed in terms of a bound on the spectral magnitude.

Thus we require that the reconstructed image belong to the constraint set

$$C_2 = \{h(x,y): \ |H(u,v)| \leq B(u,v)\}, \tag{9}$$

where $B(u,v)$ is a suitable bounding function. The tightest possible bounding function would, of course, be $|F(u,v)|$; but this is not available to us. Instead, we might choose

$$B(u,v) = \begin{cases} G(u,v), & \rho = (u^2 + v^2)^{1/2} \leq \rho_0, \\ G(\rho_0 u/\rho, \rho_0 v/\rho)\rho^{-\alpha}, & \text{else.} \end{cases} \tag{10}$$

The rationale behind this choice is that at low frequencies we may expect $G(u,v) \simeq F(u,v)$; so we use this as our bound from the origin out to radial frequency ρ_0. Beyond this point, the bound decreases according to the inverse of the α-th power of the radial frequency. To completely determine $B(u,v)$, we still need to specify the parameters ρ_0 and α.

The sets C_1 and C_2 contain all the information that we presume to know about the continuous-tone original image f. To recover f, we seek an image which lies in the intersection of C_1 and C_2. In general, this intersection will contain more than one image; so we cannot expect to uniquely reconstruct f. However, we will be content to reconstruct a continuous-tone image that is close to f, especially as perceived by a human viewer. Projection operators provide the means for finding such an image.

An operator P is said to be a *projection operator* for the constraint set C if for any image h it satisfies the following two conditions

1. $Ph \in C$,
2. $||Ph - h|| \leq ||d - h|| \; \forall \; d \in C$,

i.e. P maps any image h to the closest member of the set C. If $h \in C$, then P is an identity operator. Here the norm $||\cdot||$ is the usual L_2 norm. It is straightforward to find projection operators P_1 and P_2, respectively, for the sets C_1 and C_2 given by (8) and (9). They are

$$P_1 h(x,y) = \begin{cases} h(x,y), & 0 \leq h(x,y) < s(x,y) \text{ and } g(x,y) = 0, \\ & s(x,y) \leq h(x,y) \leq 1 \text{ and } g(x,y) = 1, \\ \\ s(x,y), & 0 \leq h(x,y) < s(x,y) \text{ and } g(x,y) = 1, \\ & s(x,y) \leq h(x,y) \leq 1 \text{ and } g(x,y) = 0, \\ \\ 0, & h(x,y) < 0, \\ \\ 1, & 1 < h(x,y), \end{cases} \qquad (11)$$

and

$$P_2 H(u,v) = \begin{cases} H(u,v), & |H(u,v)| \leq B(u,v), \\ \\ B(u,v) \dfrac{H(u,v)}{|H(u,v)|}, & \text{else.} \end{cases} \qquad (12)$$

Image recovery by alternating projections onto constraint sets then proceeds as follows. We choose an initial estimate $f_0(x,y)$ for the unknown continuous-tone image. A reasonable choice is to let $f_0(x,y) = g(x,y)$. We then carry out the iteration

$$f_{k+1}(x,y) = P_1 \mathscr{F}^{-1} P_2 \mathscr{F} f_k(x,y), \qquad (13)$$

where $f_k(x,y)$ is our estimate of the continuous-tone image after k iterations.

The iteration can be accelerated by under-relaxation. To do this, we replace each projection operator P_i in (13) by

$$\tilde{P}_i = (1 - \lambda) I + \lambda P_i. \qquad (14)$$

Here I is the identity operator; and λ is a scalar constant. If $\lambda = 1$, $\tilde{P}_i = P_i$; and $\tilde{P}_i$ is still a projection. If $0 < \lambda < 1$, $\tilde{P}_i$ maps h to a new image somewhere between h and the closest point on the boundary of C_i. As λ decreases from 1 to 0, the projection operator is increasingly relaxed. At $\lambda = 0$, it is maximally relaxed, and does nothing to the image h. On the other hand, if $\lambda > 1$, $\tilde{P}_i$ takes h beyond the closest point on the boundary of C_i to a point in its interior. We refer to this condition as being under-relaxed.

Youla and Webb [14] showed that provided the constraint sets are convex, weak convergence of the iteration (13) is assured if $0 < \lambda < 2$. A constraint set C is said to be *convex* if $\forall$ images d and h $\in$ C, and any value of the scalar $0 \leq \mu \leq 1$, the image $\mu d + (1-\mu)h$ also $\in$ C. It is straightforward to show that the sets C_1 and C_2 given by (8) and (9) are both convex.

Estimating the Screen Function. As described in the previous section, recovery of the continuous-tone original image from its halftone version by alternating projections onto constraint sets requires knowledge of the screen function. This is in contrast to image recovery by low-pass filtering which requires no such knowledge. A wide variety of screen functions are used to produce halftone images. In general, we cannot expect to know *a priori* the screen function corresponding to a particular halftone image. However, it is reasonable to expect that we could estimate the screen function directly from the halftone image.

First, let us consider how the screen function may be obtained from the dot profile. If we envision stacking the dot patterns shown in Fig. 4, starting with absorptance $b = 0$, and adding the patterns corresponding to increasing b, until we reach $b = 1$, the surface which forms the boundary between the black volume on top and the white volume underneath it corresponds to the screen function. More precisely, it follows from (2) that

$$s(x,y) = 1 - \int_0^1 p[x,y;\beta]d\beta. \tag{15}$$

Note that any dot profile function corresponding to a halftone process implemented by screening must obey the *stacking property* [15], *i.e.* for fixed (x,y), $p[x,y;b_0] = 0, \Rightarrow \quad p[x,y;b] = 0,$ $\forall \ 0 \leq b < b_0$; and $p[x,y;b_0] = 1, \Rightarrow \quad p[x,y;b] = 1, \forall \ b_0 < b \leq 1$.

Thus, given the dot profile function, we can determine the screen function. Now consider a typical halftone image such as that shown in Fig. 1. Suppose we segment it into individual halftone cells. This should be a relatively straightforward task, given the strongly periodic structure in the image. Within each halftone cell, we calculate the average absorptance b. For each absorptance $0 < b_0 < 1$, we average the dot patterns within all cells that have average absorptance b_0. This provides an estimate of one period of the dot profile function for absorptance b_0. If the continuous-tone original image had constant value within each halftone cell, this procedure would generate the dot profile function exactly. However, in general, the continuous-tone image will vary within many of the halftone cells. Within these cells, the partial dot structure will not match that of the dot profile function for the same average absorptance. Including these dot patterns in the average will degrade the estimate of the dot profile function. To eliminate the effect of these partial dots, we could calculate the median rather than the average of the dot patterns within all cells with the same average absorptance. An alternative would be to base the estimate only on dot patterns from halftone cells for which the average absorptance was the same as that in all neighboring halftone cells. This would make it very probable that the continuous-tone image was, in fact, constant over all halftone cells used in the estimation procedure.

Conclusions. Screening is a widely used method for generating halftone images. Frequency domain dispersion and subsequent aliasing limits the success of low-pass filtering as a means of recovering the continuous-tone original image from its halftone version. Alternating projections onto constraint sets is well matched to the nonlinear, periodic point-to-point screening process. In order to use alternating projections to recover the continuous-tone original image, some assumptions must be made about the spectral structure of that image. Finally, the screen function may be estimated from the halftone image itself.

Acknowledgements: The author is with the School of Electrical Engineering, Purdue University, West Lafayette, IN 47907. He wishes to thank Thyagarajan Balasubramanian and Gregory Lesher for their assistance in generating the images in this paper.

References

[1] T. N. Cornsweet, *Visual Perception*, New York: Academic Press, 1970.

[2] R. Ulichney, *Digital Halftoning*, Cambridge MA: The MIT Press, 1987.

[3] J. D. Stoffel and J. F. Moreland, "A Survey of Electronic Techniques for Pictorial Image Reproduction," *IEEE Trans. Comm.*, Vol. COM-29, pp. 1898-1925, Dec. 1981.

[4] J. P. Allebach, "Visual Model-Based Algorithms for Halftoning Images," *Proceedings of the 25th Annual International SPIE* Technical Symposium, San Diego, CA, August 24-28, 1981, Vol. 310, pp. 151-158.

[5] H. S. Black. *Modulation Theory*, New York: Van Nostrand, 1953.

[6] R. W. Floyd and L. Steinberg, "An Adaptive Algorithm for Spatial Greyscale," *Proc. SID*, Vol. 17, pp. 75-77, 2nd Quarter, 1976.

[7] R. A. Ulichney, "Dithering with Blue Noise," *Proc. IEEE*, Vol. 76, January 1988.

[8] K. D. Sauer and J. P. Allebach, "Iterative Reconstruction of Bandlimited Images from Nonuniformly Spaced Samples," *IEEE Trans. Circuits Syst*, Vol. CAS-34, pp. 1497-1506, December 1987.

[9] C. A. Weber and J. P. Allebach, "Reconstruction of Frequency-Offset Fourier Data by Alternating Projection onto Constraint Sets," *Proceedings of the 24th Annual Allerton Conference on Communication, Control, and Computing*, Monticello, IL, October 1-3, 1986, pp. 194-203.

[10] J. P. Allebach and D. W. Sweeney, "Iterative Approaches to Computer-Generated Holography," *Proceedings of the 1988 SPIE International Symposium on Optoelectronics and Laser Applications in Science and Engineering*, Los Angeles, CA, January 10-15, 1988, Vol. 884, pp. 2-9.

[11] H. Stark, ed. *Image Recovery: Theory and Application*, New York: Academic Press, 1987.

[12] D. Kermisch and P. G. Roetling, "Fourier Spectrum of Halftone Images," *J. Opt. Soc. Am.*, Vol. 65, pp. 716-720, June 1975.

[13] J. P. Allebach and B. Liu, "Analysis of Halftone Dot Profile and Aliasing in the Discrete Binary Representation of Images," *J. Opt. Soc. Am.*, Vol. 67, pp. 1147-1154, Sept. 1977.

[14] D. C. Youla and H. Webb, "Image Restoration by the Method of Convex Projections - Part 1," *IEEE Trans. Med. Imaging*, Vol. MI-1, pp. 81-94, Oct. 1982.

[15] P. D. Wendt, E. J. Coyle, and N. C. Gallagher, "Stack Filters," *IEEE Trans. Acoust. Speech Signal Process.*, Vol. ASSP-34, pp. 898-911, August 1986.

Design of Two-Dimensional FIR Digital Filters
by Using the Singular-Value Decomposition

Wu-Sheng Lu, Hui-Ping Wang, and Andreas Antoniou

University of Victoria

Introduction. The design of 2-D digital filters using the singular-value decomposition (SVD) and other similar decompositions has been investigated by a number of researchers [1] [2] [3] [4]. This design method has several advantages. First, the design can be accomplished by designing a set of 1-D subfilters and, therefore, the many well-established techniques for the design of 1-D filters can be employed; second, the resulting 2-D filter is stable as long as all the 1-D subfilters involved are stable; and third, the designs obtained can be fast since the 1-D subfilters form a parallel structure and, therefore, allow parallel processing. As pointed out in [3], either infinite-impulse response (IIR) or finite-impulse response (FIR) 2-D filters can be designed by using the SVD method. While high selectivity can be achieved by using low-order IIR designs for the parallel 1-D subfilters, zero phase is required for each subfilter. This necessitates data transpositions at the inputs and outputs of subfilters and, as a result, the usefulness of these designs is limited to nonreal-time applications, where the delay introduced in the processing is unimportant.

In this paper, the SVD method reported in [3] is applied in conjunction with 1-D FIR techniques for the design of 2-D quadrantally symmetric FIR filters. It is shown that three realizations are possible and in each case, causal, linear-phase designs can easily be obtained. The first realization is based on the direct application of the SVD to the required amplitude response, the second realization is a modified version of the first, and the third realization involves the application of the SVD in conjunction with the LU decomposition (LUD). It is shown that in the last two realizations, the approximation error introduced by the application of the SVD can be reduced to an insignificant level without increasing the number of parallel sections required.

Direct SVD Realization. A 2-D FIR digital filter with support in the rectangle defined by $-(N_i - 1)/2 < n_i < (N_i - 1)/2$, $i = 1$, 2, can be characterized by the transfer function

$$H(z_1, z_2) = \sum_{n_1=-(N_1-1)/2}^{(N_1-1)/2} \sum_{n_2=-(N_2-1)/2}^{(N_2-1)/2} h(n_1, n_2) z_1^{-n_1} z_2^{-n_2} \tag{1}$$

where $h(n_1, n_2)$ is its impulse response. If $h(n_1, n_2)$ is real and

$$h(n_1,\ n_2) = h(-n_1,\ -n_2) \tag{2}$$

then the frequency response of the filter

$$H(e^{j\omega_1 T_1},\ e^{j\omega_2 T_2}) = \sum_{n_1=-(N_1-1)/2}^{(N_1-1)/2} \sum_{n_2=-(N_2-1)/2}^{(N_2-1)/2} h(n_1,\ n_2) e^{-j\omega_1 n_1 T_1} e^{-j\omega_2 n_2 T_2}$$

$$\equiv X(\omega_1,\ \omega_2) \tag{3}$$

is a real function which is even with respect to ω_1 and ω_2.

For a quadrantally symmetric filter, (1) can be rewritten as [5]

$$H(z_1,\ z_2) = \sum_{i=1}^{K} F_i(z_1) G_i(z_2) \tag{4}$$

where $F_i(z_1)$ and $G_i(z_2)$ are transfer functions of 1-D subfilters in the z_1 and z_2 domains, respectively. If these subfilers are FIR filters with support in the rectangle defined by $-(N_i-1)/2 < n_i < (N_i-1)/2$, $i = 1,\ 2$, we have

$$F_i(z_1) = \sum_{n_1=-(N_1-1)/2}^{(N_1-1)/2} f_i(n_1) z_1^{-n_1} \tag{5}$$

$$G_i(z_2) = \sum_{n_2=-(N_2-1)/2}^{(N_2-1)/2} g_i(n_2) z_2^{-n_2} \tag{6}$$

and if $F_i(z_1)$ and $G_i(z_2)$ are assumed to represent zero-phase filters, then their frequency responses are given by

$$F_i(e^{j\omega_1 T_1}) = \sum_{n_1=-(N_1-1)/2}^{(N_1-1)/2} f_i(n_1)\ e^{-j\omega_1 n_1 T_1} \equiv \Phi_i(\omega_1) \tag{7}$$

$$G_i(e^{j\omega_2 T_2}) = \sum_{n_2=-(N_2-1)/2}^{(N_2-1)/2} g_i(n_2)\ e^{-j\omega_2 n_2 T_2} \equiv \Gamma_i(\omega_2) \tag{8}$$

where $\Phi_i(\omega_1)$ and $\Gamma_i(\omega_2)$ are real functions which are even with respect to ω_1 and ω_2, respectively. Consequently, from (4) - (8), the frequency response for a quadrantally symmetric 2-D filter can be written as

$$H(e^{j\omega_1 T_1},\ e^{j\omega_2 T_2}) = \sum_{i=1}^{K} F_i(e^{j\omega_1 T_1}) G_i(e^{j\omega_2 T_2})$$

$$= \sum_{i=1}^{K} \Phi_i(\omega_1)\ \Gamma_i(\omega_2) \tag{9}$$

On comparing (9) with (3), we obtain

$$X(\omega_1,\ \omega_2) = \sum_{i=1}^{K} \Phi_i(\omega_1)\ \Gamma_i(\omega_2) \tag{10}$$

Now assume that matrix $A = \{a_{l,m}\}$ represents a desired frequency response, i.e.

$$X(\frac{\pi\mu_l}{T_1},\ \frac{\pi\nu_m}{T_2}) = a_{l,m},\ 1 \leq l \leq L \text{ and } 1 \leq m \leq M \tag{11}$$

where μ_l and ν_m are normalized frequencies such that

$$\mu_l = \frac{l-1}{L-1}, \quad \nu_m = \frac{m-1}{M-1}$$

and $0 \leq \mu_l \leq 1, 0 \leq \nu_m \leq 1$. The SVD of A gives

$$A = \sum_{i=1}^{r} \sigma_i u_i v_i^T = \sum_{i=1}^{r} \tilde{u}_i \tilde{v}_i^T \tag{12}$$

where σ_i are the singular values of A such that $\sigma_1 \geq \sigma_2 \geq \cdots \geq \sigma_r$, u_i is the ith eigenvector of AA^T associated with the ith eigenvalue σ_i^2, v_i is the ith eigenvector of $A^T A$ associated with the ith eigenvalues σ_i^2, r is the rank of A, and $\tilde{u}_i = \sigma_i^{\frac{1}{2}} u_i$, $\tilde{v}_i = \sigma_i^{\frac{1}{2}} v_i$.

On comparing (10) with (12), and assuming that $\tilde{u}_i$ and $\tilde{v}_i$ are sampled versions of the frequency responses of the 1-D filters characterized by $F_i(z_1)$ and $G_i(z_2)$, respectively, a 2-D zero-phase FIR filter can be designed by designing two sets of 1-D zero-phase FIR subfilters characterized by $F_i(z_1)$ and $G_i(z_2)$, where $1 \leq i \leq r$. A 2-D causal linear-phase filter can readily be obtained by shifting the impulse response by $(N_1 - 1)/2$ and $(N_2 - 1)/2$ with respect to the n_1 axis and n_2 axis, respectively. This can be accomplished by multiplying $F_i(z_1)$ and $G_i(z_2)$ by $z_1^{-(N_1-1)/2}$ and $z_2^{-(N_2-1)/2}$, respectively.

In practice, it may be possible to neglect all singular values σ_i for $i > K$ without introducing a large approximation error in the design. The solution obtained in such a case is a minimal mean-square-error approximation for any given value of K in the range 1 to r, as was demonstrated in [3]. Obviously, as $K \to r$ the approximation error in the design of the 2-D filter is reduced to that introduced in the design of the 1-D FIR filter, but the number of multiplications required in the realization becomes very large.

The design of the 2-D filter can be completed by using any one of the standard methods for the design of 1-D FIR filters. Using the Fourier series method in conjunction with the window technique [6], designs can be obtained very quickly with a small amount of computational effort. These design are not optimal although the approximation error can be made arbitrarily small by increasing the order of the 1-D filters used. On the other hand, by using methods based on the Remez algorithm [7], it may be possible to obtain optimal designs although a large amount of computation would be required.

If $N_1 = N_2 = N$, the number of the multiplications needed in the direct SVD realization is $2K\bar{N}$, where

$$\bar{N} = \begin{cases} N/2 & \text{if } N \text{ is even} \\ (N+1)/2 & \text{if } N \text{ is old} \end{cases}$$

Integer K $(1 \leq K \leq r)$ is the number of singular values that must be used in the design to reduce the approximation error introduced by the SVD to an insignificant amount.

Modified SVD Realization. The modified SVD realization can be obtained by manipulating the transfer function of the 2-D filter. From (4) - (6), we can write

$$H(z_1,\ z_2) \;=\; \sum_{i=1}^{K}[\sum_{n_1=-(N-1)/2}^{(N-1)/2} f_i(n_1)z_1^{-n_1}][\sum_{n_2=-(N-1)/2}^{(N-1)/2} g_i(n_2)z_2^{-n_2}]$$

$$=\; \sum_{n_1=-(N-1)/2}^{(N-1)/2}\sum_{n_2=-(N-1)/2}^{(N-1)/2}[\sum_{i=1}^{K} f_i(n_1)g_i(n_2)]z_1^{-n_1}z_2^{-n_2} \tag{13}$$

and if we let

$$c(n_1,\ n_2) = \sum_{i=1}^{K} f_i(n_1)g_i(n_2) \tag{14}$$

we have

$$H(z_1,\ z_2) = \sum_{n_1=-(N-1)/2}^{(N-1)/2}\sum_{n_2=-(N-1)/2}^{(N-1)/2} c(n_1,\ n_2)z_1^{-n_1}z_2^{-n_2} \tag{15}$$

The SVD of matrix $C = \{c(n_1,\ n_2)\}$ can be written as

$$C = \sum_{i=1}^{r_e}\sigma_{ci}u_{ci}v_{ci}^{T} = \sum_{i=1}^{r_e}\tilde{u}_{ci}\tilde{v}_{ci}^{T} \tag{16}$$

where r_e is the rank of C. Combining (15) and (16), we have

$$H(z_1,\ z_2) = \sum_{n_1=-(N-1)/2}^{(N-1)/2}\sum_{n_2=-(N-1)/2}^{(N-1)/2}\sum_{i=1}^{r_e}\tilde{u}_{ci}(n_1)\tilde{v}_{ci}(n_2)z_1^{-n_1}z_2^{-n_2} = \sum_{i=1}^{r_e}\tilde{F}_{ci}(z_1)\tilde{G}_{ci}(z_2) \tag{17}$$

where

$$\tilde{F}_{ci}(z_1) = \sum_{n_1=-(N-1)/2}^{(N-1)/2}\tilde{u}_{ci}(n_1)z_1^{-n_1} \tag{18}$$

$$\tilde{G}_{ci}(z_2) = \sum_{n_2=-(N-1)/2}^{(N-1)/2}\tilde{v}_{ci}(n_2)z_2^{-n_2} \tag{19}$$

Therefore, a modified SVD realization can be obtained by connecting r_e 2-D zero-phase filters in parallel, where each 2-D filter consists of two cascaded 1-D FIR filters represented by $\tilde{F}_{ci}(z_1)$ and $\tilde{G}_{ci}(z_2)$.

By (14), $C = \sum_{i=1}^{K} f_i g_i^T = [f_1 \cdots f_K][g_1 \cdots g_K]^T$ where $f_i = [f_i(\frac{-(N-1)}{2}) \cdots f_i(\frac{(N-1)}{2})]^T$ and $g_i = [g_i(\frac{-(N-1)}{2}) \cdots g_i(\frac{(N-1)}{2})]^T$, and it follows that the rank of C satisfies the inequality $r_e \leq K$. Moreover, since each f_i and g_i are mirror-image symmetric, matrix C is quadrantally symmetric. Consequently, there are at most $\bar{N}$ linearly independent row (or column) vectors in C and, therefore

$$r_e \leq \min\ (\bar{N},\ K) \tag{20}$$

In the modified SVD realization, vectors $\tilde{u}_{ci}$ and $\tilde{v}_{ci}$ in (16) are all mirror-image symmetric, as shown in Appendix A.1. Consequently, $\tilde{F}_{ci}(z_1)$ and $\tilde{G}_{ci}(z_2)$ represent 1-D zero-phase FIR filters.

As a result, the number of multiplications required is $2r_e\bar{N}$, which is always less than $2K\bar{N}$, the number of multiplications required in the direct SVD realization. Also notice that by (20), $2r_e\bar{N}$ has an upper bound $2\bar{N}^2$. In other word, no matter how many singular values are used in the design, the modified SVD realization scheme described here requires fewer multiplications than the direct SVD realization. A corresponding, 2-D causal linear-phase realization can be obtained by multiplying $\tilde{F}_{ei}(z_1)$ and $\tilde{G}_{ei}(z_2)$ by $z_1^{-(N-1)/2}$ and $z_2^{-(N-1)/2}$, respectively.

SVD-LUD Realization. The realization of 2-D FIR linear-phase filters using the SVD and LUD is similar to that of the modified SVD realization scheme but, as will be shown, it is more economical in terms of the number of multiplications required. Instead of decomposing C using the SVD, the LUD is performed [8, Ch. 4] such that

$$C = L_e U_e \tag{21}$$

where L_e and $U_e \in R^{N \times N}$ are the lower- and upper-triangular matrices, respectively. Since matrix C given by (16) is quadrantally symmetric, it can be shown that L_e and U_e have the forms of

$$L_e = \begin{bmatrix} * & 0 & & \cdots & & 0 \\ * & * & 0 & & \cdots & 0 \\ * & \cdots & * & 0 & \cdots & 0 \\ \vdots & & \vdots & \vdots & & \vdots \\ * & \cdots & * & 0 & \cdots & 0 \\ * & * & 0 & & \cdots & 0 \\ * & 0 & & & \cdots & 0 \end{bmatrix} \quad \text{and} \quad U_e = \begin{bmatrix} * & * & * & \cdots & * & * & * \\ 0 & * & \vdots & & \vdots & * & 0 \\ & 0 & * & \cdots & * & 0 & \\ & & 0 & \cdots & 0 & & \\ \vdots & \vdots & \vdots & & \vdots & \vdots & \vdots \\ 0 & 0 & 0 & \cdots & 0 & 0 & 0 \end{bmatrix}$$

respectively, (see Appendix A.2) where the nonzero columns (rows) in L_e (U_e) are also mirror-image symmetric.

Now if we let $Z_i = [z_i^{(N-1)/2}, \cdots, 1, \cdots, z_i^{-(N-1)/2}]^T$, $i = 1, 2$, then (15) can be written as

$$H(z_1, z_2) = Z_1 L_e U_e Z_2^T = \sum_{i=1}^{r_e} L_{ei}(z_1) U_{ei}(z_2) \tag{22}$$

where

$$L_{ei}(z_1) = \sum_{n_1 = i-(N+1)/2}^{(N+1)/2-i} L_e(n_1, i) z_1^{-n_1} \tag{23}$$

$$U_{ei}(z_2) = \sum_{n_2 = i-(N+1)/2}^{(N+1)/2-i} U_e(i, n_2) z_2^{-n_2} \tag{24}$$

A corresponding, 2-D, causal, linear-phase realization can be obtained by multiplying $L_{ei}(z_1)$ and $U_{ei}(z_2)$ by $z_1^{-(N-1)/2}$ and $z_2^{-(N-1)/2}$, respectively.

From (22) - (24), it follows that the number of multiplications required in the SVD-LUD realization is $r_e(2\bar{N} - r_e + 1)$, which is always less that $2r_e\bar{N}$, the number of multiplications required in the modified SVD method. Also note that by (20), $r_e(2\bar{N}-r_e+1)$ has an upper bound $\bar{N}(\bar{N}+1)$ which is less than $2\bar{N}^2$, the upper bound for the modified SVD method. Consequently, the SVD-LUD method leads to the most economical realization. The numbers of multiplications required by the three realization schemes are given in Table I.

TABLE I

Numbers of Multiplications Required by the Realization Schemes

Realizations	Direct SVD	Modified SVD	SVD-LUD
No. of Multipli.	$2K\bar{N}$	$2r_e\bar{N}$	$r_e(2\bar{N} - r_e + 1)$
Upper Bound	$2r\bar{N}$	$2\bar{N}^2$	$\bar{N}(\bar{N}+1)$

Examples. In this section, the designs of a bandpass and a fan 2-D filter are presented to illustrate the effectiveness of the proposed method. The steps needed to complete the design are as follows:

- Specify the desired amplitude response and thereby obtain the corresponding sampled amplitude response matrix $\mathbf{A}$.

- Decompose matrix $\mathbf{A}$ using (12) to get $\tilde{\mathbf{u}}_i$ and $\tilde{\mathbf{v}}_i$, where $1 \leq i \leq r$.

- Design K 2-D FIR filters, each of which is obtained by designing two 1-D zero-phase FIR filters characterized by transfer functions $F_i(z_1)$ and $G_i(z_2)$ corresponding to the desired amplitude responses $\tilde{\mathbf{u}}_i$ and $\tilde{\mathbf{v}}_i$, respectively, where $1 \leq K \leq r$.

- Obtain the resulting 2-D zero-phase transfer function using (14) and (15).

- Multiply the resulting 2-D zero-phase transfer function by $z_1^{-(N-1)/2}$ and $z_2^{-(N-1)/2}$, to obtain a linear-phase causal transfer function.

The desired amplitude response of a circularly-symmetric, 2-D, bandpass FIR filter, shown in Fig. 1(a), is specified by

$$|H_1(e^{j\omega_1 T_1}, e^{j\omega_2 T_2})| = \begin{cases} 0 & 0 \leq \sqrt{\omega_1^2 + \omega_2^2} \leq \omega_{a_1} \\ 1 & \omega_{p_1} \leq \sqrt{\omega_1^2 + \omega_2^2} \leq \omega_{p_2} \\ 0 & \omega_{a_2} \leq \sqrt{\omega_1^2 + \omega_2^2} \leq \pi \end{cases}$$

where $\omega_{a_1} = 0.24\pi$, $\omega_{p_1} = 0.36\pi$, $\omega_{p_2} = 0.64\pi$ and $\omega_{a_2} = 0.76\pi$.

The corresponding sampled amplitude response $A_1 = |H_1(e^{j\pi\mu_l}, e^{j\pi\nu_m})|$ can be expresed as

$$|H_1(e^{j\pi\mu_l}, e^{j\pi\nu_m})| = \begin{cases} 0 & 0 \leq \sqrt{\mu_l^2 + \nu_m^2} < \omega_{c_1}/\pi \\ 1 & \omega_{c_1}/\pi \leq \sqrt{\mu_l^2 + \nu_m^2} \leq \omega_{c_2}/\pi \\ 0 & \omega_{c_2}/\pi < \sqrt{\mu_l^2 + \mu_m^2} \leq 1 \end{cases}$$

where $\omega_{c_1} = \frac{1}{2}(\omega_{a_1} + \omega_{p_1})$ and $\omega_{c_2} = \frac{1}{2}(\omega_{a_2} + \omega_{p_2})$.

An easy-to-use, numerically reliable software package called MATLAB, has been used to perform SVD on matrix A_1 in order to obtain the necessary data for the designs in the following steps.

The 1-D FIR filters were designed by using the Fourier series method along with the window technique, and several window functions have been tried. It appears that the Kaiser window gives the best results in terms of the approximation error for a given transition band and filter order. Furthermore, as may be expected, the higher the order of the 1-D filters, the lower the approximation error. By trial and error, it has been found that a value of 29 for N gives satisfactory result for the above example.

There are 19 non-zero singular values resulting from the SVD of matrix A_1, but only the first 9 are significant. The amplitude response of the bandpass filter with 9 sections is shown in Fig. 1(b). The maximum passband and stopband errors for 9 and 19 sections are given in Table II.

The above approach has also been applied for the design of a fan filter having an amplitude response

$$|H_2(e^{j\omega_1 T_1}, e^{j\omega_2 T_2})| = \begin{cases} 1 & \omega_2 < 0.6\omega_1 - 0.02857\pi \\ 0 & \omega_2 > 0.6\omega_1 + 0.1143\pi \end{cases}$$

as depicted in Fig. 2(a).

The corresponding sampled amplitude response $A_2 = |H_2(e^{j\pi\mu_l}, e^{j\pi\nu_m})|$ can be written as

$$|H_2(e^{j\pi\mu_l}, e^{j\pi\nu_m})| = \begin{cases} 1 & \nu_m < 0.6\mu_l + 0.0457 \\ 0 & \nu_m \geq 0.6\mu_l + 0.0457 \end{cases}$$

There are 22 non-zero singular values resulting from the SVD of matrix A_2, but only the first 9 are significant. The amplitude response achieved with 9 sections is shown in Fig. 2(b). The maximum passband and stopband errors for 9 and 22 sections are given in Table II.

The number of multiplications for the direct, modified and SVD-LUD realizations for different numbers of singular values used are given in Table III. As can be seen, a significant reduction in the number of multiplications can be achieved by using the SVD-LUD realization.

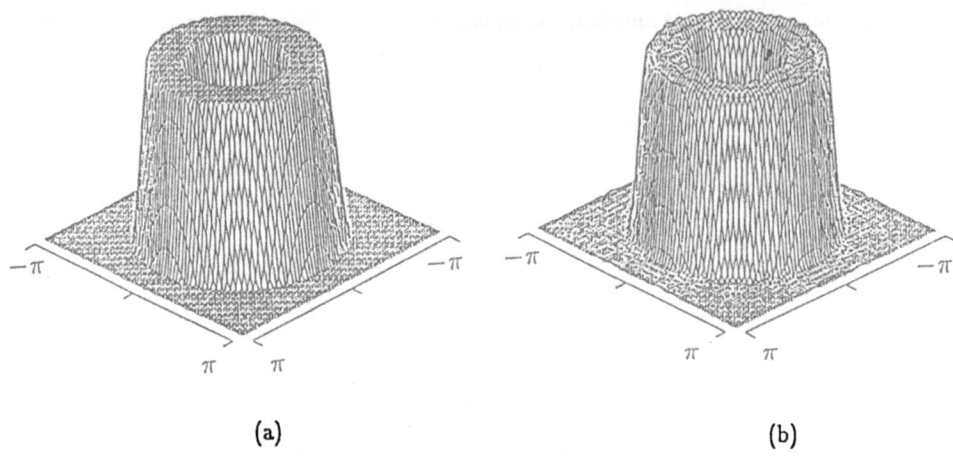

(a)   (b)

Fig. 1. Amplitude response of 2-D bandpass SVD FIR filters: (a) ideal amplitude response; (b) amplitude response obtained with 9 correction sections $N = 29$.

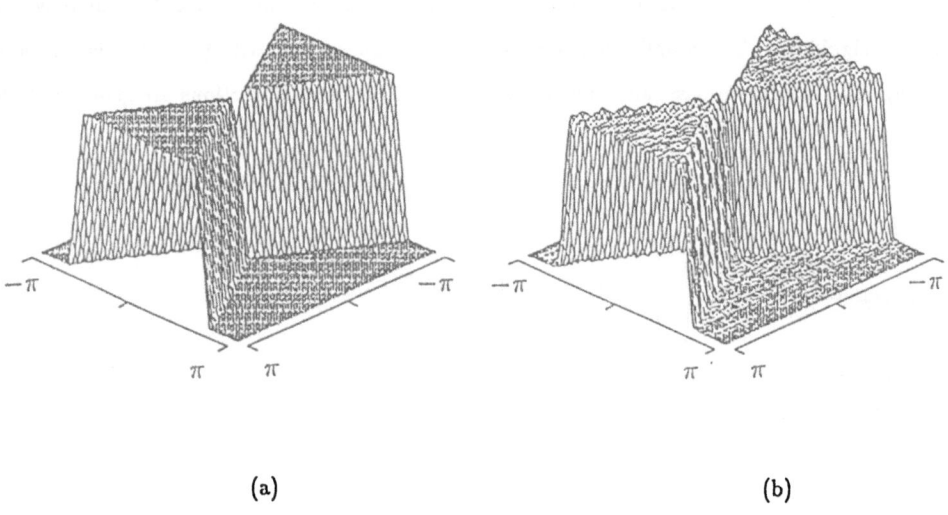

(a)   (b)

Fig. 2. Amplitude response of 2-D fan SVD FIR filters: (a) ideal amplitude response; (b) amplitude response obtained with 9 correction sections $N = 29$.

TABLE II

Approximation Errors

	Bandpass Filter		Fan Filter	
Sections Used	9	19	9	22
Max. Approx. Error in Passband	0.0399	0.0303	0.0465	0.0358
Max. Approx. Error in Stopband	0.0299	0.0263	0.0346	0.0293

TABLE III

Numbers of Multiplications

	Direct SVD	Modified SVD	SVD-LUD
9 Sections	270	270	198
19 Sections	570	450	240
22 Sections	660	450	240

Conclusions. The SVD has been applied in conjunction with 1-D FIR-filter techniques for the design of 2-D quadrantally-symmetric FIR filters. It has been shown that three related realizations are possible and in each case, causal, linear-phase designs can easily be obtained. Each realization consists of a number of parallel 1-D filters and, therefore, fast designs are possible which are suitable for real-time or quasi-real-time application.

The modified SVD realization is more efficient than the direct SVD realization and, in turn, the SVD-LUD realization is more efficient than the modified SVD realization. The last two methods have the important advantage that the number of parallel sections, and hence the number of multiplications, dose not increase as more singular values are included in the design. Consequently, the approximation error introduced by the application of the SVD can be reduced to an insignificant level.

Some examples based on the application of 1-D window techniques have shown that our method leads to very good designs with low computational effort.

Appendix A. Let $C = \{c_{i,j}, 0 \le i, j \le N-1, N \text{ odd}\}$ be an $N \times N$ quadrantally-symmetric matrix, i.e.

$$c_{i,j} = c_{i,N-1-j} = c_{N-1-i,N-1-j}$$

If a matrix $\hat{I}$ is defined by

$$\hat{I} = \begin{bmatrix} 0 & \cdots & 0 & 1 & 0 \\ 0 & \cdots & 1 & 0 & 0 \\ \vdots & \vdots & \vdots & \vdots & \vdots \\ 1 & \cdots & 0 & 0 & 0 \end{bmatrix}$$

where the size of $\hat{I}$ is $n_2 \times n_1$, and $n_1 = (N+1)/2$, $n_2 = (N-1)/2$, then matrix C can be decomposed as

$$C = \begin{bmatrix} I_{n_1} & 0 \\ \hat{I} & I_{n_2} \end{bmatrix} \begin{bmatrix} C_1 & 0 \\ 0 & 0 \end{bmatrix} \begin{bmatrix} I_{n_1} & \hat{I}^T \\ 0 & I_{n_2} \end{bmatrix} \tag{A.1}$$

where C_1 is the $N_1 \times N_1$ principal minor of C. Assume that the SVD of matrix C_1 is given by

$$C_1 = U_1 S_1 V_1^T$$

. Alternatively, matrix C can be written as

$$
\begin{aligned}
C &= \begin{bmatrix} U_1 & 0 \\ \hat{I}U_1 & 0 \end{bmatrix} \begin{bmatrix} S_1 & 0 \\ 0 & 0 \end{bmatrix} \begin{bmatrix} V_1 & 0 \\ \hat{I}V_1 & 0 \end{bmatrix}^T \\[2mm]
&= (\frac{1}{\sqrt{2}} \begin{bmatrix} U_1 & 0 \\ \hat{I}U_1 & 0 \end{bmatrix}) \begin{bmatrix} 2S_1 & 0 \\ 0 & 0 \end{bmatrix} (\frac{1}{\sqrt{2}} \begin{bmatrix} V_1 & 0 \\ \hat{I}V_1 & 0 \end{bmatrix})^T \\[2mm]
&= \begin{bmatrix} \bar{U}_1 & 0 \end{bmatrix} \begin{bmatrix} 2S_1 & 0 \\ 0 & 0 \end{bmatrix} \begin{bmatrix} \bar{V}_1 & 0 \end{bmatrix}
\end{aligned}
$$

where

$$\bar{U}_1 = \frac{1}{\sqrt{2}} \begin{bmatrix} U_1 \\ \hat{I}U_1 \end{bmatrix} \quad \text{and} \quad \bar{V}_1 = \frac{1}{\sqrt{2}} \begin{bmatrix} V_1 \\ \hat{I}V_1 \end{bmatrix} \tag{A.2}$$

If $\bar{U}_2$ and $\bar{V}_2$ are the orthonormal complements of $\bar{U}_1$ and $\bar{V}_1$, respectively, then the SVD of matrix C is given by

$$C = USV^T = \sum_{i=1}^{r_e} \sigma_i u_i v_i^T = \sum_{i=1}^{r_e} \tilde{u}_i \tilde{v}_i^T \tag{A.3}$$

where $U = [\bar{U}_1 \ \bar{U}_2]$ and $V = [\bar{V}_1 \ \bar{V}_2]$ are orthogonal matrices, u_i and v_i represent the ith column of U and V, respectively, $\tilde{u}_i = \sigma_i^{\frac{1}{2}} u_i$, $\tilde{v}_i = \sigma_i^{\frac{1}{2}} v_i$, and

$$S = \begin{bmatrix} 2S_1 & 0 \\ 0 & 0 \end{bmatrix} = \begin{bmatrix} \sigma_1 & 0 & \cdots & & 0 \\ 0 & \sigma_2 & 0 & \cdots & 0 \\ \vdots & \cdots & \sigma_{r_e} & 0 & \cdots & \vdots \\ & \cdots & \cdots & 0 & \cdots \\ & \cdots & & \vdots & 0 \\ 0 & \cdots & & & 0 \end{bmatrix}$$

By (A.2), the first r_e columns of U and V are all mirror-image symmetric and, consequently, vectors $\tilde{u}_i$ and $\tilde{v}_i$ of (A.3) are all mirror-image symmetric.

Appendix B. Equation (A.1) can be used to obtain the LUD of matrix C. As a matter of fact, if the LUD of matrix C_1 is given by

$$C_1 = L_1 U_1$$

where L_1 and U_1 are lower- and upper-triangular matrices of size N_1, then (A.1) implies that

$$C = \begin{bmatrix} I_{n_1} & 0 \\ \hat{I} & I_{n_2} \end{bmatrix} \begin{bmatrix} L_1 & 0 \\ 0 & 0 \end{bmatrix} \begin{bmatrix} U_1 & 0 \\ 0 & 0 \end{bmatrix} \begin{bmatrix} I_{n_1} & \hat{I}^T \\ 0 & I_{n_2} \end{bmatrix} = \begin{bmatrix} L_1 & 0 \\ \hat{I}L_1 & 0 \end{bmatrix} \begin{bmatrix} V_1 & U_1\hat{I}^T \\ 0 & 0 \end{bmatrix}$$

Acknowledgments. This research was supported by the Natural Sciences and Engineering Research Council of Canada.

Dept. of Electrical and Computer Engineering, University of Victoria, P.O. Box 1700, Victoria, BC, Canada, V8W 2Y2.

References

[1] S.Treitel and J. L. Shanks, "The design of multistage separable planar filters", *IEEE Trans. Geosci. Electron.*, vol. GE-9, pp. 10-27, Jan. 1971.

[2] R. E. Twogood and S. K. Mitra, "Computer-aided design of separable two-dimensional digital filters", *IEEE Trans. Acoust., Speech, Signal Processing*, vol. ASSP-25, pp. 165-169, April 1977.

[3] A. Antoniou and W.-S. Lu, "Design of two-dimensional digital filters by using the singular value decomposition", *IEEE Trans. Circuits and Syst.*, vol. CAS-34, pp. 1191-1198, Oct., 1987.

[4] A. N. Venetsanopoulos and C. L. Nikias, "Realization of two-dimensional digital filters by LU decomposition of their transfer function", *in Proc. IEEE Int. Conf. Acoust., Speech, Signal Processing*, San Diego, CA, pp. 20.4.1-20.4.4, March 1984.

[5] P. K. Rajan and M. N. S. Swamy, "Quadrantal symmetric associated with two-dimensional digital transfer functions," *IEEE Trans. Circuits and Syst.*, vol. CAS-25, pp. 340-343, June 1983.

[6] A. Antoniou, *Digital Filters: Analysis and Design*, New York: McGraw-Hill, 1979.

[7] T. W. Parks and J. H. McClellan, " Chebyshev approximation for nonrecursive digital filters with linear phase", *IEEE Trans. Circuits Theory*, vol. CT-19, pp. 189-194, March 1972.

[8] G. H. Golub and C. F. Van Loan, *Matrix Computations*, Baltimore: The Johns Hopkins University Press, 1983.

GEOMETRIC MEASURES OF ROBUSTNESS
IN SIGNAL PROCESSING

M. W. Thompson   D. R. Halverson

Colorado State University   Texas A&M University

I. Introduction

Robustness is a property which is increasingly being associated with a desirable signal processor. While such a property may present different connotations among the various constituents
of signal processing, we will confine attention here to associating robustness with a desirable
insensitivity in performance to lack of knowledge of an underlying statistical distribution function (or other important quantity) as it varies about a nominal. A traditional approach toward
imparting robustness has been to apply classical robust statistical techniques such as the saddlepoint methodology of Huber and Strassen (see, for example, [1-3]). Although tractable results
have been obtained via the application of such saddlepoint techniques (within, for example,
the areas of signal detection and estimation), there are inherent limitations associated with the
saddlepoint criterion. For example, this criterion is applicable to only certain restricted families
of noise distributions, and we might find it reasonable to doubt that nature would be so kind
as to provide a distribution within an admissible family. Moreover, the saddlepoint criterion is
inherently nonquantitative and therefore does not lend itself to the consideration of questions
concerning the quantitative tradeoff of performance and robustness. Such questions arise quite
naturally since we might intuitively expect that it would be necessary to sacrifice performance
(when the distribution takes on its nominal value) in order to gain robustness; quantitative measures of robustness as well as performance would thus be of important interest to the designer of
an appropriate signal processor. Finally, we note that the classical techniques do not readily lend
themselves to admit dependent and/or nonstationary data, a limitation of growing importance.
In the forthcoming sections we describe a new geometric approach toward measuring robustness
which transcends the aforementioned limitations.

II. The Finite Dimensional Geometric Concept

As noted above, we claim that the robustness problem may be viewed from a new perspective
which is based on geometric concepts. Let D_n denote the family of n-dimensional distribution
functions. From a geometric point of view, the performance of a signal processor is thus expressed
by considering a performance functional $P : D_n \rightarrow R$. We wish to choose a detector which
achieves a reasonably high value for P, and yet such that P does not vary much near the
nominal element of D_n. Viewing P as a height function over D_n, we could say that a desirable
robust scheme would yield a "surface" which above the nominal has an acceptable value and
is not strongly sloped. Such a perspective thus would indicate that a geometric approach to

robust signal processing might be appropriate. What would be needed would be to provide a differentiable structure to D_n so that the concept of slope would have the proper meaning. We would then be considering a height function over a differentiable manifold M which would result in a new manifold M_1 for which the Riemannian metric would yield a norm. It is the immersion of M_1 in $M \times R$, together with the Riemannian metric that offers the potential for quantifying the concept of slope, and therefore robustness.

Consider first the case where M_1 is a surface in R^{m+1}. Such a case arises, for example, when the underlying multivariate distribution function is parameterized by m parameters. In this situation the performance functional generates M_1 via a height function $h : R^m \to R$. An appropriate metric tensor $g(\cdot, \cdot)$ is inherited from the standard inner product on R^{m+1} with the obvious choice of coordinate system $\frac{\partial}{\partial y_i} = (\delta_{i1}, \delta_{i2}, \cdots, \delta_{im}, \frac{\partial h}{\partial x_i})$ leading to the components of the metric tensor given by

$$g_{ij} \triangleq g\left(\frac{\partial}{\partial y_i}, \frac{\partial}{\partial y_j}\right) = \begin{cases} \frac{\partial h}{\partial x_i} \cdot \frac{\partial h}{\partial x_j} & \text{if } i \neq j \\ 1 + \left(\frac{\partial h}{\partial x_i}\right)^2 & \text{if } i = j \end{cases} \quad .$$

Associating the slope of the surface with the cosine of the angle of the unit normal to vertical (with M_1 immersed in R^{m+1}) it is then straightforward, although somewhat lengthy, to show that at the point corresponding to the nominal distribution this cosine is given by

$$\phi_{2,m} = \cos\gamma_m = \frac{1}{\sqrt{1 + \sum_{i=1}^{m}\left(\frac{\partial h}{\partial x_i}\right)^2}} \quad .$$

Note that $\phi_{2,m}$ provides a measure of local robustness, with larger values corresponding to greater robustness. This is actually an "l_2" type measure, and suggests the alternative l_1 measure which yields

$$\phi_{1,m} = \frac{1}{1 + \sum_{i=1}^{m}\left|\frac{\partial h}{\partial x_i}\right|} \quad .$$

As an example of a specific application of this geometric concept of measuring robustness, we recall from [4] the consideration of the discrete time detection of a constant, positive signal s in dependent zero-mean Gaussian noise with the parameters being the elements of the covariance matrix Λ. Thus, we consider a hypothesis test of the following form:

$$H_0 : X_i = N_i, \quad i = 1, 2, \cdots, n$$
$$H_1 : X_i = N_i + s, \quad i = 1, 2, \cdots, n$$

where $\{N_i\}_{i=1}^{n}$ are jointly Gaussian, zero mean random variables with a nominal covariance matrix Λ. For the three sample case, we shall compare the performance and robustness of the canonical memoryless linear detector to the performance and robustness of the Huber-type detector, which is obtained by censoring the linear detector at the vertical heights $-k$ and k. Censoring of this nature is often employed in the hope of imparting robustness.

First consider the linear detector. The threshold T_L is chosen such that for a false alarm rate α,

$$\alpha = P(\sum_{i=1}^{3} N_i \geq T_L).$$

Note that the joint density of $\{N_i\}_{i=1}^{3}$ is given by

$$f_{\underline{N}}(\underline{x}) = \frac{1}{(2\pi)^{3/2}|\Lambda|^{\frac{1}{2}}}\exp\{-\frac{1}{2}\underline{x}^T\Lambda^{-1}\underline{x}\}.$$

It is then straightforward to show that

$$\alpha = \int_{-\infty}^{\infty}\int_{-\infty}^{\infty}\int_{T_L-x_1-x_2}^{\infty} f_{\underline{N}}(\underline{x})dx_1dx_2dx_3.$$

Furthermore, the detection probability is given by

$$\beta_L = \int_{-\infty}^{\infty}\int_{-\infty}^{\infty}\int_{T_L-x_1-x_2}^{\infty} f_{\underline{N}}(\underline{x}-\underline{s})dx_1dx_2dx_3,$$

where $\underline{s}^T = (s,s,s)$ and $\underline{x}^T = (x_1,x_2,x_3)$.

Lengthy but straightforward calculations then yield expressions for $\frac{\partial\beta_L}{\partial\sigma_{ij}}$, $i,j = 1,2,3$, where σ_{ij} is the i,j th element of Λ. These expressions are too cumbersome to display here, but bear a passing resemblance to those found in [5] for the parameterized Gaussian example. We then have that a convenient robustness measure for the linear detector is given by

$$\phi_{1,3}^L = \frac{1}{1 + \sum_{i=1}^{3}|\frac{\partial\beta_L}{\partial\sigma_{ij}}| + \sum_{\substack{i=1 \\ i\neq j}}^{3}\sum_{j=1}^{3}|\frac{\partial\beta_L}{\partial\sigma_{ij}}|} .$$

Evaluating this expression using the nominal parameter values $\sigma_{11} = \sigma_{22} = \sigma_{33} = 1$, $\sigma_{12} = \sigma_{21} = \sigma_{23} = \sigma_{32} = 0.5$, and $\sigma_{13} = \sigma_{31} = 0.25$, we have for a false alarm rate $\alpha = 0.05$ and signal strength $s = 0.5$ that $\beta_L = 0.157$ and $\phi_{1,3}^L = 0.83$.

A similar approach may be used when employing the Huber-type detector. For this detector, α is given by

$$\alpha = \int_{T_H-2k}^{\infty}\int_{T_H-2k}^{\infty}\int_{T_H-2k}^{\infty} f_{\underline{N}}(\underline{x})d\underline{x}$$
$$- \int_{T_H-2k}^{4k-T_H}\int_{T_H-2k}^{2k-x_1}\int_{T_H-2k}^{T_H-x_1-x_2} f_{\underline{N}}(\underline{x})d\underline{x}$$

where T_H is the threshold chosen to achieve α. It also follows that

$$\beta_H = \int_{T_H-2k}^{\infty}\int_{T_H-2k}^{\infty}\int_{T_H-2k}^{\infty} f_{\underline{N}}(\underline{x}-\underline{s})d\underline{x}$$
$$- \int_{T_H-2k}^{4k-T_H}\int_{T_H-2k}^{2k-x_1}\int_{T_H-2k}^{T_H-x_1-x_2} f_{\underline{N}}(\underline{x}-\underline{s})d\underline{x}$$

Again, lengthy calculations yield expressions for $\frac{\partial\beta_H}{\partial\sigma_{ij}}$, $i,j = 1,2,3$.

In a manner analogous to the linear detector we then may compute the robustness measure $\phi_{1,3}^H$ for the Huber-type detector. Evaluating at the aforementioned nominal parameter values, we have for $a = 0.05$, $s = 0.5$, and $k = 1$ that $\beta_{\mathrm{H}} = 0.141$ and $\phi_{1,3}^{\mathrm{H}} = 0.78$. Note that the linear detector offers both greater performance (as measured by β) and greater robustness (as measured by $\phi_{1,3}$) than the Huber-type detector, a contributing factor being the dependency present in the noise.

The presence of a parameterized noise distribution facilitates the aforementioned analysis by giving rise to a finite dimensional underlying manifold. Analogous applications for robust estimators, as well as other signal processors, could also be considered in the presence of parameterized distributions. However, for many practical problems, greater uncertainty might be expected in the available statistical knowledge, thus suggesting the need to modify our finite dimensional approach. This topic is considered in the next section.

III. The Generalized Concept

For many (if not most) applications, we would like to admit essentially arbitrary perturbations of a general nonparameterized nominal distribution. In such a situation, the locally Euclidean nature of a differentiable manifold may make difficult the identification of a differentiable structure with the family of admissible distributions. However, the performance measure for many signal processors involves integration, and in such cases a generalized concept of measuring robustness is often feasible. The basic approach is that, because of the presence of the integral, it suffices to approximate the operative distribution function with step functions (subject to a suitable partition of Euclidean space) whose levels hence parameterize the distribution. We then could define an appropriate robustness measure by taking a limit of the corresponding finite dimensional robustness measure as the norm of the partition approaches zero. Note that there is nothing "approximate" in our resultant definition; the sufficiency of the approximation is in a similar sense as that employed, for example, by Billingsley [6, p.270].

To illustrate this concept more specifically, consider first the discrete time detection of time varying (possibly random) signals in independent nonstationary noise. Then, as indicated in [4], we observe that an appropriate performance measure could be associated with either the detection probability β or the false alarm rate α, thus implying a choice of height function h which, being α or β, possesses an integral representation.

Following the approach of [4], we then choose a rectilinear partition $\mathcal{P}$ of R^n, and note that in order to extend the parameterized robustness measure to a generalized robustness measure it suffices to limit consideration to distribution functions $F_i(\cdot)$ given by step functions, i.e. functions of form

$$F_i(\cdot) = \sum_{j=1}^{m_i} a_{ij} I_{A_{ij}}(\cdot),$$

where the m_i are natural numbers, the a_{ij} are reals between zero and unity, and the A_{ij} are the bounded subsets of the reals which generate $\mathcal{P}$. Letting m denote the total number of bounded partitioning subsets of R^n, we note that the a_{ij} parameterize $F_1(\cdot), \cdots, F_n(\cdot)$, and we thus define the general robustness measure

$$\phi = \lim_{|P| \to 0} \phi_{1,m},$$

where (as is the custom for extended integrals) the existence of the limit corresponds to the same value being obtained regardless of how the union of the bounded partitioning sets approaches R^n. We then have (defining $a_{i,m_i+1} = 1$ for $i = 1, \cdots, n$)

$$h = \sum_{j_1=1}^{m_1} \cdots \sum_{j_n=1}^{m_n} \prod_{i=1}^{n} (a_{i,j_i+1} - a_{i,j_i}) I_{B_n}(x_{1,j_1}, \ldots, x_{n,j_n}) \quad,$$

where $x_{i,j_i} = \sup A_{i,j_i}$ for $i = 1, \ldots, n$ and $j_i = 1, \ldots, m_i$. Taking the appropriate partial derivatives we obtain for $k = 1, \ldots, n$ and $l = 2, \ldots, m_k$

$$\frac{\partial h}{\partial a_{k,l}} = \sum_{j_1=1}^{m_1} \cdots \sum_{j_{k-1}=1}^{m_{k-1}} \sum_{j_{k+1}=1}^{m_{k+1}} \cdots \sum_{j_n=1}^{m_n} \prod_{\substack{i=1 \\ i \neq k}}^{n} (a_{i,j_i+1} - a_{i,j_i})$$

$$\cdot [I_{B_n}(x_{1,j_1}, \ldots, x_{k-1,j_{k-1}}, x_{k,l-1}, x_{k+1,j_{k+1}}, \ldots, x_{n,j_n})$$
$$- I_{B_n}(x_{1,j_1}, \ldots, x_{k-1,j_{k-1}}, x_{k,l}, x_{k+1,j_{k+1}}, \ldots, x_{n,j_n})] \quad.$$

As in [4], we define

$$\partial_i^+ B_n = \{(x_1, x_2, \ldots, x_n) : \text{ there exists } \varepsilon > 0 \text{ such that}$$
$$(x_1, \cdots, x_{i-1}, y, x_{i+1}, \cdots, x_n) \in \overline{B_n} \text{ for } y \in (x_i - \varepsilon, x_i) \text{ and}$$
$$(x_1, \cdots, x_{i-1}, z, x_{i+1}, \cdots, x_n) \in B_n \text{ for } z \in (x_i, x_i + \varepsilon)\} \quad,$$

and

$$\partial_i^- B_n = \{(x_1, x_2, \cdots, x_n) : \text{ there exists } \varepsilon > 0 \text{ such that}$$
$$(x_1, \cdots, x_{i-1}, y, x_{i+1}, \cdots, x_n) \in B_n \text{ for } y \in (x_i - \varepsilon, x_i) \text{ and}$$
$$(x_1, \cdots, x_{i-1}, z, x_{i+1}, \cdots, x_n) \in \overline{B_n} \text{ for } z \in (x_i, x_i + \varepsilon)\} \quad.$$

Suppose that if $x \notin \partial_1^+ B_n \cap \cdots \cap \partial_n^+ B_n$ then $x \in \text{int}\,(B_n) \cup \text{int}\,(\bar{B}_n)$. For the common situation where $\partial_n^- B_n$ is empty for $i = 1, \cdots, n$, we then have

$$\phi = \frac{1}{1 + \sum_{k=1}^{n} |\int_{\partial_k^+ B_n} dF_1(x_1) \cdots dF_{k-1}(x_{k-1}) dF_{k+1}(x_{k+1}) \cdots dF(x_n)|},$$

where $F_k(\cdot)$ is the nominal distribution of X_k.

As an example of the application of the above expression for the robustness measure ϕ, consider first the employment of the linear detector, where for a set of observations $\{y_i; i = 1, 2, \cdots, n\}$ the test statistic $\sum_{i=1}^{n} y_i$ is compared to a threshold T. Note that such detector might be appropriate with identical, Gaussian nominal distributions; we are still admitting nonstationary and arbitrary perturbations about the nominal, however. This leads to a decision region B_n with the associated sets (for $i = 1, 2, \cdots, n$)

$$\partial_i^+ B_n = \{(y_1, y_2, \cdots, y_n) : \sum_{j=1}^{n} y_j = T\}$$

$$\partial_i^- B_n \quad \text{is empty} \quad.$$

We therefore obtain

$$\phi_{\text{linear}} = \frac{1}{1+n},$$

regardless of the choice of α or β as performance measure.

Note that as the number of samples n approaches infinity then ϕ_{linear} approaches zero; thus, the linear detector becomes completely "unrobust" as the number of samples approaches infinity.

On the other hand, consider the Huber-type detector which employs a censored form of the linear detector, wherein a test statistic of form $\sum_{i=1}^{n} g(y_i)$ is compared to a threshold T, with

$$g(x) = \begin{cases} x & \text{if } x \leq k \\ k & \text{if } x > k \\ -k & \text{if } x < -k \end{cases},$$

where k is a positive real number. For $n = 2$ we can succinctly specify the relevant sets

$$\partial_1^+ B_2 = \{(y_1, y_2) : y_1 = (T - k),\ y_2 > k;$$
$$\text{or } y_1 + y_2 = T,\ (T - k) \leq y_1 < k\}$$
$$\partial_2^+ B_2 = \{(y_1, y_2) : y_2 = (T - k),\ y_1 > k;$$
$$\text{or } y_1 + y_2 = T,\ (T - k) < y_1 \leq k\}$$
$$\partial_1^- B_2 \text{ and } \partial_2^- B_2 \text{ are empty.}$$

Considerably more complicated expressions can be obtained for larger values of n. Letting $\alpha_n(T)$ denote the false alarm probability for this Huber-type detector with threshold T, it follows (cf. [7]) that if the false alarm probability is chosen as the performance measure, then for a nonrandomized test,

$$\phi_{\text{Huber},\alpha} = \frac{1}{1 + n\alpha_{n-1}(T - k)}.$$

We therefore observe that $\phi_{\text{Huber}} \geq \phi_{\text{linear}}$.

Consider now the situation where we fix a desired false alarm rate $\alpha > 0$. Subscripting the threshold to indicate the sample size n, we then have for a nonrandomized test,

$$\phi_{\text{Huber},\alpha} = \frac{1}{1 + n\alpha_{n-1}(T_n - k)},$$

where, for each n, $\alpha_n(T_n) = \alpha$. Noting,

$$\alpha_{n-1}(T_n - k) = \text{Pr}_0(\sum_{i=1}^{n-1} g(X_i) + k > T_n)$$
$$\geq \text{Pr}_0(\sum_{i=1}^{n} g(X_i) > T_n) = \alpha,$$

we observe that $\lim_{n \to \infty} \phi_{\text{Huber},\alpha} = 0$, and thus, as with the linear detector, the Huber-type detector becomes completely "unrobust" as the number of samples approaches infinity.

Similar conclusions hold if detection probability is used as the performance measure. Letting $\beta_n(T)$ represent the detection probability of the Huber-type detector with n samples and threshold T, we then choose the threshold T_n to achieve a prescribed α (under the nominal distribution) and obtain for a nonrandomized test

$$\phi_{\text{Huber},\beta} = \frac{1}{1 + n\beta_{n-1}(T_n - k)}.$$

Analogous to before, we then have

$$\lim_{n\to\infty} \phi_{\text{Huber},\beta} = 0.$$

If $\partial_i^- B_n$ is not empty for some $i = 1, 2, \cdots, n$, then the preceding expression for the robustness measure ϕ must be modified. Although this can be done for various cases of interest with larger sample sizes, the procedure   is tedious to illustrate for large n and we therefore confine attention here to the two sample situation. Suppose for each $x \in R$ one of the following occurs:

(1) there does not exist y such that $(x, y) \in \partial_1^+ B_2$ or $(x, y) \in \partial_1^- B_2$, or

(2) there exists y such that $(x, y) \in \partial_1^+ B_2$ and $(x, y) \notin \partial_1^- B_2$, or

(3) there exists y such that $(x, y) \notin \partial_1^+ B_2$ and $(x, y) \in \partial_1^- B_2$, or

(4) there exists unique $l_1^+(x), l_1^-(x)$ such that $(x, l_1^+(x)) \in \partial_1^+ B_2$ and $(x, l_1^-(x)) \in \partial_1^- B_2$.

Furthermore, we denote the set of all x satisfying (4) by S_1 and assume that the functions l_1^+ and l_1^- are continuous on S_1. In addition, we let

$$S_1^+ = \{(x, y) : x \in S_1 \text{ and } (x, y) \in \partial_1^+ B_2\}$$

$$S_1^- = \{(x, y) : x \in S_1 \text{ and } (x, y) \in \partial_1^- B_2\}.$$

Analogously, we can define the functions l_2^+ and l_2^-, as well as the sets S_2^+ and S_2^-. Note that these conditions and definitions, while appearing somewhat complex, are really quite natural when viewed from a geometric perspective and apply to many decision regions of engineering interest, including a wide variety of polygonal regions in R^2. We then have

$$\phi = \frac{1}{1 + \{|\int_{(\partial_1^+ B_2 \cup \partial_1^- B_2) - (S_1^+ \cup S_1^-)} dF_2(y)| + |\int_{(\partial_2^+ B_2 \cup \partial_2^- B_2) - (S_2^+ \cup S_2^-)} dF_1(x)| \atop + \int_{S_1} |d(F_2 \circ l_1^+)(x) - d(F_2 \circ l_1^-)(x)| + \int_{S_2} |d(F_1 \circ l_2^+)(y) - d(F_1 \circ l_2^-)(y)|\}}.$$

This expression generalizes the preceding result for the robustness measure, and could be extended to larger sample sizes (with a corresponding increase in complexity).

The methods of this section can also be applied to robust parameter estimation, as illustrated in [8]. Suppose we wish to estimate the real parameter θ based on data which are realizations of the independent random variables $X_1, X_2, \cdots, X_n$ i.e., we estimate θ with $\hat{\theta} = g(X_1, X_2, \cdots, X_n)$ where $g : R^n \to R$ is Borel measurable. Letting $Q : R \to R$ be Borel measurable, we employ the performance measure $P = E\{Q(\theta - \hat{\theta})\}$. Such a general performance measure is desirable since, as noted by Hall and Wise [9], too heavy a reliance has been traditionally placed on conventional measures, such as mean square error. A "robust" estimate $\hat{\theta}$ would thus be one for which the corresponding value of P is both relatively small and insensitive to inexact knowledge of the univariate distributions of the X_i, $i = 1, 2, \cdots, n$.

To pursue the above analysis further, suppose the distribution function of X_i lies in a set containing a nominal element $F_i(\cdot)$. We are interested in analyzing what happens to P as the true distribution function varies within the set about $F_i(\cdot)$. Since P involves an expectation, it suffices to approximate the distribution function $F_i(\cdot)$ with step functions of form $\sum_{j=1}^{m_i} a_{ij} I_{A_{ij}}(\cdot)$, where the indicator functions are evaluated at sets A_{ij} generated by a partition $\mathcal{P}_i$ of the real line. The partial derivatives $\left\{\frac{\partial P}{\partial a_{ij}}\right\}_j$ then provide a measure to the degree of variation of P as the step function approximations vary about the nominal distribution $F_i(\cdot)$. We then allow the partition norm to approach zero, and consider two measures of robustness. The first is the l_2 measure

$$\phi_2 = \lim_{|\mathcal{P}| \to 0} \left[\sum_{i=1}^{n} \sum_{j=1}^{m_i} \left(\frac{\partial P}{\partial a_{ij}} \right)^2 \right]^{1/2}$$

where $|\mathcal{P}| = \max_i |\mathcal{P}_i|$.

Letting the partitioning points of $\mathcal{P}_i$ be $x_{i1}, \cdots, x_{i(m_i-1)}$, we have

$$P = \sum_{j_n=1}^{m_n-1} \cdots \sum_{j_2=1}^{m_2-1} \sum_{j_1=1}^{m_1-1} \prod_{i=1}^{n} (a_{i(j_i+1)} - a_{ij_i}) \cdot Q(\theta - g(x_{1j_1}, x_{2j_2}, \cdots, x_{nj_n})).$$

We then obtain

$$\frac{\partial P}{\partial a_{ij}} = \sum_{j_n=1}^{m_n-1} \cdots \left(\sum_{j_i=1}^{m_i-1} \right)^{\bullet} \cdots \sum_{j_1=1}^{m_1-1} \prod_{\substack{k=1 \\ k \neq i}}^{n} (a_{k(j_k+1)} - a_{kj_k})$$

$$\cdot [Q(\theta - g(x_{ij_1}, x_{2j_2}, \cdots, x_{ij_i}^{\bullet\bullet}, \cdots, x_{nj_n}))$$

$$- Q(\theta - g(x_{1j_1}, x_{2j_2}, \cdots, x_{ij_i}^{\bullet\bullet\bullet}, \cdots, x_{nj_n}))] \quad,$$

where $*$ denotes deletion, $**$ denotes replacing x_{ij_i} with $x_{i(j-1)}$, and $***$ denotes replacing x_{ij_i} with x_{ij}. This expression then implies

$$(\phi_2)^2 = 0 \cdot \sum_{i=1}^{n} \int_{-\infty}^{\infty} \left(\int_{R^{n-1}} \left(\frac{\partial Q(\theta - g(x_1, \cdots, x_n))}{\partial x_i} dF_1(x_1) \cdots dF_i(x_i)^{\bullet} \cdots dF_n(x_n) \right)^2 dx_i = 0,$$

whenever the integrals exist. We therefore observe that ϕ_2 does not lead to a useful measure because the squaring operation results in too rapid convergence to zero of the $\frac{\partial P}{\partial a_{ij}}$ as the partition norm approaches zero.

As an alternative, we consider the l_1 measure which results in

$$\phi_1 = \lim_{|\mathcal{P}| \to 0} \sum_{i=1}^{n} \sum_{j=1}^{m_i} \left| \frac{\partial P}{\partial a_{ij}} \right|.$$

Note that larger values of ϕ_1 correspond to less robustness, and smaller values of ϕ_1, correspond to greater robustness. Applying the preceding derivation, we then obtain

$$\phi_1 = \sum_{i=1}^{n} \int_{-\infty}^{\infty} \left| \int_{R^{n-1}} \frac{\partial Q(\theta - g(x_1, \cdots, x_n))}{\partial x_i} \cdot dF_1(x_1) \cdots dF_i(x_i)^{\bullet} \cdots dF_n(x_n) \right| dx_i,$$

whenever the integrals exist, where the superscript $*$ denotes omission. Note that, in particular, the integrand of ϕ_1 vanishes on sets where g is constant, thus illustrating the appeal of censoring observations of large magnitude, a method of imparting robustness often seen to arise from classical saddlepoint criteria. More generally, however, the above expression for ϕ_1 provides a quantitative way to compare the robustness of various estimators beyond those based simply on censoring. This makes possible the consideration of estimators which trade off performance and robustness in a way suitable to the designer.

For a specific application of the above expression, consider the estimation of the parameter $\mu = \sum_{j=1}^{n} \mu_j/n$, where $\mu_j = E\{X_j\}$. We employ an estimator of form

$$g(X_1, X_2, \cdots, X_n) = \sum_{j=1}^{n} h(X_j)/n,$$

where

$$h(x) = \begin{cases} x & \text{if } x \leq k \\ k & \text{if } x > k \\ -k & \text{if } x < -k \end{cases}.$$

Using first the mean square error criteria on performance, we obtain

$$\phi_1 = 2 \sum_{i=1}^{n} A_i/n^2, \quad \text{where}$$

$$A_i = \begin{cases} k^2 + \left(\sum_{\substack{j=1 \\ j \neq i}}^{n} E_N\{h(X_j)\} - n\mu \right)^2 & \text{if } \left| \sum_{\substack{j=1 \\ j \neq i}}^{n} E_N\{h(X_j)\} - n\mu \right| \leq k \\ 2k \left| \sum_{\substack{j=1 \\ j \neq i}}^{n} E_N\{h(X_j)\} - n\mu \right| & \text{otherwise} \end{cases},$$

and where $E_N\{\cdot\}$ denotes expectation computed using the nominal distributions. One consequence of this expression is that, as shown in [8], for many important situations the estimator becomes completely unrobust as the number of samples approaches infinity.

As an illustration of the employment of an alternative to the mean square error performance criterion, consider the L_1 criterion obtained by choosing

$$Q(\cdot) = |\cdot|,$$

which is almost everywhere differentiable with respect to Lebesgue measure. This choice of $Q(\cdot)$ implies

$$\phi_1 = \sum_{i=1}^{n} \int_{-\infty}^{\infty} \left| \left| \int_{R^n} \text{sgn}\left(\mu - g(x_1, x_2, \cdots, x_n)\right) I_{[-k,k]}(x_i) dF_1(x_1) \cdots dF_i(x_i)^* \cdots dF_n(x_n) \right| dx_i/n, \right.$$

where $\text{sgn}(x) = \begin{cases} 1 & \text{if } x > 0 \\ -1 & \text{if } x < 0, \end{cases}$ and thus

$$\phi_1 = \frac{1}{n} \sum_{i=1}^{n} \int_{-k}^{k} \left| \left| \int_{R^n} \text{sgn}\left(\mu - g(x_1, x_2, \cdots, x_n)\right) dF_1(x_1) \cdots dF_i(x_i)^* \cdots dF_n(x_n) \right| dx_i. \right.$$

This expression can be evaluated analytically for specific choices of the F_i, however, the analysis is tedious and numerical integration is an appealing alternative for larger sample sizes.

As a final note to this section, we remark that the aforementioned geometric methods have been local in nature, with the corresponding results applying to anticipated perturbations in the underlying distribution which are confined to a small neighborhood about the nominal. As indicated in [4], these methods can be extended in a natural way to the nonlocal situation.

IV. Conclusions

We have presented a newly originated geometric approach toward measuring the robustness of a variety of signal processors. This approach admits nonstationary and, in certain cases, dependent data in a manner which reflects the effects of essentially arbitrary perturbations in an underlying distribution about a nominal. In particular, we include extensive results for robust signal detection and parameter estimation, and observe that our expressions reveal important limitations in robustness schemes based on classical saddlepoint techniques, thus encouraging further investigations into appropriate alternatives. For example, such alternative signal processors could be designed with the goal of optimizing a weighted combination of performance and robustness, a design criterion which is of important practical interest and is not readily addressed by the nonquantitative classical saddlepoint criteria.

Acknowledgement

This research was supported by the Air Force Office of Scientific Research under Grant AFOSR-87-0087.

References

[1] P. J. Huber, "Robust estimation of a local parameter," *Ann. Math. Stat.*, vol. 35, pp. 73–101, March 1964.

[2] P. J. Huber, "A robust version of the probability ratio test," *Ann. Math. Stat.*, vol. 36, pp. 1753-58, December 1965.

[3] P. J. Huber, *Robust Statistics*, New York: Wiley, 1981.

[4] M. W. Thompson and D. R. Halverson, "A differential geometric approach toward robust signal detection in nonstationary and/or dependent noise," *Proc. Twenty-Fifth Annual Allerton Conf. on Communications, Control, and Computing*, Monticello, Illinois, September 30–October 2, 1987, pp. 163–170.

[5] M. W. Thompson and D. R. Halverson, "A novel differential geometric approach toward robust signal detection," *Proc. 1986 Conf. on Information Sciences and Systems*, Princeton, New Jersey, March 19–21, 1986, pp. 11–16.

[6] P. Billingsley, *Convergence of Probability Measures*. New York: Wiley, 1968.

[7] M. W. Thompson "Applications of robustness measures in signal detection", to appear in *Proc. 1988 International Symposium on Communications and Controls*, Baton Rouge, Louisiana, October 19–21, 1988.

[8] M. W. Thompson and D. R. Halverson, "Robust estimation of signal parameters with nonstationary data," *Proc. 1988 Conf. on Information Sciences and Systems*, Princeton, New Jersey, March 16–18, 1988, pp. 180–184.

[9] E. B. Hall and G. L. Wise, "Simultaneous optimal estimation over a family of fidelity criteria," *Proc. 1987 Conf. on Information Sciences and Systems*, Baltimore, Maryland, March 25–27, 1987, pp. 65–70.

STATE-DEPENDENT ROUTING FOR A MULTI-SERVICE NETWORK

Zbigniew Dziong   Lorne Mason
INRS-Telecommunications   INRS-Telecommunications

Introduction.   A widely recognized objective in the telecommunications industry is the design and deployment of integrated services digital networks. These networks should be capable of handling a broad range of services with heterogeneous bandwidths, holding times and set-up objectives. In the paper we deal with circuit switched multi-slot loss networks. A major problem of current interest concerns the synthesis of control procedures for these multi-service networks, which ensure efficient operation while maintaining prescribed service levels for the different traffic classes, in the face of uncertain and variable traffic demand and unpredictable facility failures. One promising approach to this problem involves the synthesis of adaptive control algorithms for call set-up, (flow control and routing), which respond on-line to measured network conditions. A number of approaches to the adaptive routing of calls in telephone networks have been proposed in the literature. These methods, which are surveyed in [1] and [2], range from decentralized adaptive schemes employing learning automata [3], through centralized time-variable but non-adaptive schemes [4] to centralized adaptive routing procedures [5]. These schemes have proven to be more efficient and robust than the fixed hierarchy, long used in telephone networks throughout the world. This has led the recent introduction of centralized schemes into the North American telecommunications network. In spite of their demonstrated success, the routing policies in place have been developed on a somewhat ad hoc basis. Important exceptions to the ad hoc approach are the papers [6], [7] and [8]. In [6] and [7], Markov decision theory was employed to compute a state dependent routing policy off-line by executing a single step of policy iteration. It was shown in [7] that this approach can produce a routing policy marginally more efficient than the DNHR centralized time variable routing scheme currently implemented in the AT&T network. In [8], Kelly has applied the notion of link shadow prices from linear programming to the problem of decentralized adaptive load sharing in a telephone network. The link shadow price is interpreted as a price paid for carrying a call on the link. Recently, in [2] and [9], it was shown how this approach can be generalized to state-dependent link shadow prices, and was demonstrated by way of simulation how this approach leads to improved network performance in the case of a single traffic class. The maximum revenue criterion was introduced in [2] and [9], which can be viewed as a generalization of the standard maximum throughput criterion, as the latter case corresponds to setting the revenue parameter to unity. The call revenue parameter provides an important additional vehicle for controlling the grade of service for different traffic classes. Note that the last feature is a central issue in the control of multi-service networks. This feature provided motivation to develop a flow control and routing method, for such networks, on the basis of revenue maximization and the link shadow price concept.

In the first part of the paper (Sections 1,2), the main control model is derived within the framework of Markov decision theory. Benes [10] originally applied Markov decision theory to circuit switched networks, however the exact state model he employed led to enormous complexity, putting computation of an optimal control beyond reach. In the approach described here, we approximate the true state distribution by assuming state dependent Poisson link arrivals and link independence. The ensuing simplification enables decomposition of the Markov decision problem into a set of link analysis problems involving the evaluation of link shadow prices. This suboptimal solution provides a significant state space and memory requirement reduction. Moreover the recursive procedure used for policy iteration, based on real time traffic measurements, also implies that the network control will track or adapt to time variable traffic demand.

In Section 3, the model for link shadow price evaluation is described. The approach is based on the limiting solution for the equivalent system with discounting. In case of a one link network this solution provides the near optimal call access control. Moreover it is shown that, under the additional assumption that the mean holding time of wideband calls is significantly greater than that for narrowband calls, the link analysis problem can be decomposed into three subproblems: evaluation of shadow prices for narrowband traffic, evaluation of shadow prices for wideband traffic and dynamic bandwidth allocation. This decomposed model provides additional reduction of numerical computation.

1 Problem formulation.

We describe the network as a set of nodes and a set of trunk groups connecting the nodes. The network is offered many classes of calls. The j-th class is characterized by the following:

- origin-destination (OD) node pair,

- number of required channels: d_j,

- intensity of arrival process (assumed to be Poissonian): λ_j,

- mean holding time (assumed to be exponentially distributed): μ_j^{-1},

- set of alternative paths (in general this set can contain all feasible paths): W_j,

- parameter r_j with a value from $(0, \infty)$ which can be interpreted as the revenue parameter of the j-th class call carried on the network (it should be emphasized that r_j need not have anything to do with the real charge for the call), the revenue from a carried call is defined by the revenue rate $q_j = r_j \mu_j$.

The network operates in a lost call mode. This means that under routing policy, π, a new call can be either carried on the path recommended by the routing policy or lost in case there is no available path or it is not efficient to carry this call. In the case where the call is accepted we assume instantaneous call set up.

The problem addressed in the paper can be formulated as follows:

- find the optimal routing policy π^* which maximizes the mean value of revenue from the network defined as:

$$\overline{R}(\pi) = \sum_j r_j \overline{\lambda}_j \tag{1}$$

where $\overline{\lambda}_j$ denotes the mean number of j-th class calls carried in the network (the process is assumed to be stationary).

In general this problem can be solved within the framework of the theory of continuous-time Markov Decision Processes (MDP). The state of the system can be described by a matrix $z = \{z_j^k\}$ where z_j^k denotes the number of j-th class calls carried on the k-th path from the set W_j. From the theory of MDP [11], it follows that since our system is ergodic, the optimal policy π^* is deterministic and can be found by the policy iteration procedure which can be stated as follows:

- for a given policy π solve the set of network equations

$$g = q(z) + \sum_j \lambda_j\, v(z_P, \pi) + \sum_j \sum_k z_j^k\, \mu_j\, v(z_k, \pi) \qquad ;\ z \in Z \tag{2}$$

for all relative values $v(z, \pi)$ and gain g by setting the relative value for a certain z to zero. $q(z)$ denotes the rate of revenue in state z given by $q(z) = \sum_j \sum_k r_j z_j^k \mu_j$, z_P denotes the new network state after j-th class call acceptance on path P in state z and z_k denotes the new network state after j-th class call departure on path k in state z.

- define the improved routing policy, π', using the relative values from the previous policy in the following way: if a new call arrives in state z, carry the call on the path P which provides maximum relative value $v(z_P, \pi)$ on condition that this relative value is larger than that of the current state z, otherwise reject the call. This improved policy π' is to be used again in the first step.

The theory of MDP ensures that starting from an arbitrary initial policy this procedure converges to π^* in a finite number of iterations.

It is important to note that the relative value can be interpreted as the expected difference between the revenue from the network in the interval (t_0, ∞), assuming state z in t_0, and the revenue in the same interval but assuming a certain arbitrary reference state z_r in t_0 (see [11]). This can be stated as follows

$$v(z, \pi) = \lim_{t \to \infty} [R(z, \pi, t) - R(z_r, \pi, t)] \tag{3}$$

where $R(z, \pi, t)$ denotes the expected revenue from the network in the interval $(t_0, t_0 + t)$, assuming state z in t_0. Then let us notice that the second step of the algorithm provides that the call, with revenue parameter r_j, is carried on the path providing the maximum increase of the expected revenue from the network (compared to the call rejection). This expected increase, for each path $P \in W_j$, can be defined as

$$g_P(z, r_j, \pi) = v(z_P, \pi) - v(z, \pi) \tag{4}$$

and henceforth we call it net-gain. Using eq. 3, the net-gain is given by

$$g_P(z, r_j, \pi) = \lim_{t \to \infty} [R(z_P, \pi, t) - R(z, \pi, t)] \tag{5}$$

For further considerations it is convenient to express net-gain in the equivalent form as

$$g_P(z, r_j, \pi) = \lim_{t \to \infty} [R_P(z, r_j, \pi, t) - R(z, \pi, t)] \tag{6}$$

where $R_P(z, r_j, \pi, t)$ denotes the expected revenue from the network in the interval $(t_0, t_0 + t)$, assuming that in t_0 the j-th type call with revenue parameter r_j was accepted on the path P when the network was in state z.

The path net gains, which depend upon the state, call revenue parameter and policy enable one to determine the optimal control policy. Unfortunately the exact solution is intractable due to the enormous cardinality of the state and policy spaces. In the following sections we describe an approach which approximates the optimal policy and reduces computational and memory requirements to manageable levels.

2 Decomposition of Markov Decision Problem.

One approach to simplify complex network problems involves decomposition into the set of link analysis problems. First we assume that link arrivals are state dependent Poisson streams and that link states are statistically independent. We observe that these assumptions are commonly made in network performance analysis and design and have proven to be one of the most powerful approaches which dramatically reduce complexity while maintaining an adequate quality of accuracy. In particular, these assumptions imply that a call connected on a path consisting of l links is decomposed into l independent link calls characterized by the same mean holding time as the original call. Then the Markov process for a given policy π can be described separately for each link in terms of the link state $x^s = \{x_j^s\}$, where x_j^s denotes the number of j-th class calls carried on link s, and the transition rates defined by the intensities $\lambda_j^s(x^s, \pi)$ and departure rates μ_j (to simplify notation, we assume that alternative paths for j-th class calls have no common links, although it should be stressed that the presented approach also covers cases with common links). The global network state is given by $y = \{x^s\}$.

Although the above mentioned assumptions provide a decomposition of the Markov process, it is not sufficient to decompose the Markov Decision problem. To do that we must also decompose the path net-gain into a set of separable link net-gains. This can be done by dividing the revenue parameter of a call offered to a multilink path between the link calls, so each link call is characterized by the link call revenue parameter evaluated according to a certain division rule. Then the decision, whether to accept or reject the divided calls, is made separately for each link call with the objective of maximizing the mean value of revenue from the link.

As it will be shown later, the division of call revenue depends on the state of the entire path. As a consequence, the link call revenue parameter can have different values for the cases where its link state is the same but the state of other links on the path is different. Thus, the link call revenue parameter for the j-th class call and each link state $\mathbf{x}^s$ is the random variable, $r_j^s(\mathbf{x}^s, \pi)$, with mean value $\bar{r}_j^s(\mathbf{x}^s, \pi)$. To facilitate analysis we assume that $r_j^s(\mathbf{x}^s, \pi)$ and $r_j^s(\mathbf{x}^{ls}, \pi)$ are statistically independent for $\mathbf{x}^s \neq \mathbf{x}^{ls}$. This approximation is justified by the fact that usually the intensity of multilink calls, from the j-th class, offered to the link s is very small compared to the total intensity of calls offered to this link.

Now, for a given routing policy, each link revenue process can be described independently enabling definition of the link net-gain, $g_j^s(\mathbf{x}^s, r_j^s(\mathbf{x}^s, \pi), \pi)$, as the expected gain from accepting the j-th class link call with revenue parameter $r_j^s(\mathbf{x}^s, \pi)$. The link net-gain can be expressed in an analogous way to eq. 6 as

$$g_j^s(\mathbf{x}^s, r_j^s(\mathbf{x}^s, \pi), \pi) = \lim_{t \to \infty} [R_j^s(\mathbf{x}^s, r_j^s(\mathbf{x}^s, \pi), \pi, t) - R^s(\mathbf{x}^s, \pi, t)] \tag{7}$$

where $R_j^s(\mathbf{x}^s, r_j^s(\mathbf{x}^s, \pi), \pi, t)$ denotes the expected revenue from the link in the interval $(t_0, t_0 + t)$, assuming that in t_0 the j-th class link call with revenue parameter $r_j^s(\mathbf{x}^s, \pi)$ was accepted when the link was in state $\mathbf{x}^s$ and $R^s(\mathbf{x}^s, \pi, t)$ denotes the expected revenue from the link in the interval $(t_0, t_0 + t)$, assuming state $\mathbf{x}^s$ in t_0.

Finally, the separable form of the path net-gain for the decomposed model is given by

$$g_P(\mathbf{y}, r_j, \pi) = \sum_{s \in P} g_j^s(\mathbf{x}^s, r_j^s(\mathbf{x}^s, \pi), \pi) \tag{8}$$

It is convenient to introduce the notion of link shadow price $p_j^s(\mathbf{x}^s, \pi)$ which is related to the link net gain and the link call revenue by the equation

$$p_j^s(\mathbf{x}^s, \pi) = r_j^s(\mathbf{x}^s, \pi) - g_j^s(\mathbf{x}^s, r_j^s(\mathbf{x}^s, \pi), \pi) \tag{9}$$

This value can be interpreted as the price for accepting j-th type call on link s. Notice that the link shadow prices are independent of the current link call revenue parameter $r_j^s(\mathbf{x}^s, \pi)$ and that the path net-gain from carrying the j-th type call on the path P can be expressed as

$$g_P(\mathbf{y}, r_j, \pi) = r_j - \sum_{s \in P} p_j^s(\mathbf{x}^s, \pi) \tag{10}$$

Note that in the decomposed model the value of mean revenue from the network, $\overline{R}'$, under the optimal routing policy depends on the revenue parameter division rule. To make this model equivalent to the original one (assuming that the used approximations are exact), first, the revenue parameter division rule should be such that all link calls offered to a path are either accepted or rejected. This condition is fulfilled if

$$sgn[g_P(\mathbf{y}, r_j, \pi)] = sgn[g_j^s(\mathbf{x}^s, r_j^s(\mathbf{x}^s, \pi), \pi)] \qquad ; \ s \in P \tag{11}$$

Also the division rule should maximize the mean value of revenue from the network. It seems difficult to find the optimal division rule directly from this argument. Instead we impose the natural condition that in all cases where a divided call is offered to a link, the revenue parameter assigned to this link should only depend on the call revenue parameter r_j, the link shadow price $p_j^s(\mathbf{x}^s, \pi)$ and the path net-gain $g_P(\mathbf{y}, r_j, \pi)$. The only general division rule that fulfills both conditions is given by

$$r_j^s(\mathbf{x}^s, \pi) = r_j \frac{p_j^s(\mathbf{x}^s, \pi)}{r_j - g_P(\mathbf{y}, r_j, \pi)} \tag{12}$$

By using eq. 10 the rule can be rewritten as

$$r_j^s(\mathbf{x}^s, \pi) = r_j \frac{p_j^s(\mathbf{x}^s, \pi)}{\sum_{s \in P} p_j^s(\mathbf{x}^s, \pi)} \tag{13}$$

It is worth noting that this rule has the very natural interpretation that the revenue parameter assigned to the s-th link should be proportional to the price, $p_j^s(\mathbf{x}^s, \pi)$, paid for carrying the link call on this link.

In the next section we define a model for evaluating the link shadow prices, assuming that the link intensities $\lambda_j^s(\mathbf{x}^s, \pi)$ and mean values of link call revenue parameters $\bar{r}_j^s(\mathbf{x}^s, \pi)$ are given. Concerning the evaluation of $\lambda_j^s(\mathbf{x}^s, \pi)$ and $\bar{r}_j^s(\mathbf{x}^s, \pi)$ this can be done in two ways. One is to develop an analytical model for performance analysis of the network with the given routing policy π. This model should be based on the Markov process decomposition approach. The second possibility is to install a measurement system in the network providing statistics which enable estimation of $\lambda_j^s(\mathbf{x}^s, \pi)$ and $\bar{r}_j^s(\mathbf{x}^s, \pi)$. Since, at the moment, the analytical model is not available, we focus on the second possibility.

Using the concept of link shadow prices, the policy iteration procedure from the previous section can be rewritten in the following steps:

- collect the statistics in the network operating under a given routing policy π based on link shadow prices, then evaluate $\lambda_j^s(\mathbf{x}^s, \pi)$, $\bar{r}_j^s(\mathbf{x}^s, \pi)$ and, based on these values, compute the new values of the link shadow prices $p_j^s(\mathbf{x}_s, \pi)$,

- implement in the network the improved routing policy π' by executing the following algorithm for each arrival of j-th class call:

  - find the maximum net-gain over all feasible paths

$$g_{max} = \max_{P \in W_j} [r_j - \sum_{s \in P} p_j^s(\mathbf{x}^s, \pi)] \tag{14}$$

  using the new values of link shadow prices,

  - if g_{max} is positive, carry the call on the path giving the maximum net-gain, otherwise reject the call.

Assuming that starting from an arbitrary policy this procedure converges to a certain policy π^*, we could treat this solution as the optimal policy for original problem if all used assumptions are exact. Since these are approximations, policy π^* is suboptimal.

To summarize this section let us point out the two main advantages of the presented decomposition approach compared to the exact model:

- evaluation of the relative values is decomposed into link analysis problems (state space reduction),

- the policy need not to be stored for all network states but instead the decisions can be easily computed at the instant of call arrivals on the basis of link shadow prices (memory requirement reduction).

In addition, this procedure also implies that in a real implementation, the routing policy will track or adapt to a time variable traffic demand.

3 Link Shadow Price Evaluation.

Under the link independence assumption, the problem of link shadow price evaluation is restricted to the analysis of separate links. Thus to simplify notation, we omit the policy and link indices and formulate the problem as follows:

- given link intensities $\lambda_j(x) = \lambda_j^*(x^*, \pi)$ and mean values of link call revenue $\bar{r}_j(x) = \bar{r}_j^*(x^*, \pi)$ find (for each pair x, j) the link shadow price defined by analog of eq. 9:

$$p_j(x) = r_j(x) - g_j(x, r_j(x)) \tag{15}$$

Since link shadow price does not depend on the value of the link call revenue, we can put $r_j(x) = \bar{r}_j(x)$ in eq. 15. Then by using eq.7 we arrive at

$$p_j(x) = \bar{r}_j(x) - \lim_{t \to \infty} [R_j(x, \bar{r}_j(x), t) - R(x, t)] \tag{16}$$

According to the definition of $R_j(\cdot, \cdot, \cdot)$ and $R(\cdot, \cdot)$ we have

$$R_j(x, \bar{r}_j(x), t) = R(x + \delta_j, t) \tag{17}$$

where δ_j is J-vector with 1 in the position j and zeros in all other positions. Thus finally link net gain can be expressed as

$$p_j(x) = \bar{r}_j(x) - \lim_{t \to \infty} [R(x + \delta_j, t) - R(x, t)] \tag{18}$$

Now let us consider the value of $R(x, t)$ for large t. To do that it is useful to express this value as

$$R(x, t) = \bar{R} \cdot t + v(x) + v(x, t) \tag{19}$$

where $\bar{R}$ denotes mean value of revenue from the link in stationary conditions and $v(x)$ is a state dependent constant. It can be shown [11] that $v(x, t)$ is an exponential transient component that vanishes for large t. Thus the value

$$\bar{r}_j(x) - [R(x + \delta_j, t) - R(x, t)] \tag{20}$$

is as close as needed to $p_j(x)$ for sufficiently large t. This conclusion is intuitively obvious since it is well known that for ergodic systems, the influence of the initial state is negligible for large t.

Unfortunately, even for a finite value of t, evaluation of $R(x, t)$ is very difficult. One could start from differential-difference equations but this approach is tractable only for some special cases. To overcome this problem, consider the revenue defined as

$$R_\alpha(x, \alpha^{-1}) = \int_0^\infty \alpha e^{-\alpha t} R(x, t) dt \tag{21}$$

In fact this is the expected value of revenue in the exponentially distributed time interval $(t_0, t_0 + T)$ with mean α^{-1} assuming state x in t_0. $R_\alpha(x, \alpha^{-1})$ can also be interpreted as the link revenue with continuous compounding at the discount rate α (see [11]). It is obvious that for the limit we have

$$\lim_{\alpha \to 0} [R_\alpha(x + \delta_j, \alpha^{-1}) - R_\alpha(x, \alpha^{-1})] = \lim_{t \to \infty} [R(x + \delta_j, t) - R(x, t)] \tag{22}$$

Thus the value

$$\bar{r}_j(x) - [R_\alpha(x + \delta_j, \alpha^{-1}) - R_\alpha(x, \alpha^{-1})] \tag{23}$$

is as close as needed to $p_j(x)$, for sufficiently small α. In fact this approach is equivalent to the use of the optimal policy π_α^* for the system with discount rate $\alpha \to 0$, as the optimal policy π^* for the system without discounting. The theory of MDP supports such an approach since it can be shown that

$$\pi^* = \lim_{\alpha \to 0} \pi_\alpha^* \tag{24}$$

The advantage of the proposed approach lies in that while analyzing the system in the time interval $(t_0, t_0 + T)$, we can use the memoryless property of the distribution of T.

In the sequel we concentrate on the evaluation of $R_\alpha(x, \alpha^{-1})$. First consider two subproblems:

- evaluation of the probability of entering state $\mathbf{x}$ in the exponentially distributed period $(t_0, t_0 + T)$ with mean α^{-1}, assuming state $\mathbf{x}_0$ in t_0: $H_\mathbf{x}(\mathbf{x}_0, \alpha^{-1})$,

- evaluation of the expected value of the time that system spends in state $\mathbf{x}$ in the exponentially distributed period $(t_0, t_0 + T)$ with mean α^{-1}, assuming state $\mathbf{x}$ in t_0: $G(\mathbf{x}, \alpha^{-1})$.

Based on the memoryless property one can derive the following set of linear equations for each state $\mathbf{x} \in X$ $(X = \{\mathbf{x}\})$:

$$H_\mathbf{x}(\mathbf{x}_0, \alpha^{-1}) = \sum_j \frac{\lambda_j(\mathbf{x}_0)}{\sigma(\mathbf{x}_0) + \alpha} H_\mathbf{x}(\mathbf{x}_0 + \delta_j, \alpha^{-1}) + \sum_j \frac{x_j \mu_j}{\sigma(\mathbf{x}_0) + \alpha} H_\mathbf{x}(\mathbf{x}_0 - \delta_j, \alpha^{-1}) \qquad ; \mathbf{x}_0 \in X \qquad (25)$$

where

$$\sigma(\mathbf{x}_0) = \sum_j [\lambda_j(\mathbf{x}_0) + x_j \mu_j] \qquad (26)$$

denotes the intensity of departure from state $\mathbf{x}_0$. This set can be solved by adding the normalizing equation

$$H_\mathbf{x}(\mathbf{x}, \alpha^{-1}) = 1. \qquad (27)$$

To solve the second subproblem, we approximate the time interval between departure from the state $\mathbf{x}$ and the following entry to this state, as being exponentially distributed with mean value $\beta(\mathbf{x})^{-1}$. Under this assumption it can be shown that the probability of being in state $\mathbf{x}$, at the moment $(t_0 + t)$, assuming state $\mathbf{x}$ in t_0, is given by

$$\Pi(\mathbf{x}, t) = 1 - \frac{\sigma(\mathbf{x})}{\sigma(\mathbf{x}) + \beta(\mathbf{x})} (1 - e^{-(\sigma(\mathbf{x}) + \beta(\mathbf{x}))t}) \qquad (28)$$

The intensity of entering state $\mathbf{x}$ can be evaluated from the stationary state probability $\Pi(\mathbf{x})$:

$$\beta(\mathbf{x}) = \frac{\sigma(\mathbf{x})}{\frac{1}{\Pi(\mathbf{x})} - 1} \qquad (29)$$

and the state probabilities can be found by solving the set of balance equations:

$$\Pi(\mathbf{x}) = \sum_j \frac{\lambda_j(\mathbf{x} - \delta_j)}{\sigma(\mathbf{x})} \Pi(\mathbf{x} - \delta_j) + \sum_j \frac{(x_j + 1)\mu_j}{\sigma(\mathbf{x})} \Pi(\mathbf{x} + \delta_j) \qquad ; \mathbf{x} \in X \qquad (30)$$

together with the normalization condition

$$\sum_\mathbf{x} \Pi(\mathbf{x}) = 1 \qquad (31)$$

Now we can evaluate $G(\mathbf{x}, \alpha^{-1})$ from the definition

$$G(\mathbf{x}, \alpha^{-1}) = \int_0^\infty \alpha e^{-\alpha t} \int_0^T \Pi(\mathbf{x}, t) \, dt \, dT \qquad (32)$$

by using equation 28. This finally gives

$$G(\mathbf{x}, \alpha^{-1}) = [\alpha(\frac{\sigma(\mathbf{x})}{\alpha + \beta(\mathbf{x})} + 1]^{-1} \qquad (33)$$

Having formulae for $H_\mathbf{x}(\mathbf{x}_0, \alpha^{-1})$ and $G(\mathbf{x}, \alpha^{-1})$ we can directly attack the problem of $R_\alpha(\mathbf{x}, \alpha^{-1})$ evaluation. Namely, if the rate of revenue from the link in state $\mathbf{x}$, $q(\mathbf{x})$, is steady for each entry into state x, we have

$$R_\alpha(\mathbf{x}_0, \alpha^{-1}) = \sum_\mathbf{x} H_\mathbf{x}(\mathbf{x}_0, \alpha^{-1}) G(\mathbf{x}, \alpha^{-1}) q(\mathbf{x}) \qquad (34)$$

(by using the memoryless property of the time interval T). This case is equivalent to the one link network. In case of multilink network, the value of $q(\mathbf{x})$ can be different for each different entry into state $\mathbf{x}$. In fact $q(\mathbf{x})$ depends on the path states at the time of arrival of all calls currently in progress. To overcome the problem we applied an assumption that $q(\mathbf{x})$ can be treated as an independent random variable with mean value $\bar{q}(\mathbf{x})$ defined as

$$\bar{q}(\mathbf{x}) = \sum_j x_j \bar{r}_j \tag{35}$$

where $\bar{r}_j$ is the mean value of revenue parameters of the j-th class link calls to be evaluated from link distribution. This assumption is justified by the fact that if we look at random instant, the current link state is not strongly correlated with that at set up time of a call being in progress (since the mean holding time is exponentially distributed and usually the total link arrival intensity is much higher than single call departure intensity).

Under this assumption we have

$$R_\alpha(\mathbf{x}_0, \alpha^{-1}) = \sum_{\mathbf{x}} H_{\mathbf{x}}(\mathbf{x}_0, \alpha^{-1}) G(\mathbf{x}, \alpha^{-1}) \bar{q}(\mathbf{x}) \tag{36}$$

This procedure enables the link shadow price to be evaluated from eq. 23 to the required accuracy (on the condition that the used assumption holds), by using a sufficiently small value of α. From a theoretical point of view, the solution method does not impose any limitations on this value, but in practice, due to the limited accuracy of computer calculations, one can expect some numerical problems for very low values of α since we have

$$\lim_{\alpha \to 0} H_{\mathbf{x}}(\mathbf{x}_0, \alpha^{-1}) = 1 \tag{37}$$

and

$$\lim_{\alpha \to 0} \alpha G(\mathbf{x}, \alpha^{-1}) = \Pi(\mathbf{x}) \tag{38}$$

To check whether there exists a practical range of α and whether the assumption that the time $\beta(\mathbf{x})^{-1}$ is exponentially distributed is reasonable, we investigated the relative error function for shadow price evaluation

$$f_j(\mathbf{x}, \alpha) = \{r_j - [R_\alpha(\mathbf{x} + \delta_j, t) - R_\alpha(\mathbf{x}, t)]\}/p_j(\mathbf{x}) - 1 \tag{39}$$

for the one link network offered single class of calls with steady intensity $\lambda_1(\mathbf{x}) = \lambda$, one channel requirement $d_1 = 1$, departure intensity $\mu_1 = 1$ and constant revenue parameter $r_1 = 1$. The advantage of this system is that we can calculate the value of the shadow price with a very high accuracy since it can be shown [7], [9] that

$$p_1(x) = H_N(x, \mu_1^{-1}) \tag{40}$$

where N denotes link dimension. In fig.1 we display typical values of $f_1(\mathbf{x}, \alpha)$ versus α for sets $\{N = 48, \lambda = 42, x = 42, 46\}$ and $\{N = 240, \lambda = 230, x = 230, 238\}$. These results were obtained by a Fortran program (run on the VAX 8600 computer) with single precision variables. As we can see in both cases the optimum is achieved for the interval $\alpha \cong [10^{-3}\mu_1, 10^{-2}\mu_1]$ where the relative error is negligible. Further decreasing α causes numerical errors which stabilize when limits 37, 38 are achieved. It is worth noting that in most investigated cases the optimal value of shadow price was achieved for $\alpha \cong [10^{-3}\mu_1, 10^{-2}\mu_1]$. One would expect such results since the autocorrelation of link occupancy distribution depends strongly on μ_1. This feature can be helpful in choosing a value of α for different systems. It should be noted that in this experiment, nothing was done to reduce numerical errors. One can expect that if such an effort was undertaken the practical range of α would be greater.

The results also show that in case of a one link network the presented approach provides a solution very close to the optimal call access control.

Another numerical problem can be encountered during the solution of the equations 25 and 30 (for $H_{\mathbf{x}}(\mathbf{x}_0, \alpha^{-1})$ and $\Pi(\mathbf{x})$ respectively) if the link state space X is very large. In the following we present two approximations that can significantly reduce the cardinality of the link state space.

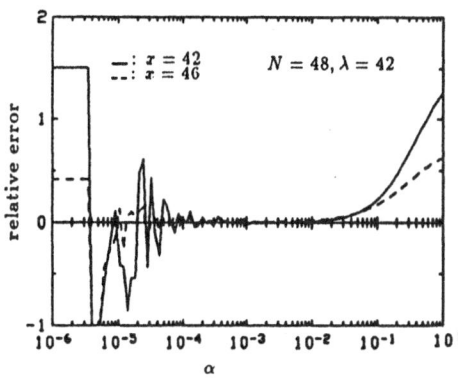 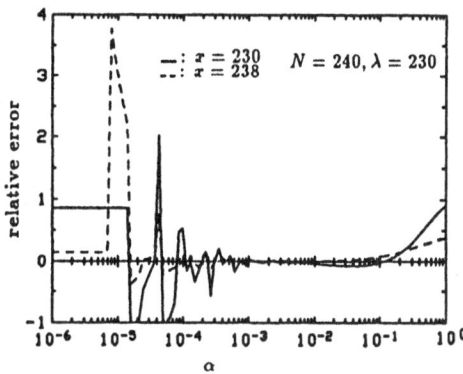

Fig. 1   Relative error of the shadow price approximation versus α

3.1 Reduction of the Number of Link Call Classes.

The first approximation described in this subsection is based on the fact that since the link shadow prices can be interpreted as the price for using d_j channels during a call holding time, this price should be roughly the same for all call classes demanding the same number of servers and having the same mean holding time. This approach is exact when accepting the j-th class call with $d_j = d_i$, has the same influence on link call intensities as accepting a call from any other call class characterized by d_i and the same mean holding time. By using this assumption we can aggregate the call classes with the same values of d_j and μ_j into the new i-th class characterized by $\bar{r}_i(\mathbf{x})$ being the expected average of respective values of aggregated call classes.

The second approximation is based on the fact that for call classes with the same channel requirements, the shadow price is roughly proportional to the mean holding time. Then we can couple the classes with the same values of d_j into a new i-th class characterized by μ_i and $\bar{r}_i(\mathbf{x})$ being the expected average of respective values of aggregated call classes.

Finally, by applying both approximations one can reduce the number of call classes to the number of different server requirements. This can facilitate solution of eq. 25 and 30.

The approximations can be used in the same policy iteration procedure as before with the exception that the maximum gain is defined by

$$g_{max} = \max_{P \in W_j}[r_j - \sum_{s \in P} \frac{\mu_i}{\mu_j} p_i^s(\mathbf{x}^s, \pi)] \tag{41}$$

where i denotes link call class comprising the j-th call class and the fraction $\frac{\mu_i}{\mu_j}$ takes into account the relation between the mean holding time of j-th class call and the i-th class call.

It is important to note that in the particular case of one type of server requirement (e.g. telephone network) the sets of eq. 25 and 30 can be rewritten in the form of recurrence relations (presented in [9]) which provide a very efficient solution. Namely, for probability $H_z(x_0, \alpha^{-1})$ and $x_0 \leq x$ we have

$$H_z(x_0, \alpha^{-1}) = \frac{\sigma(x_0 - 1) + \alpha}{\lambda(x_0 - 1)} H_z(x_0 - 1, \alpha^{-1}) - \frac{x\mu}{\lambda(x_0 - 1)} H_z(x_0 - 2, \alpha^{-1}) \qquad ; x_0 = 2, ..., x \tag{42}$$

and

$$H_z(1, \alpha^{-1}) = \frac{\lambda(0) + \alpha}{\lambda(0)} H_z(0, \alpha^{-1}) \tag{43}$$

thus by solving this recurrence for $x = N$ with any initial value of $H_N(0, \alpha^{-1})$ we can obtain values for all pairs x_0, x such that $x_0 \leq x$ by using the normalizing equation $H_z(x_0, \alpha^{-1}) = 1$ in each step of the recurrence. It can be easily shown that an analogous recurrence exists for cases $x_0 \geq x$.

For the evaluation of $\Pi(x)$ the recurrence is trivial:

$$x \mu \Pi(x) = \lambda(x-1)\Pi(x-1) \tag{44}$$

and provides the solution together with normalizing equation $\sum_x \Pi(x) = 1$.

3.2 Decomposition of the Link Analysis.

For the purpose of this section let us divide the call classes into two groups. The first one consists of classes with $d_j = 1$ and can be transformed to the one class with $d_1 = 1$ and μ_1 by using the approximation presented in the previous section. This class is called narrow-band calls. The second group consists of call classes with $d_j > 1$ and after unification can be transformed to I classes each described by d_i and μ_i'. From now on we call this group as the wide-band calls.

It is expected that in the future ISDN based on circuit switching the mean holding times of wide-band calls will be significantly longer than that for narrow-band calls. This suggests that while the state of wide-band calls remains steady, the number of narrow-band calls reaches the steady state distribution. Based on this premiss, one can develop an analytical model for performance evaluation using the approximation that narrow-band calls reach the steady state distribution instantaneously for each state of wide-band calls. This means that the Markov process is analysed separately for narrow-band calls (for each state of wide-band calls) and wide-band calls. In fact such models were investigated in [12] for delayed call set up, where it was shown that the approach is accurate and very efficient.

The decomposition of link performance analysis is also very attractive in the case of shadow price evaluation since for the narrow-band calls the shadow price can be evaluated by using the very efficient recurrence relations 42, 44 and, what is perhaps more important, the state space for wide-band calls is significantly reduced compared to the full state description.

In general the link analysis for shadow price evaluation can be decomposed into three separate problems:
• **Evaluation of shadow prices for narrow-band calls:** Since the state of wide-band calls is considered to be steady, we can analyse a system with one call class offered to the link with $N_1 = N - \sum_i x_i \, d_i$ channels. Although these channels are available for narrow-band calls, it may happen that for some states of wide-band calls $\mathbf{x}_w$ it is more efficient to reserve some space for wide-band calls to maximize mean revenue. It is obvious that in the case of link analysis decomposition, the number of reserved channels $h(\mathbf{x}_w)$ should be steady for given $\mathbf{x}_w$. The problem of $h(\mathbf{x}_w)$ evaluation, which can be also called dynamic bandwidth allocation, is described as a separate problem. Assuming that $h(\mathbf{x}_w)$ is given, we can determine the link system parameters for the narrow-band calls by the link dimension $N_1' = N - \sum_i x_i \, d_i - h(\mathbf{x}_w)$, state dependent call intensity $\lambda_1(x_1, \mathbf{x}_w)$, $d_1 = 1$ and μ_1. This is sufficient to evaluate shadow prices from eq. 23, 36 by using recurrence relations 42, 44.

• **Evaluation of shadow prices for wide-band calls:** In general the method is the same as for the link with I classes of wide-band calls offered to the link with N channels except for the fact that we have to account for the influence of narrow-band calls. Namely, the narrow-band calls have an influence on the state-dependent intensity of wide-band calls and on the revenue from the link. The first factor follows from the fact that in the case of direct wide-band calls offered to the link in state $(\mathbf{x}_w)$ some of them can be lost due to narrow-band calls if $d_i > N - \sum_i x_i - h(\mathbf{x}_w) - x_1$. This mechanism can be taken into account by reducing the intensity of direct wide-band calls by the amount

$$\lambda_i'(\mathbf{x}_w) = \lambda_i \, B_i(\mathbf{x}_w) \tag{45}$$

where

$$B_i(\mathbf{x}_w) = P\{x_1 > N - \sum_i x_i - h(\mathbf{x}_w) - d_i\} \tag{46}$$

denotes the probability of blocking an i-th type direct call in state $\mathbf{x}_w$. This value can be easily obtained from the state distribution of narrow-band calls in state $\mathbf{x}_w$. Concerning the influence of narrow-band calls on revenue from the link, this can be taken into account by modifying the mean rate of revenue in state $\mathbf{x}_w$ as follows:

$$\bar{q}(\mathbf{x}_w)' = \sum_i \bar{q}(\mathbf{x}_w) + \bar{r}_1 \bar{\lambda}_1(\mathbf{x}_w) \mu_1 \tag{47}$$

where $\bar{\lambda}_1(\mathbf{x}_w)$ denotes the mean number of narrow-band calls carried in state $\mathbf{x}_w$. After introducing the presented modifications we can evaluate shadow prices for wide-band calls from eq. 23, 36.

 • **Dynamic bandwidth allocation:** The problem of dynamic bandwidth allocation is a Markov decision problem and can be formulated as follows: find the optimal values of $h(\mathbf{x}_w)$; $\mathbf{x}_w \in X_w$ that maximize the mean value of revenue from the link $\overline{R}$. Again the problem could be solved in the framework of MDP but since this approach could significantly delay the convergence of the routing policy a more efficient approach was developed. It is based on the assumption that for practical range of link parameters the value of $h(\mathbf{x}_w)$ should be equal either to zero or to the one of the channel requirements. The detailed model of this approach for the case of one wide-band call class will be presented in a forthcoming publication.

To illustrate the advantage of the presented link analysis decomposition approach consider two examples of possible state space reduction. The first concerns a link with 240 channels and three types of service with $d_1 = 1$, $d_2 = 12$ and $d_3 = 24$. The second differs in that there is no third type of service. In the first case the number of states is reduced from 10021 to 241 for narrow-band calls and to 121 for wide-band calls. In the second case the reduction is from 2541 to 241 and 21 for narrow-band and wide-band calls respectively, and moreover for both types of traffic the recurrence solution can be applied.

Conclusions. In the paper we have synthesized a state-dependent routing policy for a multi-service circuit-switched network. To meet different requirements, the objective function was defined as the mean value of revenue from the network. The theory of MDP was applied to find the optimal routing policy. It was shown that under the link independence assumption the problem can be decomposed into the set of link analysis problems. In this approach the routing decision has a very natural interpretation. Namely, a new call should be routed on the path with maximum positive net-gain defined as the difference between the expected revenue from the network in case the call is carried on the path and the expected revenue in case the call is rejected. The net-gain is a function of link shadow prices being interpreted as prices for using each link from the path. It was shown that the values of shadow prices can be evaluated from the limiting solution for the equivalent system with discounting, separately for each link and that this solution provides the near optimal call access control in case of a one link network. Moreover it was shown that the presented approach is implementable even for large systems if certain approximations are used.

The general aim of this paper was to show that the addressed problem is tractable in the framework of Markov decision process theory. The practical implementation of these results and numerical results are to be presented in a forthcoming publication.

Another goal of further studies is to examine whether the idea of routing decisions, based on link shadow prices, can be applied to networks with delayed set up and to networks with fast packet switching with virtual paths.

Acknowledgments: The research was supported by the NSERC Strategic Grant No. G1585. The authors are with INRS-Telecommunications, 3 place du Commerce, Verdun, Quebec H3E 1H6, Canada. Zbigniew Dziong is on sabbatical leave from Warsaw University of Technology, Poland.

Bibliography:

[1] Mason L.G., Girard A., "Control Techniques and Performance Models for Circuit Switched Networks", Proceedings of 21th CDC, Orlando, USA, Dec. 1982.

[2] Dziong, Z., Pioro M. and Körner U., "An Adaptive Call Routing Strategy for Circuit Switched Networks", NTS7, Lund, 1987.

[3] Narendra, K.S., Wright E.A., Mason L.G., "Application of Learning Automata to Telephone Traffic Routing and Control" IEEE Trans. Systems, Man and Cybernetics, Vol. SMC-7, No.11, Nov. 1977

[4] Ash, G.R., Cardwell, R.H., Murray, R.P., "Design and optimization of networks with Dynamic Routing", BSTJ, vol.60, no.8, October, 1981.

[5] Szybicki, E., Lavigne, M.E., "The introduction of an advanced routing system into local digital networks and its impact on networks' economy, reliability and grade of service", ISS, Paris, 1979.

[6] Lazarev W.G., Starobinets S.M., "The use of dynamic programing for optimization of control in networks of commutations of channels", Engineering Cybernetics (Academy of Sciences, USSR), No.3, 1977.

[7] Krishnan K.R., Ott T.J., "State dependent routing for telephone traffic: theory and results ", Proceedings of 25th CDC, Athens, Greece, December 1986.

[8] Kelly, F.P., "Adaptive routing in circuit switched networks", Technical report, University of Cambridge, 1985.

[9] Dziong Z., Pioro M., Körner U., Wickberg T., "On Adaptive Call Routing Strategies in Circuit Switched Networks - Maximum Revenue Approach ", ITC12, Torino, 1988.

[10] Benes V.E., "Programming and Control Problems Arising from Optimal Routing in Telephone Networks", BSTJ, NOV. 1966.

[11] Howard R. A., "Dynamic Programing and Markov Process", The M.I.T. Press, Cambridge, Massachusetts 1960.

[12] Mason L.G., Liao K.-Q., Fortier L., DeSerres Y., "Performance Models for an Integrated Services System", ITC12, Torino, 1988.

ESTIMATION OF THE PARAMETERS OF A SECOND ORDER NONLINEAR SYSTEM

S.A. Dianat   M.R. Raghuveer

Rochester Institute of Technology

1. INTRODUCTION

Memoryless quadratic transformations occur in communication and control systems (e.g. amplitude modulation and phase locked loops) and other applications. In this paper we present an approach to estimate the parameters of such transformations when the input is white Gaussian noise and we observe only a noise corrupted version of the output. For this purpose we use the third order cumulant sequence. For details of cumulants and higher order spectra the reader is referred to [1,2]. For a zero-mean discrete stationary real process $x(n)$, the third order cumulant sequence $c(k, \ell)$ is defined as

$$c(k, \ell) = E\{x(n)x(n + k)x(n + \ell)\} \tag{1}$$

In other words it is identical to the centered third moment sequence. If $x(n)$ is Gaussian it follows that $c(k, \ell)$ is identically equal to zero. Also the cumulant sequence of the sum of two independent signals is equal to the sum of the individual cumulant sequences. Cumulants arise naturally in the statistical description of polynomial systems [3]. In particular the third order cumulant is useful when quadratic nonlinearities are involved. For these reasons the third order cumulant sequence and its spectrum (bispectrum) have been applied to problems such as quadratic phase coupling [2]-[10], system identification parameter estimation and signal reconstruction [11]-[17]. In the next section we provide a statement of the problem and develop the parameter estimation approach.

2. PROBLEM STATEMENT AND SOLUTION

Let $y(n)$ be a discrete random signal generated as the output of a memoryless quadratic nonlinear system driven by zero mean white Gaussian noise $x(n)$ i.e

$$y(n) = ax(n) + bx^2(n) \tag{2}$$

where a and b are the parameters of the transformation. Let p denote the variance of $x(n)$. Assume further that we observe only $z(n)$ a noise corrupted version of $y(n)$ as

$$z(n) = y(n) + w(n) \tag{3}$$

where $w(n)$ is zero mean Gaussian noise independent of the driving noise $x(n)$. Let $E\{w^2(n)\} = Q$. The problem is to estimate a, b, p and Q from a finite realization of $z(n)$ i.e. when we are given samples $z(0), z(1), \ldots, z(N-1)$. The parameter vector $\underline{\theta} = [a^2, b, p]^T$ is estimable to a constant factor from the mean and the third order cumulant sequence of $z(n)$ and this estimate can in turn be used along with the autocorrelation of $z(n)$ to estimate Q. The system in (2) is a special case of second order Volterra system. Hinich and Patterson [18] consider another special case of a Volterra system. The memoryless quadratic system considered in our paper is *not* a special case of the formulation in [18]. Now

$$c_z(k, \ell) = c_y(k, \ell) + c_w(k, \ell) \tag{4}$$

Since $w(n)$ is Gaussian $c_w(k, \ell) = 0$. Note that $w(n)$ *need not be white*. In fact even though we have made the Gaussian assumption all we actually need is that $w(n)$ have a probability density function which is symmetric about zero.

$$c_z(k, \ell) = c_y(k, \ell) \tag{5}$$

But (see appendix)

$$c_y(k, \ell) = [8b^3 p^3 + 6a^2 bp^2]\delta(k, \ell) \tag{6}$$

where $\delta(k, \ell)$ is the 2-D discrete impulse sequence. In deriving the above we have not made the assumption of weak nonlinearity ($b \ll a$), an assumption made by some authors (see for example [19]). Also denoting $E\{z(n)\}$ by $\bar{z}$ we get

$$\bar{z} = E\{y(n)\} + E\{w(n)\} = E\{y(n)\} = bp \tag{7}$$

From (6) and (7) bp and $a^2 p$ can be recovered as

$$bp = \bar{z} \tag{8}$$

$$a^2 p = [c_z(0,0) - 8\bar{z}^3]/6\bar{z}$$

Estimate $\hat{\underline{\theta}}$ of the parameter vector $\underline{\theta}$ can now be formed as

$$\hat{\underline{\theta}} = \alpha \begin{bmatrix} [c_z(0,0) - 8\bar{z}^3]/(6\alpha\bar{z}) \\ \bar{z}/\alpha \\ 1 \end{bmatrix} \tag{9}$$

where α is a strictly positive scale factor. Also (see appendix)

$$Q = r_z(0) - \frac{10\bar{z}^3 + c_z(0,0)}{6\bar{z}} \tag{10}$$

where $r_z(k) = E\{z(n)z(n+k)\}$. Equations (8) and (10) can be used when we have the true

mean, autocorrelation and third order cumulant sequences of $z(n)$. If we are given finite samples

of $z(n)$ as mentioned earlier, we can use estimates of these quantities in place of the true values.

We form these estimates as

$$\hat{\bar{z}} = \frac{1}{N} \sum_{n=0}^{N-1} z(n)$$

$$\hat{r}_z(0) = \frac{1}{N} \sum_{n=0}^{N-1} z^2(n) \tag{11}$$

$$\hat{c}_z(0,0) = \frac{1}{N} \sum_{n=0}^{N-1} [z(n) - \hat{\bar{z}}]^3$$

Since we make a very simple assumption about the additive noise namely that it have a sym-

metric probability density function about zero, the approach of the paper is likely to be more

robust than methods which impose greater restrictions on the form of $w(n)$.

3.SIMULATION RESULT

Forty separate realizations of 4000 samples each of the signal

$$z(n) = 1.29x(n) + 3x^2(n) + w(n) \tag{12}$$

were generated. $x(n) \sim N(0,1)$ and white and $w(n)$ is also normal with mean zero and variance

Q. $\hat{\theta}$ was estimated using the procedure of the paper. Results are given below

SNR= 35 dB	Sample Bias	Sample Variance
bp	0.0084	0.0027
a^2p	−0.067	0.601

SNR= 15 dB	Sample Bias	Sample variance
bp	0.011	0.003
a^2p	−0.26	0.77

As can be observed estimates bp show lower bias and variance than those of a^2p. This can perhaps be attributed to the fact that the estimate of the mean is of lower variance than that of the third order cumulant. Also the SNR for the two cases does not appear to influence the estimates.

4.CONCLUSION

We have presented an approach to estimate the parameters of a memoryless quadratic transformation based on the third order cumulant sequence sample of noisy measurement. The derivations are valid for zero mean measurement noise which is symmetrically distributed and is independent of the nonlinear process. We exploit the property of third order cumulants that they are identically zero for symmetric distributions. What we have here are just preliminary results. Comparisons have to be made — both in terms of theory and experiments — with techniques such as the least squares and maximum likelihood estimators with different contaminating noise characteristics. Extensions to dynamic second order nonlinear systems need to be made. Work related to these problems is in progress.

Appendix A

Detailed derivations of the mathematical results of the paper are presented here. From (2)

$$E\{y(n)\} = E\{ax(n) + bx^2(n)\} = bE\{x^2(n)\} = bp \tag{13}$$

$$E\{y^2(n)\} = E\{a^2x^2(n) + b^2x^4(n) + 2abx^3(n)\} = a^2p + 3b^2p^2 \tag{14}$$

where we have used the fact that $x(n)$ is gaussian and therefore [20]

$$E\{x^k(n)\} = \begin{cases} 0 & k \text{ odd} \\ 1 \times 3 \times 5 \times \ldots \times (k-1)p^{k/2} & k \text{ even} \end{cases} \tag{15}$$

Also

$$E\{y^3(n)\} = 9a^2bp^2 + 15b^3p^3 \tag{16}$$

where again we have used (15). Since $x(n)$ in (2) is Gaussian white noise it follows that $y(n)$, $n = 0, \pm 1, \ldots$ are non-Gaussian, independent and identically distributed random variables. Thus $c_y(k, \ell)$ the cumulant sequence of $y(n)$ is equal to zero for all (k, ℓ) except at $(k, \ell) = (0, 0)$.

$$c_y(0,0) = E\{[y(n) - E\{y(n)\}]^3\} = E\{[y(n) - bp]^3\} = 8b^3p^3 + 6a^2bp^2 \tag{17}$$

Here we have used (13) through (16).

Now for $z(n)$ in (3)

$$r_z(0) \equiv E\{z^2(n)\} = E\{y^2(n)\} + E\{w^2(n)\} = a^2p + 3b^2p^2 + Q \tag{18}$$

The above equation along with (5) and (8) yields (10).

References

[1] D.R. Brillinger, "An introduction to polyspectra," *Ann. Math. Stat.*, vol.36, pp.1351-1374, 1965.

[2] C.L. Nikias and M.R. Raghuveer, "Bispectrum estimation: a digital signal processing framework," *Proc. IEEE*, vol.75, pp.869-891, July 1987.

[3] D.R. Brillinger, "The identification of polynomial systems by means of higher order spectra," *Jour. Sound Vib.*, vol.12, pp.301-313, 1970.

[4] K. Hasselmann, W. Munk and G. Macdonald, "Bispectrum of ocean waves," in *Time Series Analysis*, M. Rosenblatt, ed., Wiley, New York 1963.

[5] M.J. Hinich and C.S. Clay, "The application of discrete Fourier transform in the estimation of power spectra, coherence and bispectra of geophysical data," *Rev. Geophysics*, vol.6, pp.347, 1968.

[6] M.R. Raghuveer, "Bispectrum and multidimensional power spectrum estimation algorithms based on parametric models with applications to the analysis of ECG data," *Ph.D. Dissertation*, Univ. of Connecticut, December 1984.

[7] M.R. Raghuveer and C.L. Nikias, "Bispectrum estimation: a parametric approach," *IEEE Trans. on ASSP*, vol.33, pp.1213-1230, October 1985.

[8] M.R. Raghuveer and C.L. Nikias, "Bispectrum estimation via AR modeling," *Signal Processing*, vol.10, pp.35-48, January 1986.

[9] C.K. An, S.B. Kim and E.J. Powers, "Optimized parametric bispectrum estimation," in *IEEE Int. Conf. on Acoust. Speech and Sig. Proc.*, pp.2392-2395, New York, April 1988.

[10] M.R. Raghuveer, "High resolution estimation of quadratic phase coupling in nonlinear systems," in *Proc. American Control Conf.*, pp.2124-2128, Atlanta, GA, June 1988.

[11] K.S. Lii and M. Rosenblatt, "Deconvolution and estimation of transfer function phase and coefficients for non-Gaussian linear processes," *Ann. Stat.*, vol.10, pp.1195-1208, 1982.

[12] T. Subba Rao and M.M. Gabr, "A test for linearity of a stationary time series," *Jour. Time Series Analysis*, vol.1, no.2, pp.145-158, 1980.

[13] T. Matsuoka and T.J. Ulrych, "Phase estimation using the bispectrum," *Proc. IEEE*, vol.72, pp.1403-1411, October 1984.

[14] C.L. Nikias, "ARMA bispectrum approach to non-minimum phase system identification," *IEEE Trans. on ASSP*, vol.36, pp.513-524, April 1988.

[15] G.B. Giannakis and J.M. Mendel, "Identification of non-minimum phase systems using higher order statistics," to be published in *IEEE Trans. on ASSP*.

[16] J.K. Tugnait, "On selection of maximum cumulant lags for non-causal autoregressive model fitting," in *Proc. Int. Conf. on Acoustics, Speech and Sig. Proc.*, pp.2372-2375, New York, NY 1988.

[17] M.R Raghuveer, S.A. Dianat and G. Sundaramoorthy, "Reconstruction of non-minimum phase multidimensional signals using the bispectrum," in *Proc. Third SPIE Conf. on Visual Communications and Image Processing*, Cambridge, MA November 1988.

[18] M.J. Hinich and D.M. Patterson, "Identification of the coefficients in a non-linear time series of the quadratic type," *Jour. of Econometrics*, vol.30, pp.269-288, 1985.

[19] V.Z. Marmarelis and D. Sheby, "Bispectral analysis of weakly nonlinear quadratic systems," in *Proc. IEEE ASSP Spectrum Estimation and Modeling Workshop III*, pp.14-16, Boston, MA November 1986.

[20] A. Papoulis, *Probability, Random Variables and Stochastic Processes,* McGraw-Hill, 1984.

Lecture Notes in Control and Information Sciences

Edited by M. Thoma and A. Wyner

Lecture Notes in Control and Information Sciences

Edited by M. Thoma and A. Wyner

Lecture Notes in Control and Information Sciences

Edited by M. Thoma and A. Wyner